Some other ELBS low-priced editions

Barry	THE CONSTRUCTION OF BUILDINGS, Volumes I–IV	*Crosby Lockwood Staples*
Bishop and Henkel	THE MEASUREMENT OF SOIL PROPERTIES IN THE TRIAXIAL TEST	*Arnold*
Case and Chilver	STRENGTH OF MATERIALS AND STRUCTURES	*Arnold*
Chapman	ELEMENTARY WORKSHOP CALCULATIONS	*Arnold*
Chapman	SENIOR WORKSHOP CALCULATIONS	*Arnold*
Chapman	WORKSHOP TECHNOLOGY, Parts I–III	*Arnold*
Coates, Coutie and Kong	STRUCTURAL ANALYSIS	*Nelson*
Fisher	WALLS	*Macmillan*
Hart	ENGINEERING DRAWING WITH PROBLEMS AND SOLUTIONS	*Hodder & Stoughton*
Haslehurst	MANUFACTURING TECHNOLOGY	*Hodder & Stoughton*
Herbert	THE STRUCTURE OF TECHNICAL ENGLISH	*Longman*
Jeffrey	MATHEMATICS FOR ENGINEERS AND SCIENTISTS	*Nelson*
John	INTRODUCTION TO ENGINEERING MATERIALS	*Macmillan*
Launder	FOUNDATIONS	*Macmillan*
Marshall and Nelson	STRUCTURES	*Pitman*
Maxwell	PLANE PROJECTIVE GEOMETRY	*Cambridge University Press*
Nuttall	HIGHER NATIONAL CERTIFICATE WORKSHOP TECHNOLOGY	*Hodder & Stoughton*
Owen	ROOFS	*Macmillan*
Page	PRINCIPLES OF MECHANICAL ENGINEERING SCIENCE IN S.I. UNITS	*Hodder & Stoughton*
Pritchard	T3 WORKSHOP TECHNOLOGY FOR MECHANICAL ENGINEERING TECHNICIANS	*Hodder & Stoughton*
Pritchard	WORKSHOP PROCESSES FOR MECHANICAL ENGINEERING TECHNICIANS	*Hodder & Stoughton*
Smith	CONSTRUCTION SCIENCE, Volume II	*Longman*
Thomson	EXERCISES IN GRAPHIC COMMUNICATION	*Nelson*
Thomson	SOLUTIONS TO EXERCISES IN GRAPHIC COMMUNICATION	*Nelson*

Technical Drawing

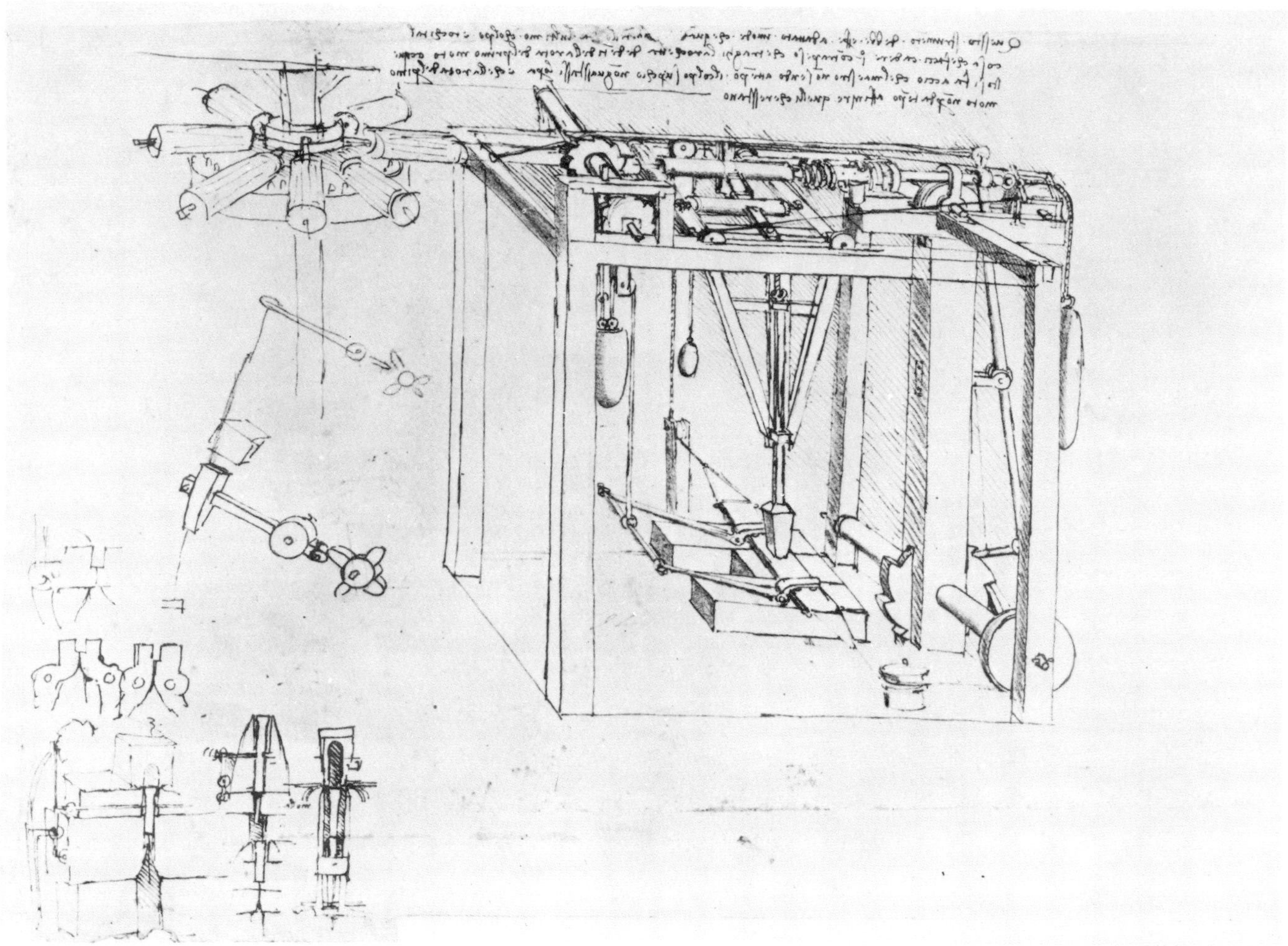

Sketch by Leonardo da Vinci (1452-1519) of an Automatic Machine for Hammering Gold Bars for the Mint in Rome

By courtesy of the Director of the Science Museum South Kensington

Technical Drawing

FOURTH EDITION

W. ABBOTT

Revised by **T. H. Hewitt**

English Language Book Society
and
Blackie & Son Limited

ISBN 0 216 90210 X (Blackie edition)
ISBN 0 216 90213 4 (ELBS edition)

First published in one volume 1958. Second edition 1962
Third edition 1970. Fourth edition 1976. Reprinted 1978

ELBS edition first published 1965
ELBS edition of Third edition 1973. Reprinted 1974
ELBS edition of Fourth edition 1976. Reprinted 1978

Published by Blackie & Son Limited
Bishopbriggs, Glasgow; and 450/452 Edgware Road
London W2 1EG

Printed in Great Britain by Robert MacLehose & Co. Ltd
Printers to the University of Glasgow

Preface

This book is arranged in three parts. It provides a complete course in Technical Drawing suitable for students whose careers will lie in one or other of the branches of the engineering industry. Part I provides an introductory course. Part II deals exclusively with applied geometry; while Part III applies the constructions and the geometry of Parts I and II to the preparation of engineering drawings. The complete book provides a course which covers the requirements of the General Certificate of Education at the Ordinary Level, and equivalent engineering technician course units.

No practical engineering knowledge on the student's part is assumed; for this reason, the technical examples have been limited to the common fastenings and simpler engineering parts. The author assumes that the subject of Technical Drawing has to justify its place in the curriculum, both by its vocational relevance and by the demands it makes upon the intelligence of the student. Hence the work in the book calls for something more than the acquisition of drafting skills and a knowledge of conventional practice, important though these are. The work in Part III asks more and more of the student in the settlement of details; and it calls for an appreciation of the problems of dimensioning—although in an elementary way.

On the vexed question of projection, inquiry of a large number of industrial organizations has shown that 3rd Angle Projection is now overwhelmingly used. So widespread a practice must be reflected in texts and in teaching, and the system predominates in this book.

This fourth edition features a substantial number of revisions, the changes being of three general types: the incorporation of the most recent recommendations of B.S. 308 and other sources of standard practice, the adoption of current usage as regards preferred sizes and other results of metrication, and a considerable expansion of the sections of exercises.

Contents

Part One **Introduction**

	Page
Technical Drawings	14
Straight Lines: Parallel, Perpendicular	16
Straight Lines: Parallel, Perpendicular, Inclined	18
Use of the Compasses. Drawing Errors	20
Division of Lines	22
Measurement and Transfer of Angles	24
Exercises	26
Drawings of Solids. The Cube	28
Rectangular Prism. Composite Solid	30
Importance of Labelling Views	32
Three-View Arrangements	34
Interpretation of Drawings	36
Hidden Lines	38
Hexagonal Prism	40
True Shape of Surfaces: Arrangement of Views	42
Hexagonal Pyramid	44
Cylinder and Cone	46
Simple Sectional Views	48
Half-Sectional Views	50
Pictorial Drawings	52
Isometric Drawings: Reference Frame	54
Isometric Drawings: Curves and Circles	56
Isometric Drawings: Circles	58
Isometric Drawings: Variety of Views	60
Dimensioning. Lettering and Numerals. Line Thicknesses	62
Freehand Sketching	64
Interpretation of Isometric Drawings. Exercises for Freehand Sketches	66
Interpretation of Drawings. Conversion to Pictorial Views	68
Solutions	70

Contents

Part Two **Applied Geometry**

	Page
Accuracy in Drawings	72
Important Ratios	74
The Rectangle: The Golden Rule	76
Triangles and Polygons	78
Polygons	80, 82
Circles	84
Parabola, Ellipse, Hyperbola	86, 88
Parallel Curves, Cams	90
Loci of Points on Moving Mechanisms	92
Approximate Constructions	94
Roulettes, Involutes and Cycloids	96
Involutes, Wheel Teeth	98
Spiral and Helix	100
Areas of Irregular Figures	102

Principles of Projection	104, 106
The Straight Line	108, 110
Auxiliary Projection of Solids: Cube	112
Pyramid	114
Cylinder and Cone	116
Sections of Solids: Cube and Pyramid	118
Cone (Conic Sections)	120
Intersections of Surfaces: Cylinders	122
Cone and Cylinder	124
Developments	126
Isometric Scale and Isometric Projection	128
Oblique Projection (Cavalier, Cabinet)	130
Horizontal Projection or Indexed Plans	132
Miscellaneous Examples	134

Contents

Part Three **Engineering Practice**

	Page
Standard Practice. Projection	138
Choice of Views. Reading Drawings	140
Drawing Procedure	142
Curves of Intersection	144
Letters and Numbers. Drawing Headings	146
Sectional Views—General Rules	148
Sectional Views: Revolved Sections, Conventions	150
Common Engineering Names. Parts not Sectioned	152
Dimensioning: Rules; Size and Position Dimensions	154
Dimensioning: Systems of Dimensioning	156
Dimensioning: Special Cases	158
Abbreviations; Machining Symbols	160
Dimensioned Drawing for Reference	162
Rivets and Riveted Joints	164
Screw Threads and Screws	166
Nuts and Bolt Heads	168
Screw Heads. Set Screws. Locking Devices	170
Freehand Sketching	172, 174
Rolled-steel Sections	176
Developments in Sheet Metal	178
Fabrication by Welding	180
Keys. Shafts and Couplings	182
Cottered Joints	184
Knuckle Joints	186
Simple Bearings	188
Brackets	190
Simple Valves	192
Toleranced Dimensions	194
Graphical Symbols	196
Circuit Diagrams	198
Exercises	200-203
Drawings in Ink	204
Solutions	206
Index	207

Part One
Introduction

Technical Drawings

The Art of Drawing is very old. Before he developed writing man tried to describe the things he saw by drawing them. Notable examples are the cave drawings in France.
Technical Drawing is a graphic language used by those concerned with the constructive arts: in the manufacture of machinery, structures, ships, aircraft, and so on. It is used when words alone will not convey all the information required. This language of lines, signs, and symbols has to be learned before it can be expressed on paper. When so expressed, it has to be understood by others. In short, the student has not only to write the language, but also to read it. The reading is often more difficult than the writing. This is because the reader has to imagine solidness, from a flat drawing made without any of the aids which the artist uses to produce the illusion of depth: perspective, colour, light and shade.

Examples

The three drawings in fig. 1 represent the object in fig. 2. They might be understood readily by a person not trained to read such drawings. But the same person would have more difficulty in giving a meaning to the drawing in fig. 3 because the conventions used will not be familiar to him.
Another purpose of a technical drawing is to give a graphical solution to a problem which would be tedious to solve by calculation. A simple loaded structure is represented in fig. 4. The forces in the various members of the structure are required. These forces can be obtained by drawing the diagram shown and actually measuring the lengths of the lines, using the scale shown. This use of the drawing board calls for very accurate work if the results are to be acceptable.

Requirements

The subject of technical drawing requires therefore three skills:
(1) The ability to make drawings of the outline and features of an object and to arrange the drawings so that the person 'reading' them may know what the object is like.
(2) A knowledge of a large number of conventions which are a kind of shorthand, used to save time and to convey exact information.
(3) The ability to work very accurately with drawing instruments and to calculate by measuring the length of lines.
In order to become expert, it is necessary for the student to accept a good deal of training in practical geometry; for an engineering part is usually composed of a number of common solids, such as the cube and the cylinder.

Equipment

It is difficult to do satisfactory work in technical drawing without good instruments and equipment. Advice on the choice and care of instruments is best given by the teacher. All that it is necessary to say here is that, for the work in this book, pencil only is required and that the following equipment will suffice: compasses, about 130 mm long, preferably needle pointed; pencils HH H HB; drawing board, tee square, set squares, scales, protractor (see later); also, if possible, spring compasses for small circles, and dividers.

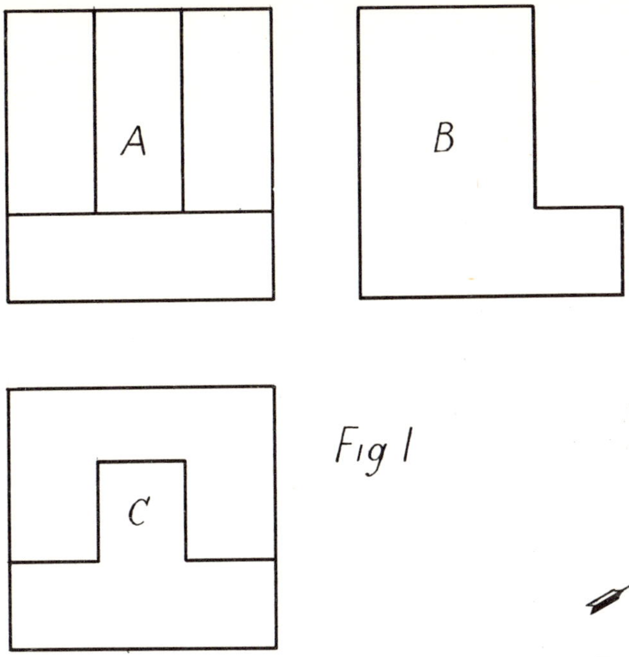

Fig 1

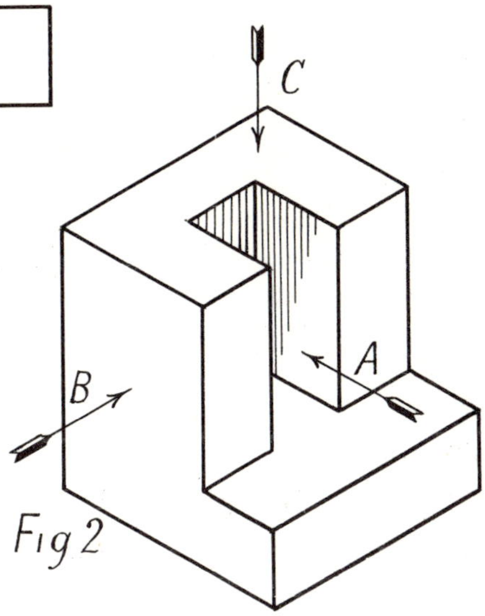

Fig 2

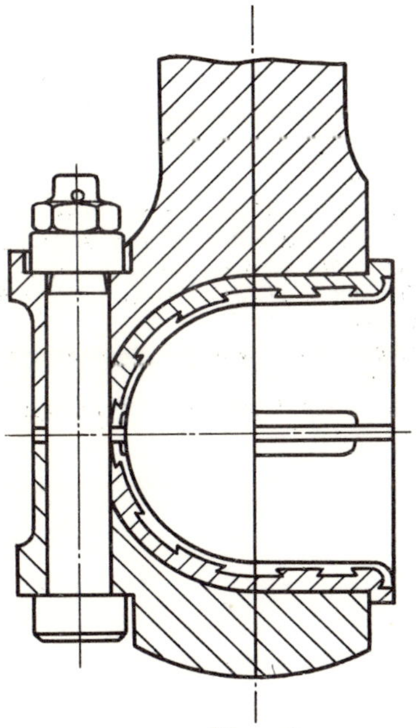
Fig 3

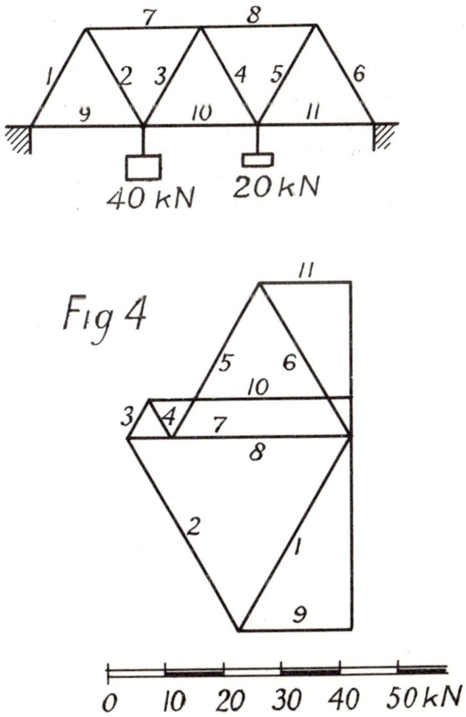

Fig 4

Straight Lines: Parallel, Perpendicular

By using a drawing board, tee square and set squares we may draw, without using geometrical constructions, lines which are parallel and perpendicular to each other. We may also draw lines which are inclined to these at various angles, as shown on the next page.

The drawing board size is a matter of choice among sizes which have been standardized. The A2 size is quite suitable for the exercises in this book. The essential things required of a drawing board are that it should be flat, that it should not warp, and that *one edge* should be straight. In practice all four edges are straight and at right angles to each other when the board is new, but this property ought not to be relied upon. Only one edge should be used, normally the one on the left if the user is right-handed.

The tee square has two important edges, one on the stock and one on the blade, straight and at right angles to each other. The stock is pressed against the straight edge of the drawing board and is held in this position when lines are drawn.

The drawing paper may be secured to the board by drawing pins (or thumb tacks), by strips of gummed paper, or by clips.

Parallel lines

Holding the tee square and pencil as in fig. 1, and keeping the stock of the tee square against the edge of the drawing board, we may draw lines which are all parallel to each other.

Perpendicular lines

Perpendicular lines may be drawn by using a set square with the tee square. *The set square* is simply a right-angled triangle. It is called a triangle in North America. Two set squares are in common use: one with angles of 90°, 60° and 30°; and the other with angles of 90°, 45° and 45°.

Using the left hand to ensure not only that the tee square is held against the edge of the board but that the set square is held against the edge of the blade, as in fig. 2, we may draw a succession of lines which will all be at right angles to those previously drawn and will therefore all be parallel.

Exercises

1. **Draw lines *of uniform thickness* at right angles to each other and spaced about 20 mm apart. Draw some of these lines accurately through points previously marked on your paper.**
2. **Draw lines at 45° to the horizontal, 5 mm apart and of uniform thickness.**
3. **Repeat Ex. 2 without measuring the distances between the lines.**
4. **Repeat Ex. 2 using each of the types of line shown on page 34.**

Fig 1

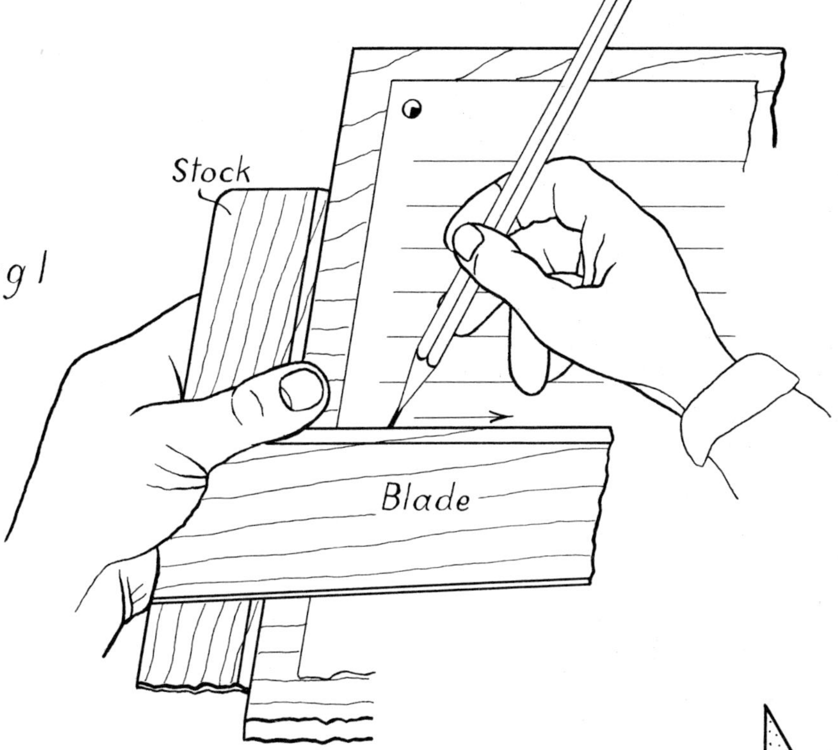

Fig 2

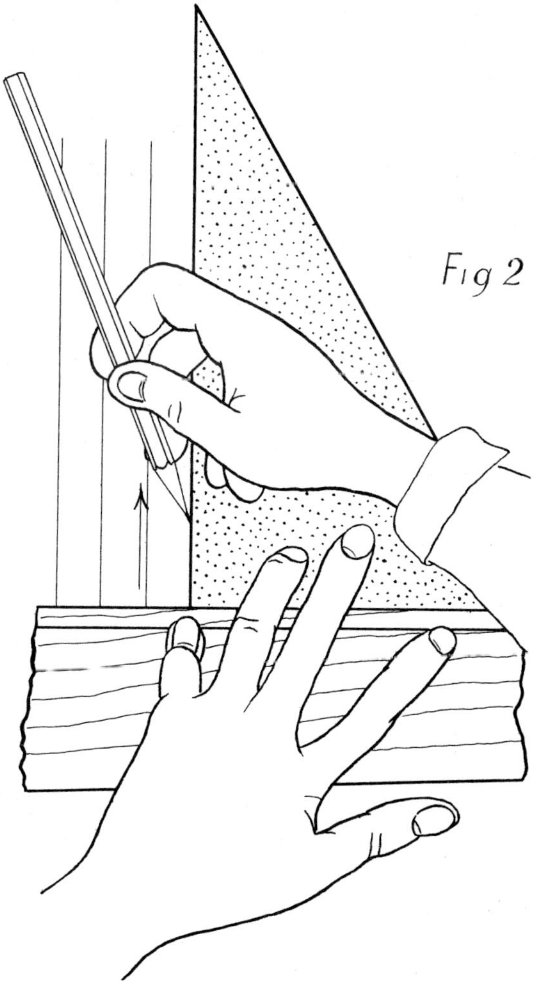

Straight Lines: Parallel, Perpendicular, Inclined

Lines which are parallel or perpendicular to given lines may be drawn by using simple geometrical constructions. These are dealt with in Part 2; they are of use for checking constructions. Here however we shall draw parallels and perpendiculars by using the tee square and a set square in combination. The constructions are therefore only as accurate as the equipment. We rely on the straightness of the working edges and the exactness of the angles of the set squares. These qualities must therefore be occasionally checked.

Parallels to a given line

Hold one of the edges of a set square against the blade of the tee square, and slide both set square and tee square so that another edge of the set square lies along the given line, as at fig. 1, I (dotted). Press the tee square on the drawing board to prevent movement and slide the set square along it to the required position for the parallel line; in the figure, at fig. 1, II, the set square is in position for a parallel to be drawn through the given point.

Perpendicular to a given line at a given point

Slide the longest edge of a set square along the blade of the tee square and move them, together, until another edge of the set square lies along the given line, as at fig. 2, I (dotted). Hold the tee square firmly in position on the drawing board and slide the set square along the blade until the third edge of the set square lies through the given point, as at fig. 2, II. We can now draw the perpendicular along this third edge of the set square.
An alternative method is shown in fig. 3.
Do not use the set square alone (i.e. without the tee square) to draw perpendiculars.

Lines at various angles to a given line

A horizontal line passing through a point P is shown in fig. 4. By using the tee square and each of the set squares we may draw lines through P inclined at angles of 30°, 45° and 60° to this horizontal, as will be clear from the figure. A complete series of inclined lines may be drawn as shown in fig. 5.
By using the two set squares *in combination*, as shown in fig. 6, we may draw lines inclined at 15° and 75° to the horizontal—or to the vertical. By reversing the position of the 30°–60° set square, other pairs of inclined lines may be drawn. All the lines shown in fig. 7 may be drawn mechanically; it will be seen that they divide the circle into 24 equal parts.

Exercises

1. Draw any line and mark several points outside the line. Through these points draw lines parallel to the original line.
2. Mark a few points on the lines you have drawn and through these points draw lines perpendicular to the original lines.
3. Draw a circle of diameter about 100 mm. Using only the tee square and set squares subdivide the circle as in fig. 7. Check the accuracy of your work by comparing the lengths along the circumference.

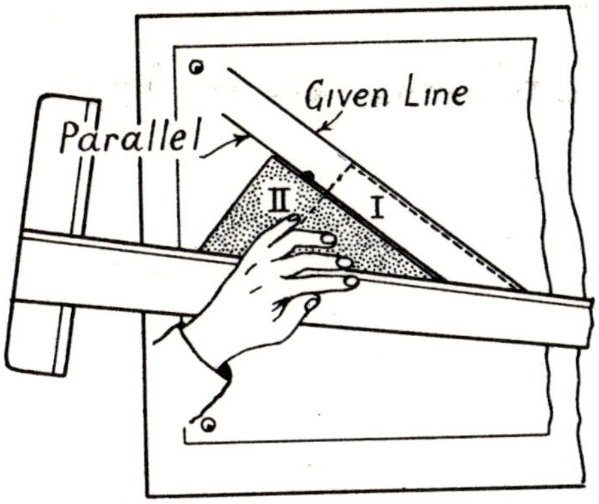

Fig 1

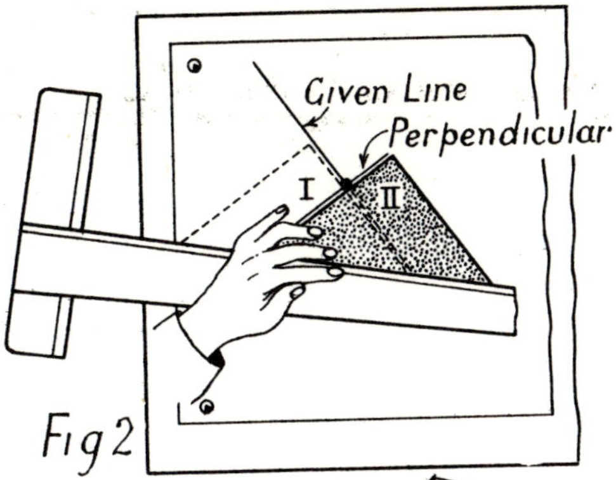

Fig 2

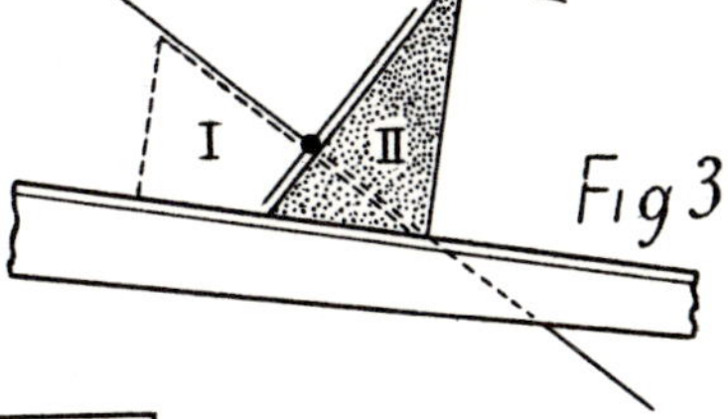

Fig 3

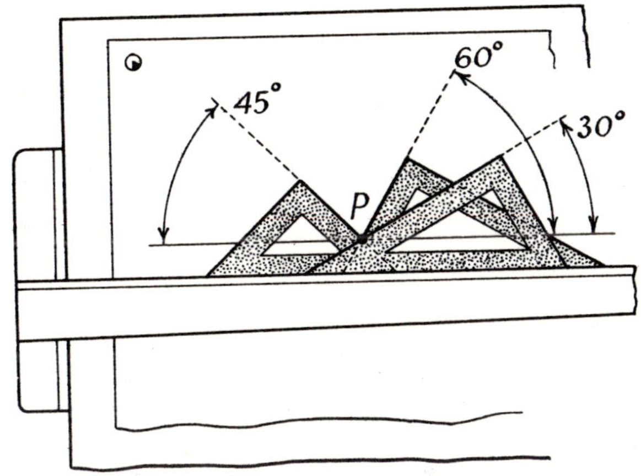

Fig 4

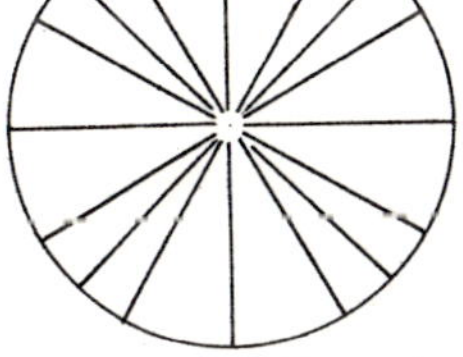

Fig 5

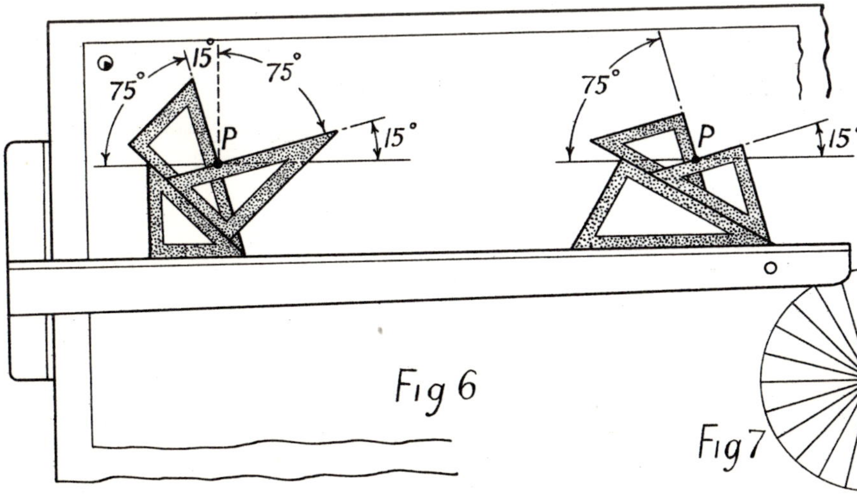

Fig 6

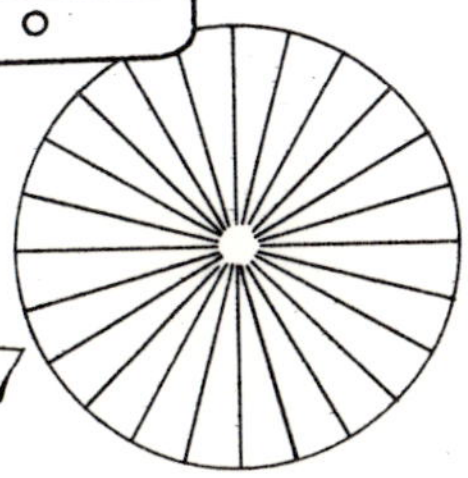

Fig 7

The Use of the Compasses

The needle point of the compasses should be sharp, otherwise it may slip when the instrument is used. Again, the pencil tip should be *just short* of the needle point when they are brought together, as in fig. 1. If the pencil point projects beyond the needle point it will be found that small circles are difficult to draw.

Because the pressure applied to the compass lead is usually less than that applied to the lead of the hand pencil it is good to have the compass lead a little softer than that of the hand pencil. The combination giving the most satisfactory results is: H for compass and 2H for hand pencil.

Very common operations in technical drawing are the construction of circular arcs in contact with other arcs or touching straight lines tangentially. Geometrical constructions are available (and will be considered later) for the positioning of the centres of the required arcs, but the experienced draughtsman usually locates these centres by trial. Consider fig. 2 (which is to scale): given circles have radii of 50 mm and 40 mm, centres 125 mm. The circles are to be joined by arcs of 75 mm and 25 mm radius. We can easily find the positions of the centres of these arcs by adding together the radii and drawing the short intersecting arcs, but it will be found that this construction must be very carefully carried out.

Common drawing errors

The most common errors among beginners are: the failure to draw lines of uniform thickness; the over-running of lines which are to meet; and the bad joining of straight lines and arcs of circles, or of arcs with other arcs. These are illustrated in fig. 3; in addition to the faults named, there are also the following: dividing marks unnecessarily heavy; construction and centre lines too thick; dotted lines irregular; and the construction for the centres of the arcs too obvious. Fig. 4 shows the correct drawing. Where straight lines join arcs they should do so tangentially and *the arcs should always be drawn first*—it is easier to draw straight lines to join curves than vice versa.

Until the student is sufficiently practiced in drawing lines of a uniform thickness, reference to the types of lines on page 34 should be made repeatedly. The exercises listed below should be made to the thickness of line (*a*).

Exercises

1. Use the construction, and make the drawing, shown in fig. 2, full size.
2. Draw a horizontal line 252 mm long, as in fig. 5, and construct the continuous curve shown, with radii descending from 36 mm to 6 mm by 6 mm steps. Make all parts of the curve uniformly thick.
3. Draw the rectangle in fig. 6, full size. Insert radii of 25 mm at the corners, using the method shown, and thicken in, as in the figure.
4. Draw the parallelogram in fig. 7, full size. Using the construction indicated, insert radii of 20 mm at the corners and thicken in, as shown.

Fig 1

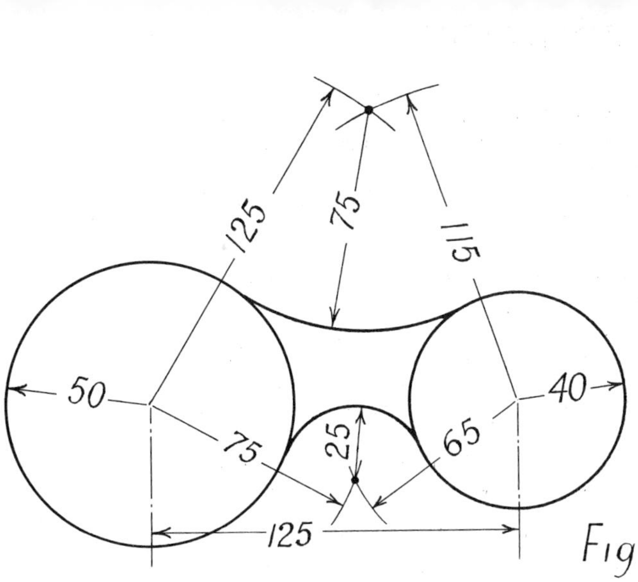

Fig 2

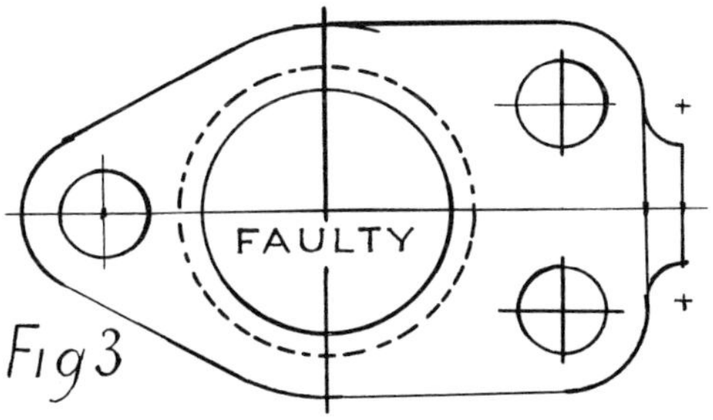

Fig 3

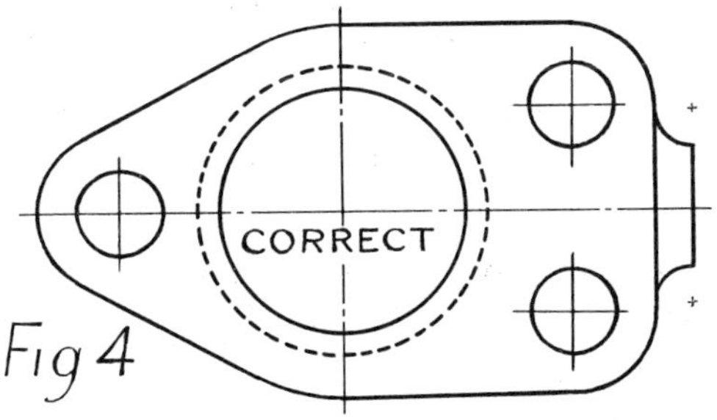

Fig 4

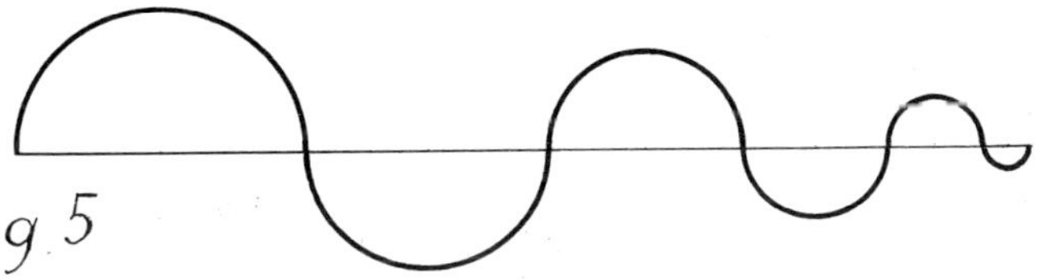

Fig 5

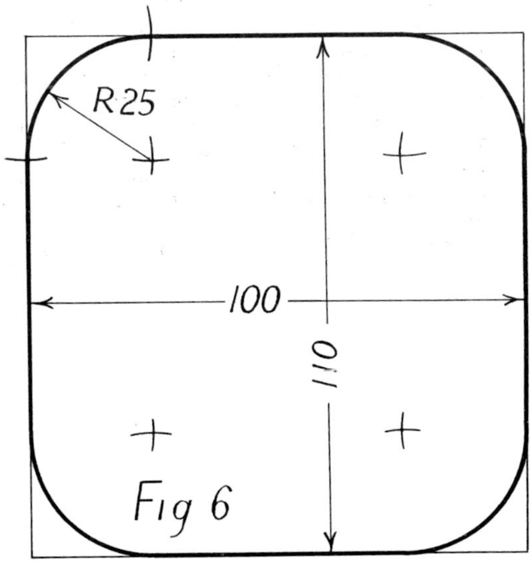

Fig 6

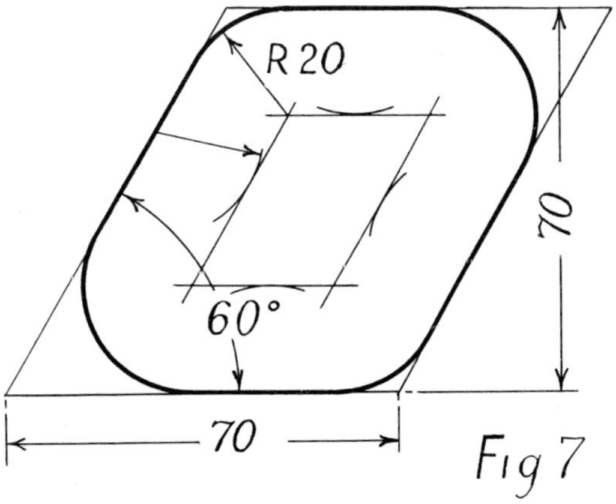

Fig 7

Division of Lines

A very common drawing operation is the division of a given straight line into a number of equal parts. This may of course be done by trial, using the dividers, but this is a method which is to be deprecated because it is time-consuming and had little or no geometrical construction.

The usual method however is a direct one, requiring the use of dividers (or scale), tee square, and finely pointed pencil. Although the operation is a simple one, very accurate working is essential if the results are to be of any value. Inaccurate work is often quite obvious to the naked eye, and the student should criticise his own work severely. Accuracy is achieved: (*a*) by using fine lines (see exercise 1 below): and (*b*) by exactly locating the points of intersection of lines.

Let the given line be AB, fig. 1, and suppose that it is to be divided into five equal parts. Proceed as follows: draw another straight line through A, fig. 2, at approximately 45° and, using either the dividers or a good scale, mark off along this line from A five equal lengths, giving points 1, 2, 3, 4 and 5. Arrange the tee square and set square as in fig. 3, so that an edge of the set square passes through points 5 and B. Slide the set square along the tee square and draw parallels through 4, 3, 2 and 1 to give the required points on AB.

Accurate working is aided if the two angles marked in fig. 3 are large. Consider the two pairs of very thick lines in fig. 4. It will be seen that it is much easier to locate the point of intersection in (*a*) than in (*b*) where the area of intersection is broad. Again, it is possible to err in drawing the parallels from points 5 to 1 not only in slightly missing the points but in failing to keep the pencil at a constant angle with the paper; if the pencil is held at (*a*), fig. 5, to begin with and as at (*b*) finally, the error of position for the point in AB may be appreciable.

Technical drawings should preferably show an object *full size;* that is to say, a part which has a dimension of 50 mm should be represented by a distance of 50 mm on the drawing.

There are however many parts, or assemblies, which are so large that a full-size drawing would be impracticable to make or use. For example, a drawing of a building, a ship, or an aeroplane could not very well be made full size. In this situation we select a distance, usually a millimetre, to represent 2, 5, 10 mm, etc. Many drawings are made larger than full size, e.g. the parts of a watch or the components of a radio.

Exercises

1. Draw two vertical lines 25 mm apart and, by using a sharp H pencil and the tee square and set square as in fig. 6, draw as many distinct and separate lines as you can, counting as you draw them. Ensure that the set square does not slip.

   You may be able to draw about 100 lines which are clear and separated by the white of the paper. If so, then each line is about one eighth of a millimetre thick, and $\frac{1}{8}$ mm = 0·125 mm. The accuracy of any drawing or graphical solution is governed to this extent by the relative thickness of the lines.

2. Take a line four times as long as AB in fig. 1. Divide it into 7 equal parts. Compare the lengths of the parts, using your dividers, and find where you have permitted an error to creep in.

   Now subdivide the first of the one-seventh parts into 10 equal divisions.

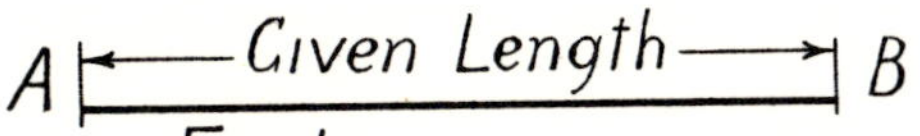

Fig 1

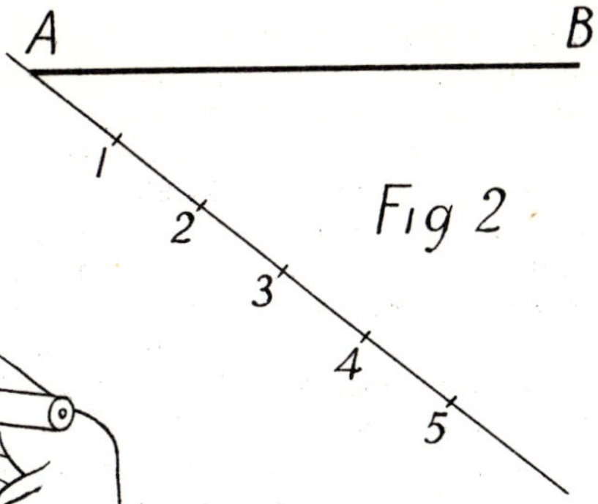

Fig 2

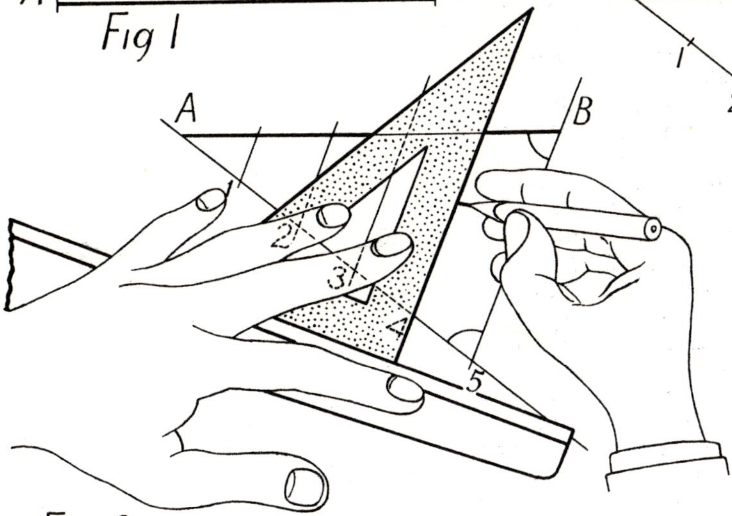

Fig 3

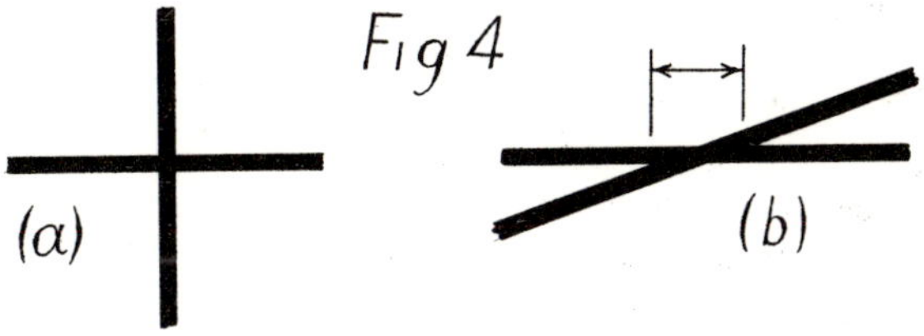

Fig 4

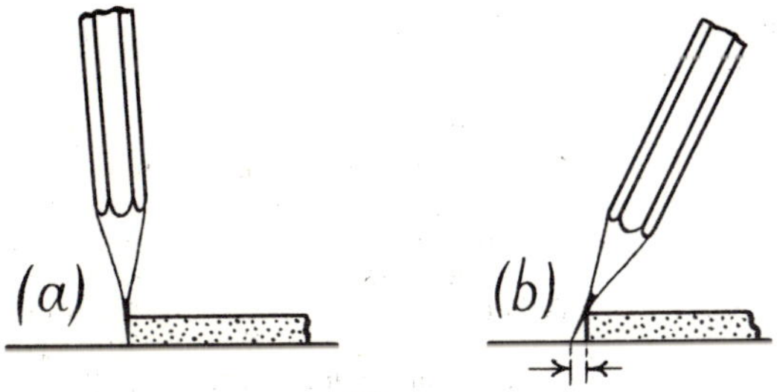

Fig 5

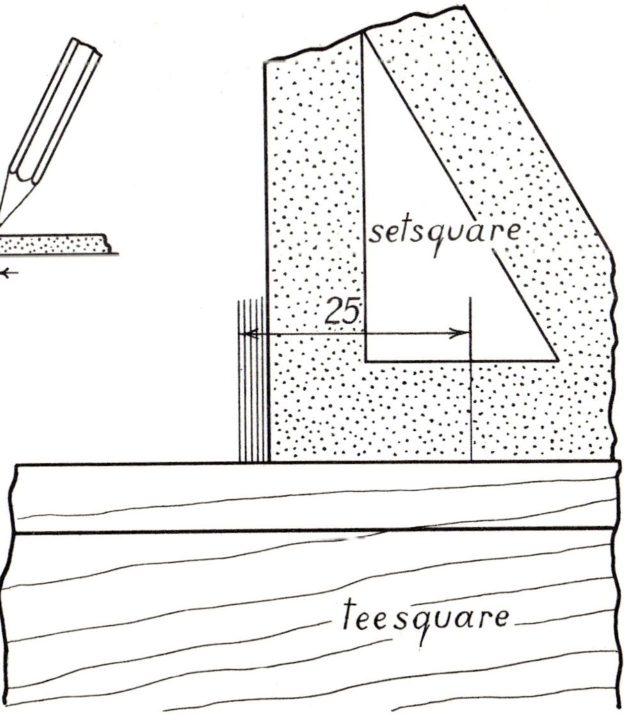

Fig 6

Measurement of Angles

If we have an arm free to rotate about one end, as in fig. 1, then in moving around once, the arm will trace out 360°, passing through four right angles.

We have seen that by using the tee square and set squares we can obtain angles of 15°, 30°, 45°, 60°, 75°, 90°, and so on up to 360°, separated by the common interval of 15°, as shown in fig. 2. In this construction we are relying upon the accuracy of the angles of our set squares.

If we require other angular measurements, we can subdivide any of the 15° angles; we do this by drawing a circular arc, fig. 4, about the point of intersection of the lines enclosing the angle, and dividing the portion of the arc between the lines, *by trial*, into 15 equal parts, using dividers. This is not an easy task and the larger the radius of the arc the more exact is likely to be the value of the degree. We are now in a position to make angular measurements in degrees.

All these angles are made available to us in the protractor, fig. 3, which in its simplest form has a range of 180°. In the figure, only one 10° interval has been divided into degrees. The base line on the protractor must always be used, not the bottom edge. The student will find that great care must be exercised in correctly placing the centre of the protractor and its base line in position, and in marking angles, if accurate work is to result. Better protractors are made with a vernier device to enable angles to be marked to one tenth of a degree.

Transfer of angles

It is often necessary to mark off, at a point in a line, an angle equal to a given angle. This is simply done, without the use of any measuring instrument such as the protractor, by using the compasses. Let the given angle be *A*, fig. 5 (I). We are to mark off an angle equal to *A*, from P on the line PQ, fig. 5 (II).

Draw any arc BC, in fig. 5 (I), cutting the lines including the angle. With the same radius, but with P as centre, draw the arc B_1C_1, fig. 5 (II), and obtain the point of intersection C_1. The line PC_1 gives the required angle.

Exercises

1. Construct, preferably on transparent paper of good quality, a semi-circular protractor as in fig. 3 having a diameter of 150 mm. Limit the number of 1 degree divisions to the first 15 degrees.
2. Set off 6 lines at 60 degree intervals, as in fig. 6, and draw concentric circles of radii 40 and 60 mm. Insert seven circles of 20 mm radius as in fig. 7. If your circles do not touch each other and the outer circle tangentially, that is, without intersecting, locate the source of your errors.

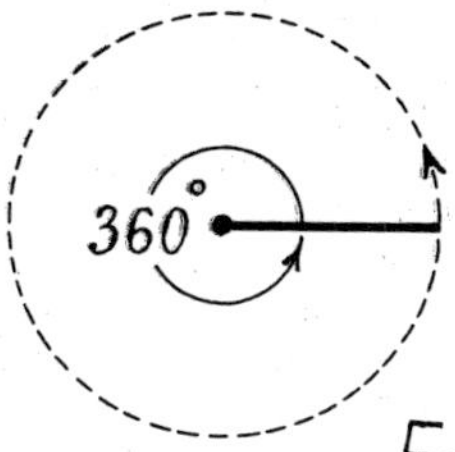

Fig 1

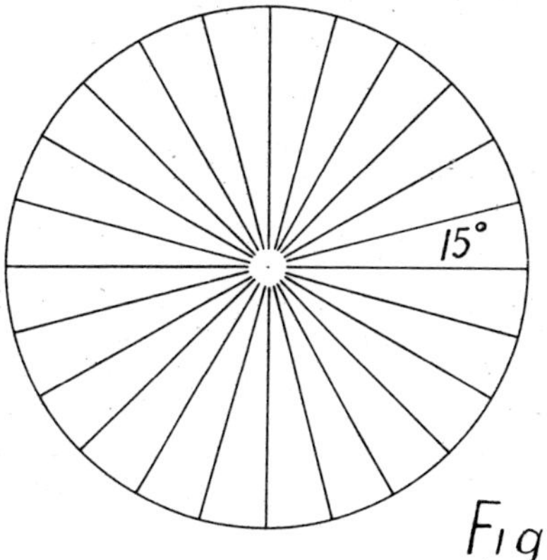

Fig 2

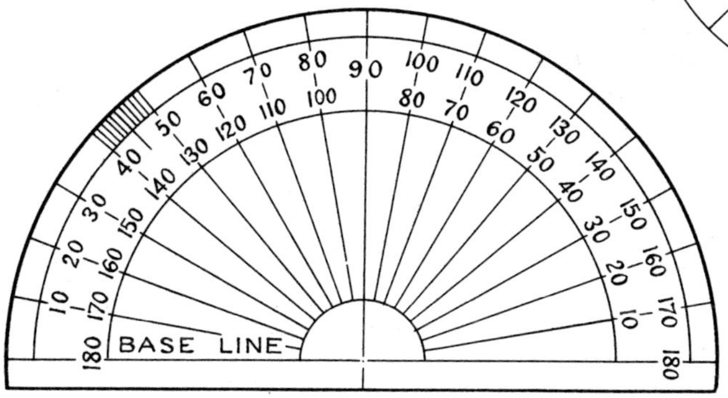

Fig 3

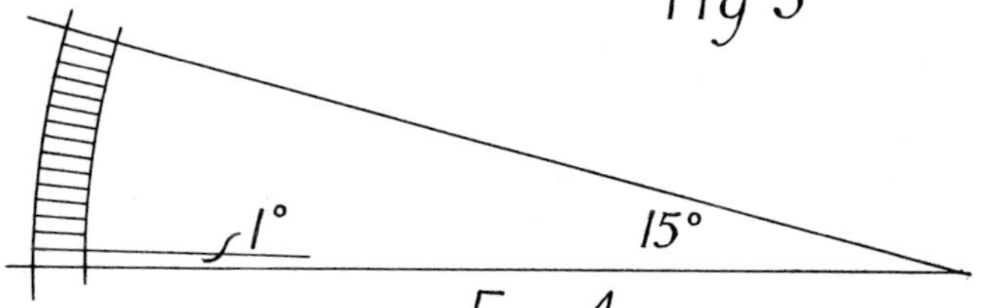

Fig 4

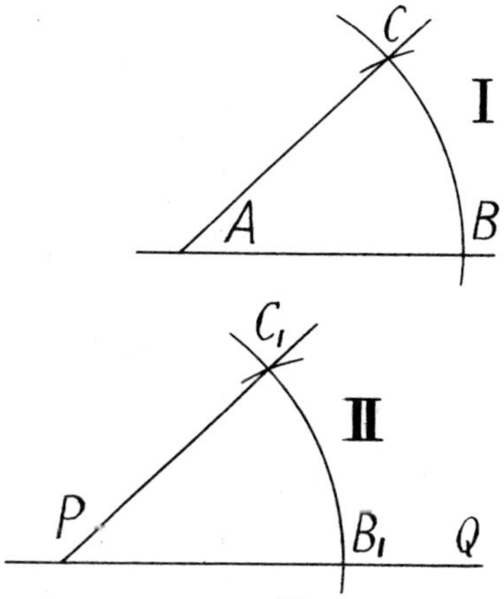

Fig 5

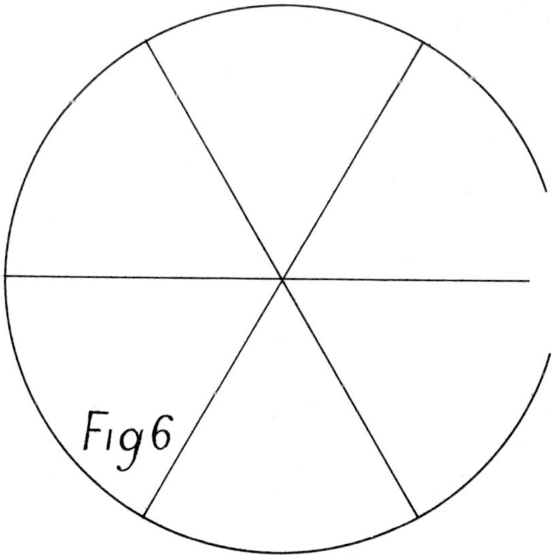

Fig 6

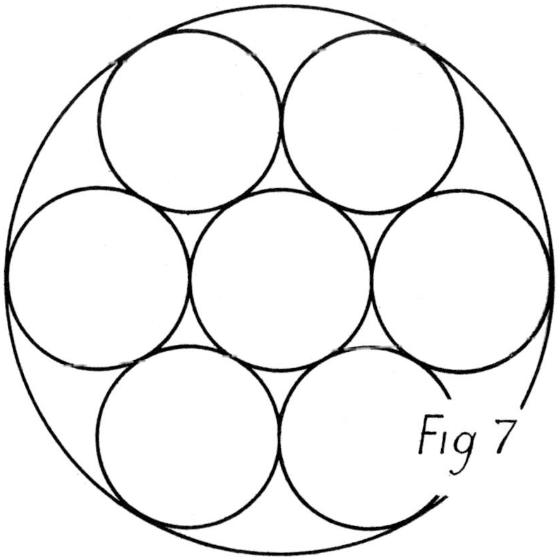

Fig 7

1. Six lines are drawn in fig. 1, some to scale and some full size. Measure and record their lengths.
   A. Full size.
   B. Full size.
   C. Scale. $^1/_5$ full size.
   D. Scale, $^1/_{20}$ full size.
   E. Scale, $^1/_{10}$ full size.
   F. Scale, 5 mm to 1 m.

2. Draw a rectangle 135 mm by 75 mm as in fig. 2. Divide the long side into three equal parts and the short side into five equal parts. Draw the horizontals and verticals. The drawing represents part of a brick wall. If the bricks measure 225 mm by 75 mm, what is the scale of your drawing? (Answer: 1:5).
3. Repeat the drawing of fig. 2 but show 10 mm of mortar between the bricks. What is now the overall size of the brickwork? (Answer: 695 mm by 415 mm).
4. A section revealed by a sawcut through a moulding is shown in fig. 3. Make a full-size drawing of the moulding.
5. Make a full-size drawing of the section of a weather board for a door shown in fig. 4.
6. Fig. 5 shows the shape of a masonry cornice which overhangs the top of a wall. The cornice is extended on the right but this part is not shown. Make a full-size drawing of the part of the cornice given in fig. 5.

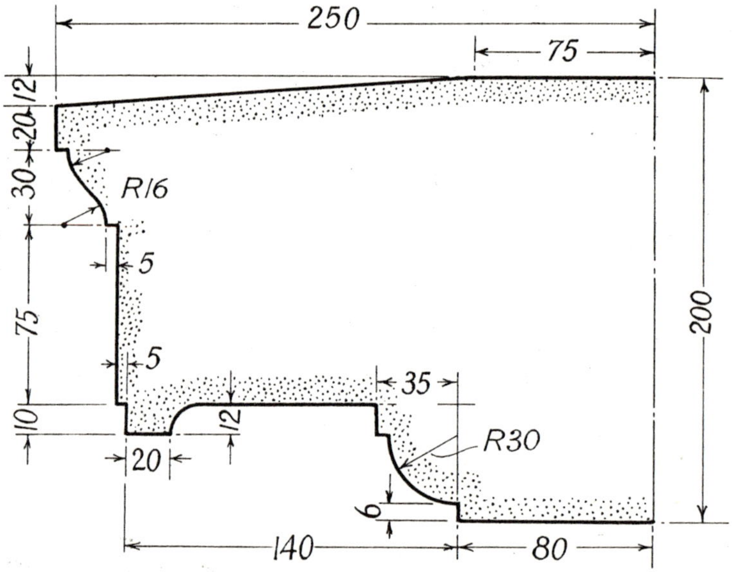

Fig 5

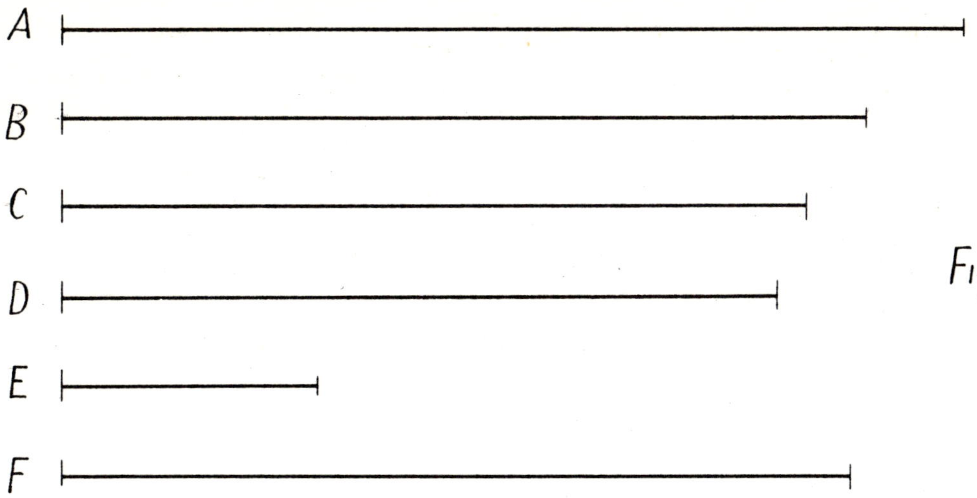

Fig 1

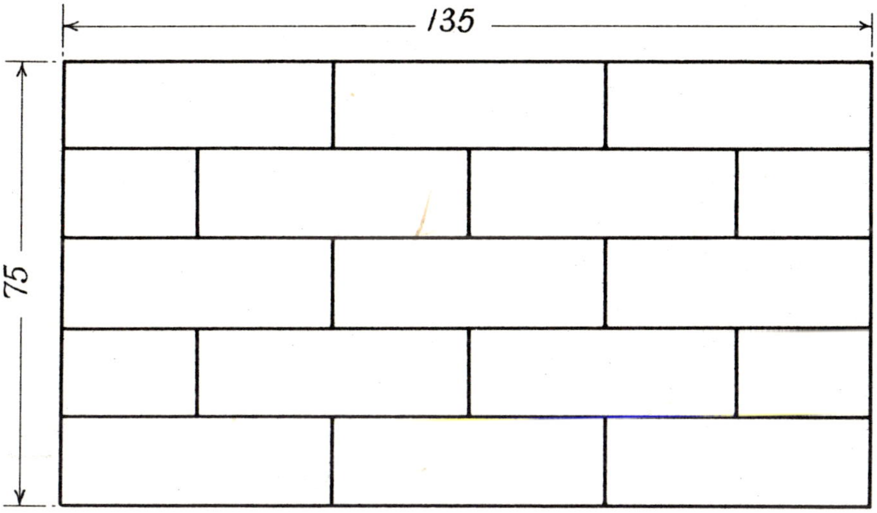

Fig 2

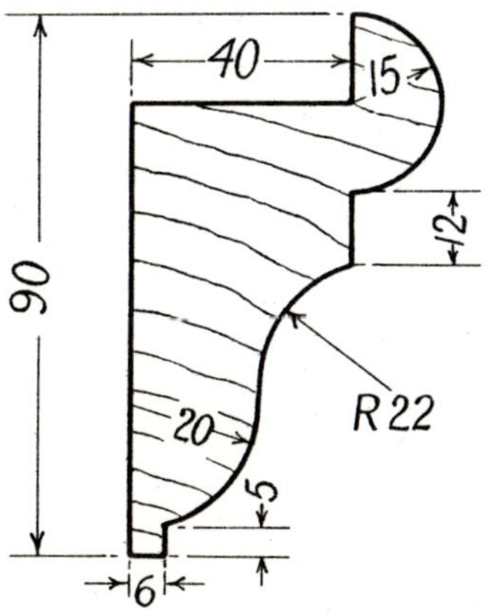

Fig 3

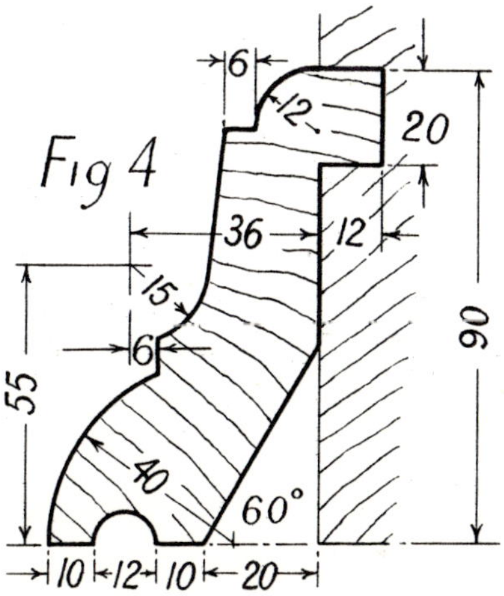

Fig 4

Drawings of Solids. The Cube

A solid object has dimensions in three directions: it has length, breadth and depth. A drawing can have only two dimensions: length and breadth; and it is not easy to describe, in these two dimensions, an article which has three dimensions, unless the drawing is like a photograph and creates the illusion of depth. The view in fig. 1 is of this kind, and it does convey the effect of solidity. Such views however are not used in normal engineering practice because they take long to prepare and are not always easy to dimension. The usual drawing is one which gives the outline of each main face of the object, together with the principal features of the face. From these flat views the observer has to conjure up a mental picture of the article they represent. This is always a little difficult; sometimes it is very difficult. In order 'to think in three dimensions' the student must constantly exercise himself in 'reading drawings', and a number of test examples are included in the pages which follow.

In times long ago it was customary to have models made, particularly of structures or fortifications, to enable designs and shapes to be understood and appreciated. Sculptors such as Michael Angelo were so employed; and it was not until about 1800, in the time of Napoleon, that Gaspard Monge devised the system of making drawings which we now use.

Solid shapes

Most engineering parts are composed of regular solids which are common in geometry. Consider the simple bracket shown in fig. 1. It carries a shaft, having a tapered end, secured by a nut. The various types of solids which go to make up the whole are shown separately. It will be seen that they are simple in form, having either flat sides, or surfaces which can be obtained by turning on a lathe. Usually the geometrical solids are linked by 'fillets' such as that at F.

Because technical drawings are composed so frequently of such combinations of geometrical shapes, it is essential that the student should have a thorough training in the drawing of the commoner solids in varying positions.

The cube and rectangular prism

The student should handle these geometrical solids and he should have no difficulty in actually making them. A supply of squared paper is required with 5 mm squares. If this not is available, 5 mm squares may be drawn on plain paper. Some adhesive tape is also required.

The cube

Draw the shape shown in fig. 2; cut it out and fold upwards along the thick lines. Apply adhesive tape and construct the cube. The pictorial view shows three faces only and these are numbered in fig. 2.

The shape cut out is called the *development* of the cube. It gives the surface areas of the solid. Notice that the cube has six equal square faces and twelve equal edges.

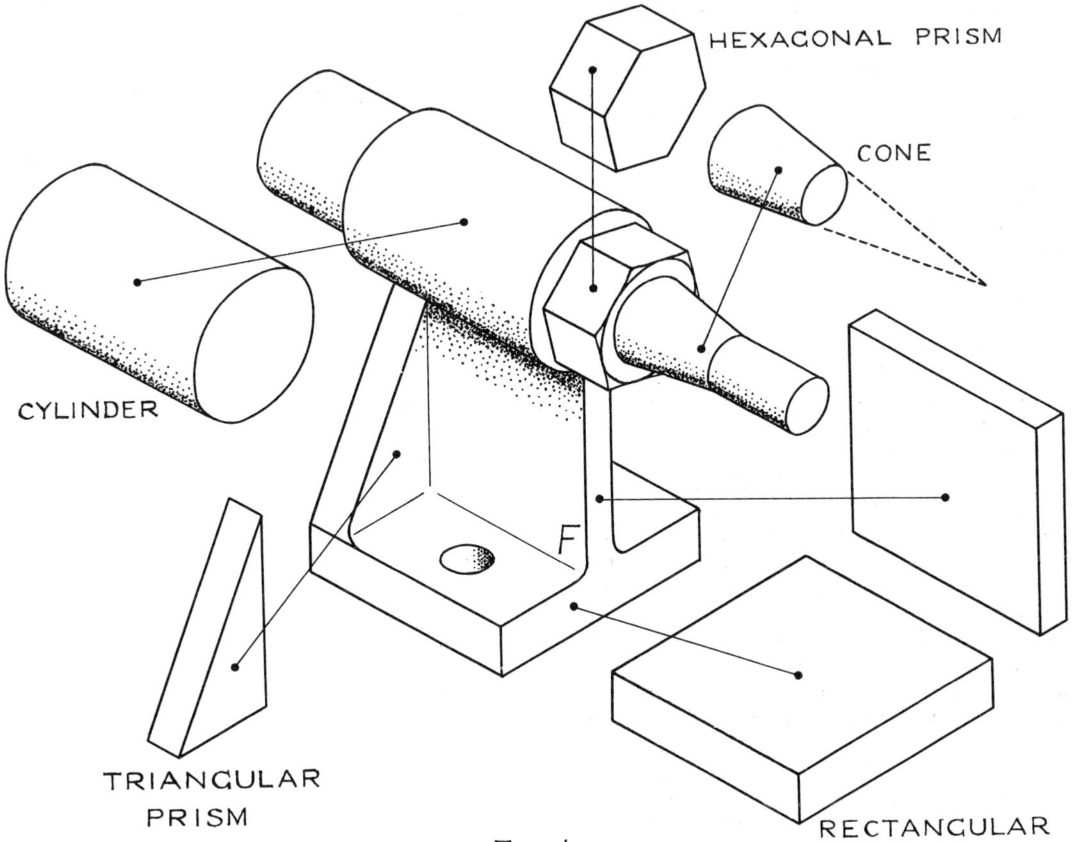
HEXAGONAL PRISM
CONE
CYLINDER
F
TRIANGULAR PRISM
RECTANGULAR PRISMS

Fig 1

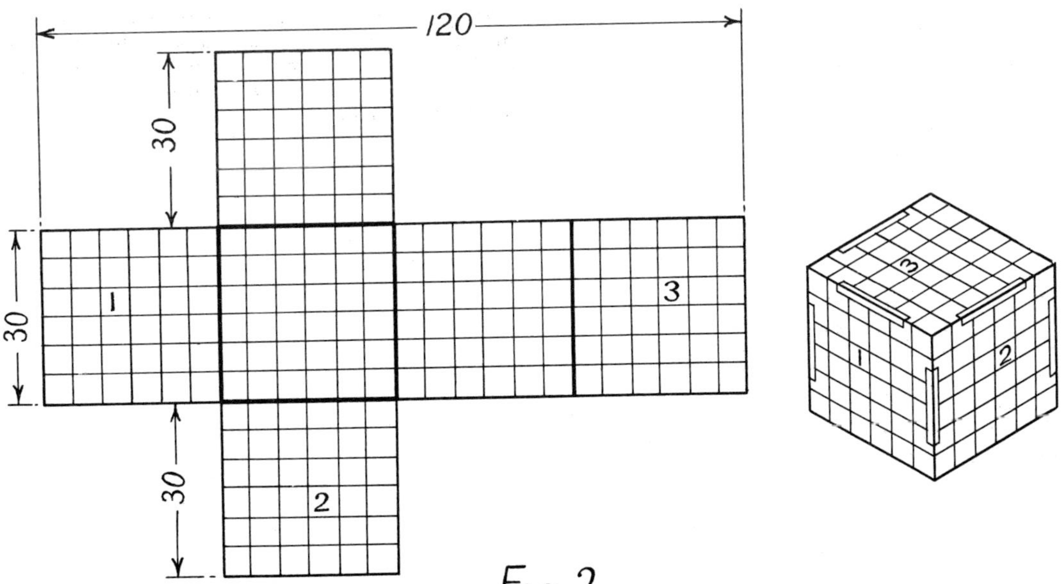
120
30
30
30
1
2
3
3
1
2

Fig 2

The Rectangular Prism

This example should be treated in the same way as the cube on the previous page. Cut out the shape shown in fig. 1, using squared paper (5 mm squares) and fold upwards along the thick lines to form the box-like form shown in the pictorial view. The slightly distorted appearance of the pictorial view will be obvious. It is unavoidable in this kind of projection and is referred to later. Notice that the solid has six rectangular faces, equal in pairs, and twelve edges, of which groups of four are equal.

Composite solid. Projections, or descriptive views

Place the cube over the prism, arranging the cube centrally at an end, as in figs. 2 and 3. Hold the object on the palm of the left hand and raise it almost to eye level, fig. 2. Make a drawing of the object viewed in this position, that is, a view in the direction of the arrow A, fig. 3. This drawing is shown in fig. 5.

Now mark on the object all the edges which the lines in fig. 5 represent. This is shown by the ticks in fig. 4—although, strictly speaking, the lines *at the back* are not shown in fig. 5. It will be clear that the edges marked O in fig. 4 are not represented by any lines in fig. 5, only by points; hence we do not know the lengths of these edges. We could not make the object from fig. 5 alone. One more drawing is required, and this could be either that given by a top view or a side view, but not an underneath view—because the larger base would obscure the smaller cube.

Draw therefore a side view in the direction of the arrow B. This is shown in fig. 6; notice that it gives the lengths of all the lines marked O.

We should naturally wish to place these two views, figs. 5 and 6, in line: but should we arrange fig. 6 on the *right* of fig. 5, as shown, or on the *left?* Where would *you* rather place it? It really does not matter where it is placed; *but it does matter that we know which view of the object the drawing represents.* The importance of this matter is emphasised by the example dealt with fully on the next page.

Exercise

Arrange the composite figure not as in fig. 3 but with the cube moved away from the short edge of the prism so that it is at the same distance from three edges. Draw four views (front, side, top, underneath) and arrange them in the positions you think most reasonable.

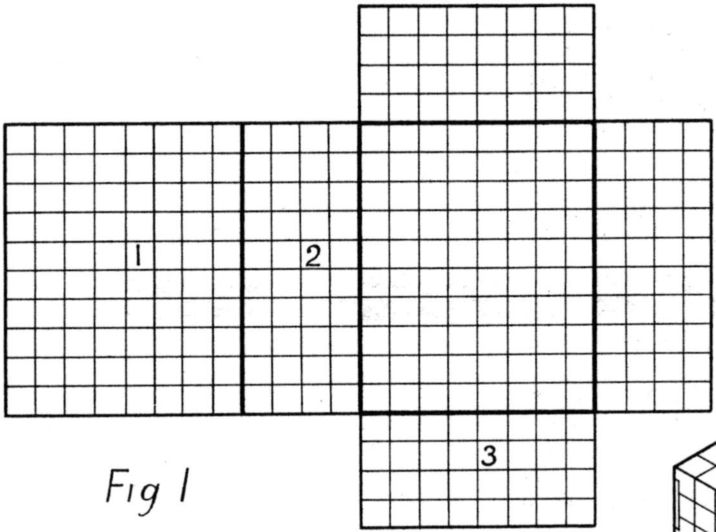

Fig 1

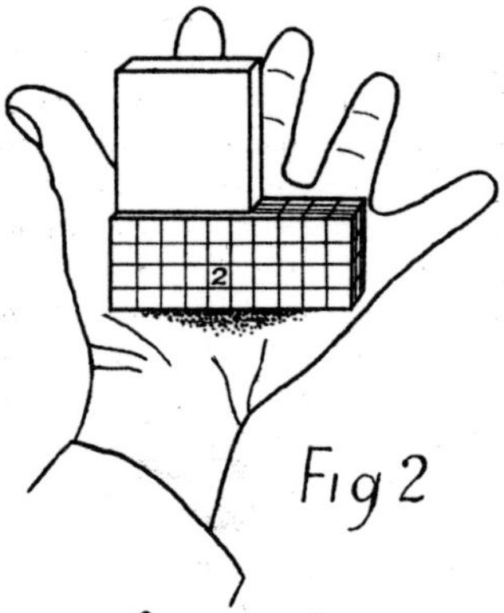

Fig 2

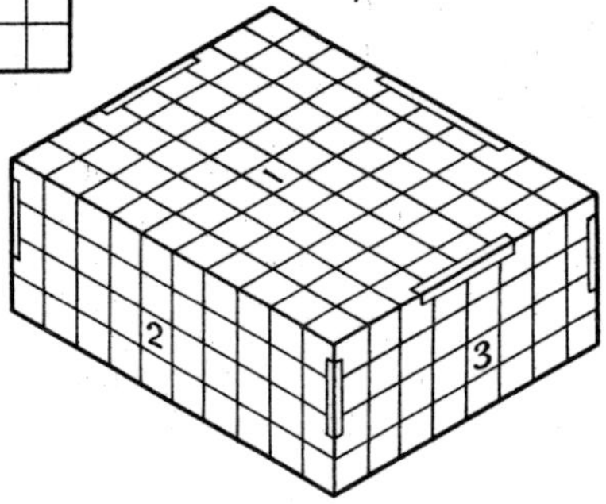

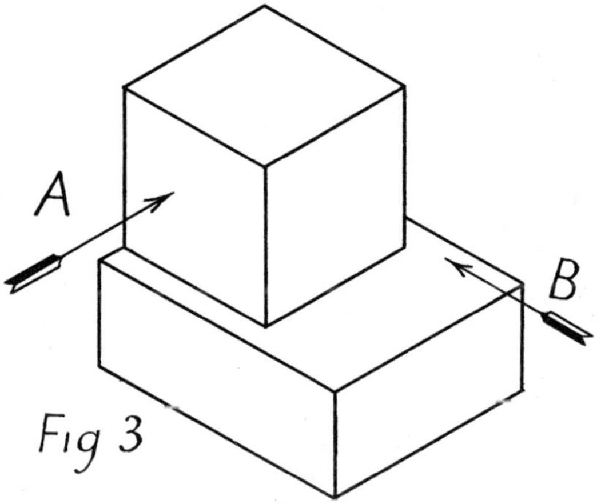

Fig 3

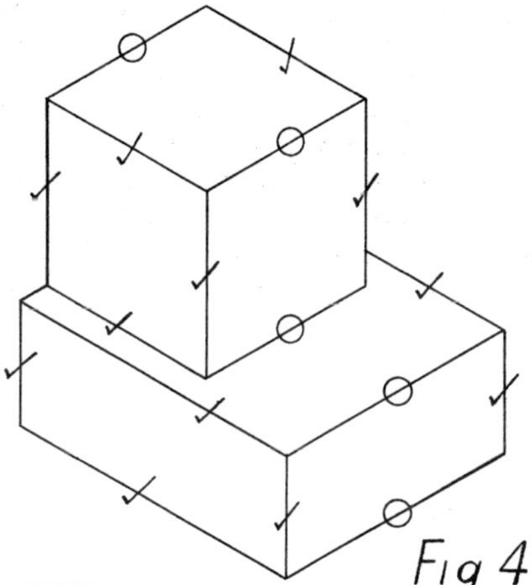

Fig 4

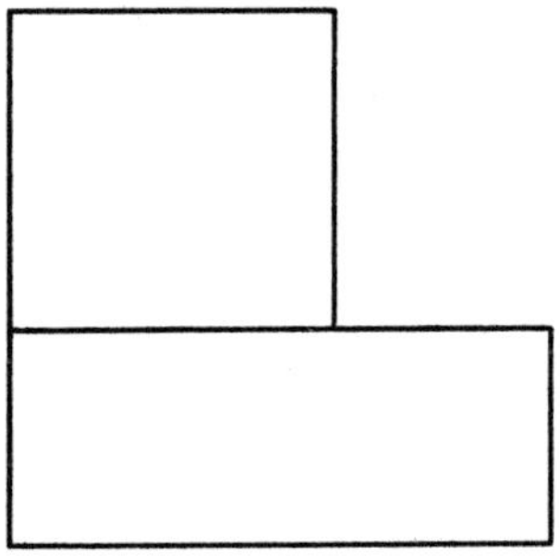

Fig 5

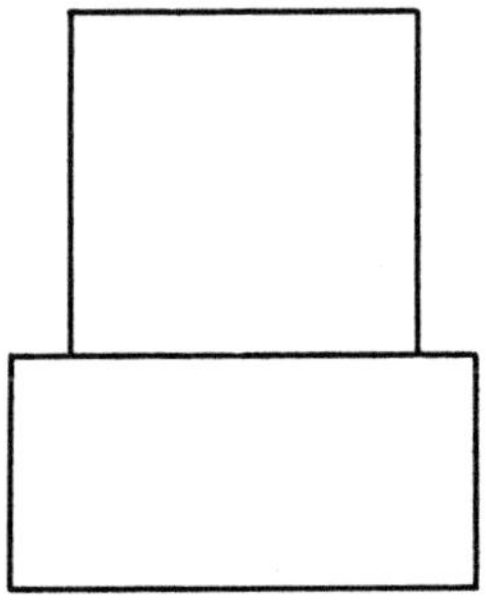

Fig 6

Importance of Labelling Views

The need for labelling some of the views in the drawing of an object will now be illustrated. Two views of a composite object are shown in figs. 1 and 2. The cube and the prism are again used, but this time the cube is displaced from the centre line of the prism and is nearer one side than the other. The two drawings do however completely describe the object.

Evidently, fig. 2 can be taken either as a view of the object in the direction of the arrow A, or as a view of the object in the direction of the arrow B. As there is no statement as to which it is, we are equally entitled to assume the one as the other.

Suppose, now, that we are to make the object in wood in the workshop; and suppose first, that we regard fig. 2 as a view of fig. 1 in the direction of the arrow A. Then the object we should make would be that shown in fig. 3.

Now suppose that we regard fig. 2 as a view of fig. 1 in the reverse direction, that shown by the arrow B. Then the object we should make would be that shown in fig. 4.

Consider figs. 3 and 4 carefully. They do not represent the same object at all. You cannot get one by turning the other around. If we require fig. 3, and an object as in fig. 4 is made, then it would be useless.

This kind of confusion is always likely to arise in dealing with unsymmetrical or 'handed' objects. Even when an additional view is provided which would settle the matter (in this case a top view) the craftsman using the drawing is confused if the arrangement is not that to which he has been used.

It may rightly be asked —*'Surely, in such an important matter, why cannot the engineering world agree on a common usage which would prevent any possibility of misunderstanding?'* To this question the answer is, that the engineering world, for reasons which will be discussed in the next part, is still divided on the matter, Europe favouring one method, North America the other, and the United Kingdom using both.

To prevent misunderstanding therefore, views should be labelled whenever there is the possibility of confusion. If the student is in doubt about the matter he should label! Suitable words are: 'View in the direction of the arrow A', or simply 'View on A'.

Exercise

Using the sizes from the previous pages, make composite objects corresponding to figs. 3 and 4 and satisfy yourself that they arise from using the alternative interpretations of fig. 2.

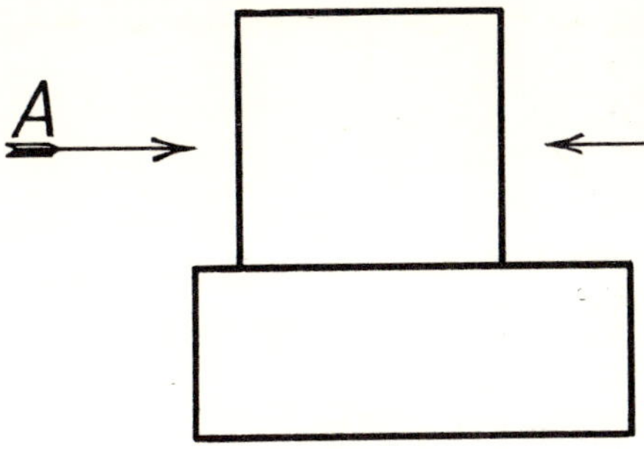

Fig 1

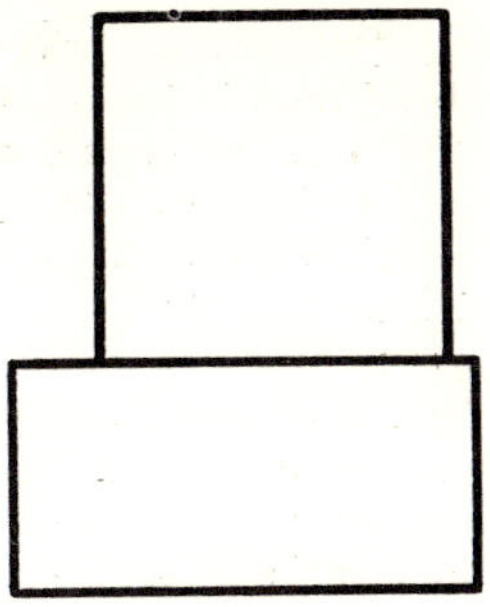

Fig 2

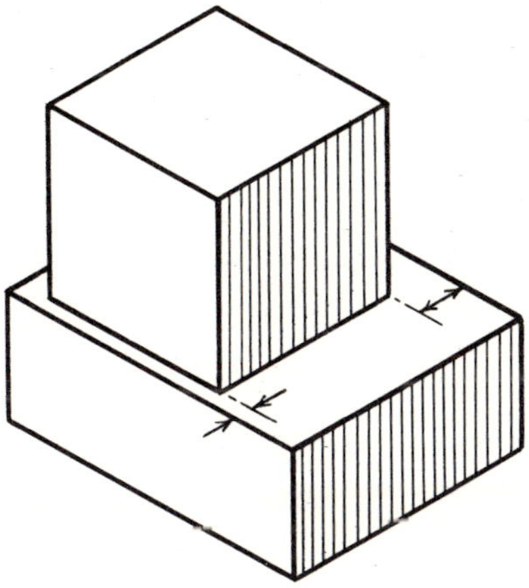

Fig 3

RESULTING OBJECT IF FIG. 2 IS TAKEN AS A VIEW IN DIRECTION A

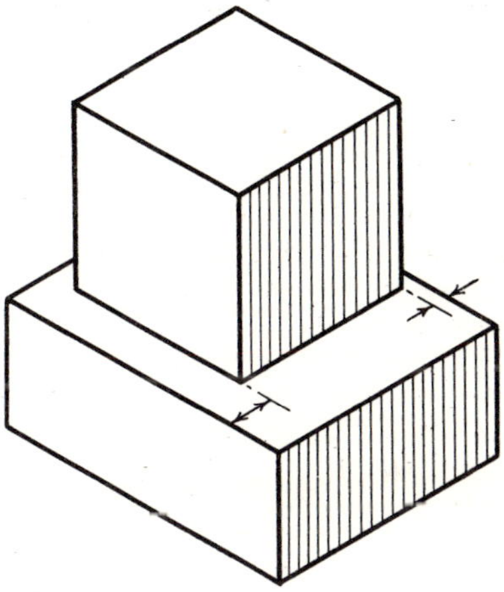

Fig 4

RESULTING OBJECT IF FIG. 2 IS TAKEN AS A VIEW IN DIRECTION B

Three-view Arrangements

The difficulty in deciding on the meaning of a two-view drawing is often cleared away by the provision of an additional view. Such three-view drawings are commonly employed, and we must now consider the best arrangement of the views. For convenience, let us again use the composite solid, fig. 1, in which the cube has been arranged to one side on the prism.
The drawing in thick lines, fig. 2, represents the view in the direction of the arrow A, fig. 1.
Let us first agree to place the other two views *as near as possible* to the faces which they represent. This means that the view in the direction of the arrow B is to be placed on the right of the view of A, and that the top view, in the direction of the arrow C, is to be placed *above* the view of A. This system, shown in fig. 2, is consistent, and is one which is in use and is known as third angle projection.*
But we could equally well agree to place the other two views on the sides *remote* from the faces they represent, as in fig. 3. This also is a consistent system and is one which is used and is known as first angle projection.*
Whichever arrangement we choose, the views should normally be labelled.
A mixture of the two systems as shown in fig. 4 is not acceptable in any circumstances.
The three views should be drawn *in line with each other*, as shown by the projection lines in fig. 2.
There are other considerations which enter into the placing of views but a discussion of these is deferred.
It should be noted that, in a three-view drawing, the same dimension will never appear in two views.
Very often the views are given names—for example: front elevation, side elevation, and plan. These terms have come to us from the architect, and they apply well enough to buildings. But where an engineering article has no obvious front or side, and where an underneath view is sometimes required, it seems undesirable to use these terms, but rather to describe the views by means of letters and arrows. Again, the terms length, breadth and height apply to drawings of buildings; but width, thickness and depth are sometimes more applicable to the engineering part.
For many years the British Standards Institution has been responsible for establishing standards of practice which British industries, colleges and schools are recommended to adopt. In this way quality and safety standards are established, and certain articles, machines, etc., are manufactured to standards of interchangeability. The specification known as BS 308 sets out the recommendations for 'Engineering Drawing Practice'. It is adopted throughout industry and in educational establishments in the making of engineering drawings. It has been modified several times to incorporate features of European and American practice deemed to be applicable and desirable. Thus engineering drawing is not only the language of the British engineer, but is becoming an international language.
Fig. 7 shows the "Types of Lines" which BS 308 recommends should be used for drawing purposes.

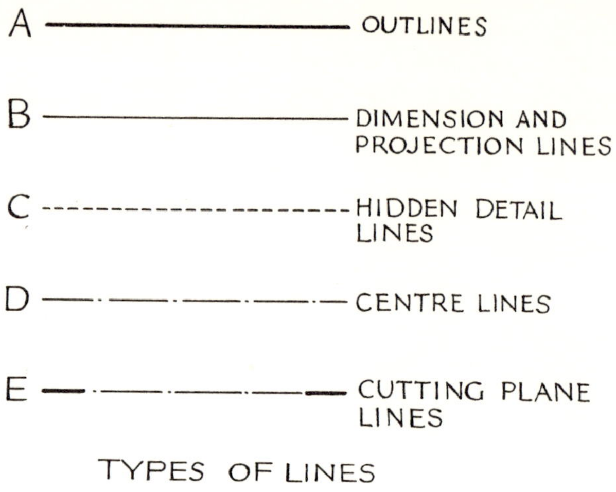

Fig 7

Exercises

1. Draw, full size, three views, arranged as in fig. 2, of the article shown in fig. 5. All dimensions are in millimetres.
2. Draw full size, three views, arranged as in fig. 3, of the vee block shown in fig. 6. All dimensions are in millimetres. (Figs. 5 and 6 are drawn to different scales.)

* A more fundamental study of the two systems is given in Part 2.

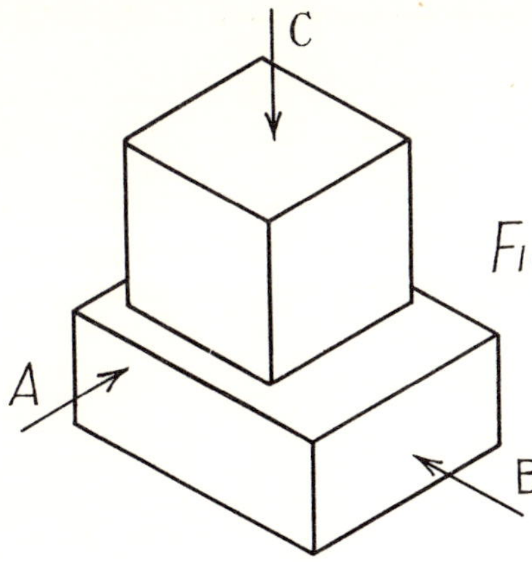

Fig 1

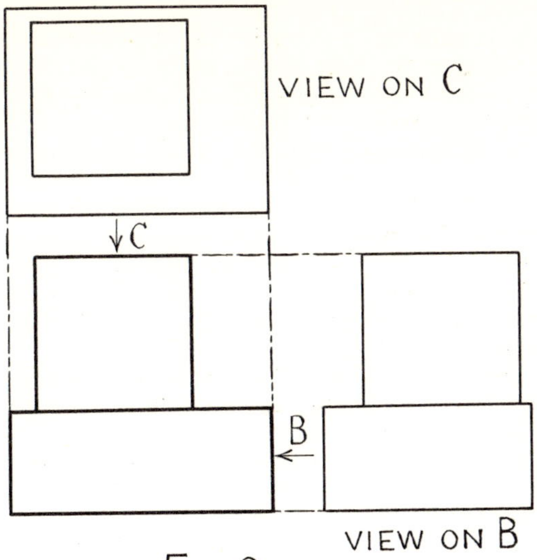

Fig 2

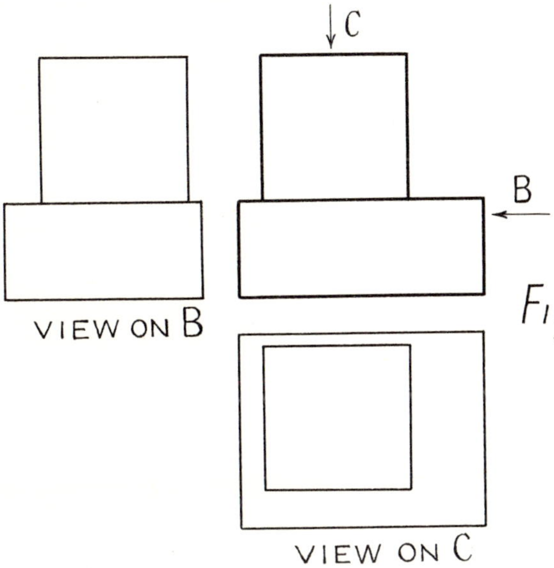

Fig 3

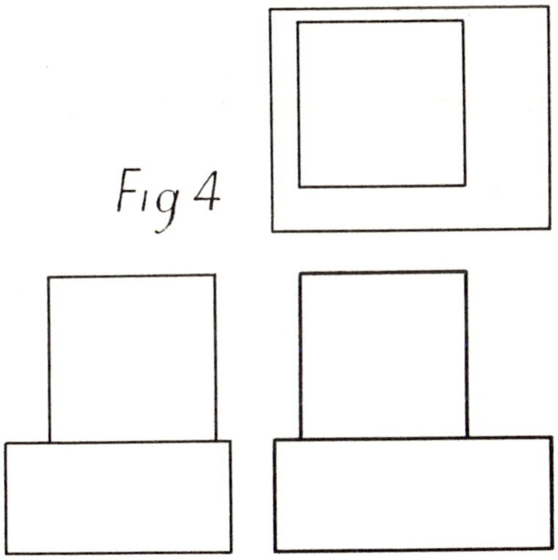
Fig 4

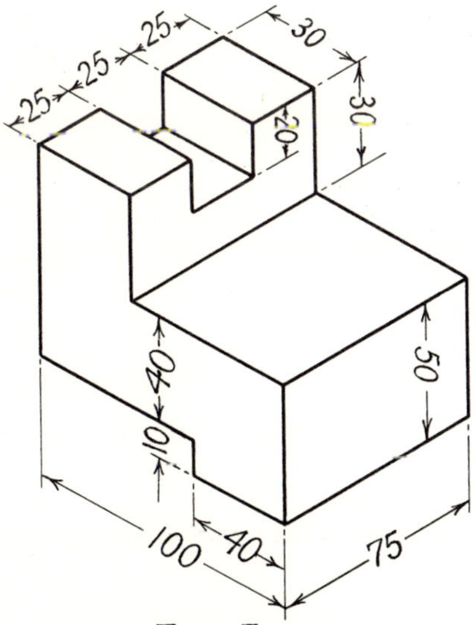

Fig 5

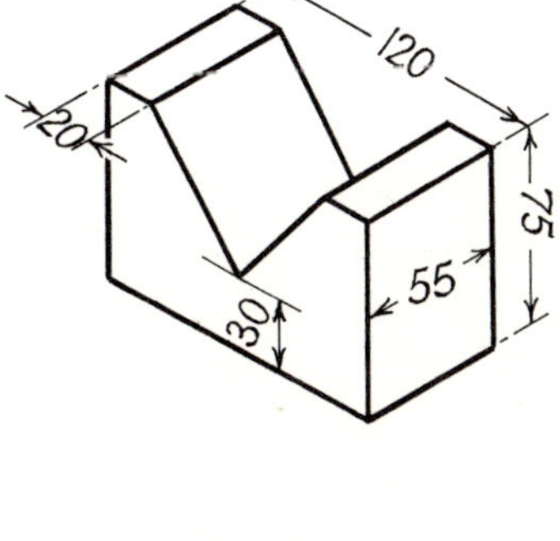

Fig 6

Interpretation of Drawings

Some important lessons can be learnt from the preparation, and subsequent interpretation, of drawings describing the object shown in fig. 1. This is a rectangular prism with one corner cut off along lines which are diagonals on three faces.
Suppose we require a view in the direction A. We might be tempted to show this as a triangle, fig. 4; but this would be wrong. A moment's thought tells us that there are lines beyond this triangular face which will be seen, and which must be shown on the drawing. The correct view is that given in fig. 5. Hence: *in preparing the drawing of a view on the face of an object we must show not only the outline of the face but also lines beyond the face which are visible.*
Views from the top and from the side are similarly rectangles with a diagonal. One set of three views is shown in fig. 2, and an alternative set is shown in fig. 3; in both, the view in direction A is shown in thicker lines. In fig. 2, each of the other two views has been placed adjacent to the face represented; in fig. 3, the views have been placed on the far side of the face represented. The student should compare both arrangements and try to decide which gives him the clearer picture. It may help him to a decision if he draws and cuts out each set of views, removes the part marked 'cut out' and folds along the dotted lines, folding backwards in fig. 2 and frontwards in fig. 3. The object may be imagined to be in the space formed by the folded paper. This conception is dealt with much more fully in the next Part.
With fig. 1 in mind it is quite easy to draw the views in fig. 2 or fig. 3. *But it is much more difficult to visualize the object in fig.* 1 *when we are given only the three views in fig.* 2 *or fig.* 3. This is what the craftsman commonly has to do. He is given a drawing such as fig. 2 and has to make the object. It will be seen that in very difficult cases it is well worth while to provide in addition a small sketch, such as fig. 1, to aid comprehension.
The development of the surface of the solid in fig. 1 is an interesting exercise which the student should attempt (see Ex.1). The solution is shown in fig. 7.

Exercises

1. Mark out the shape of the development of the surface of the object in fig. 1 taking the size of the object as 60 mm by 40 mm by 30 mm. Do this without referring to the solution in fig. 7. Cut out and fold the shape. Fix with adhesive tape to give the model of the object. Use the model and draw views corresponding to those in fig. 2.
2. In fig. 6 the same rectangular object is shown but the sliced corner extends only half-way along one side. Mark out the shape of the development of the surface, cut and fix with adhesive tape. Use the model and draw three views of the object arranged as in fig. 3.

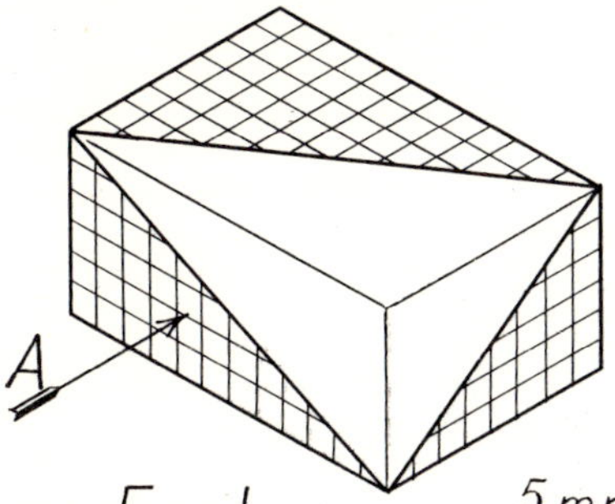

Fig 1

5 mm squares

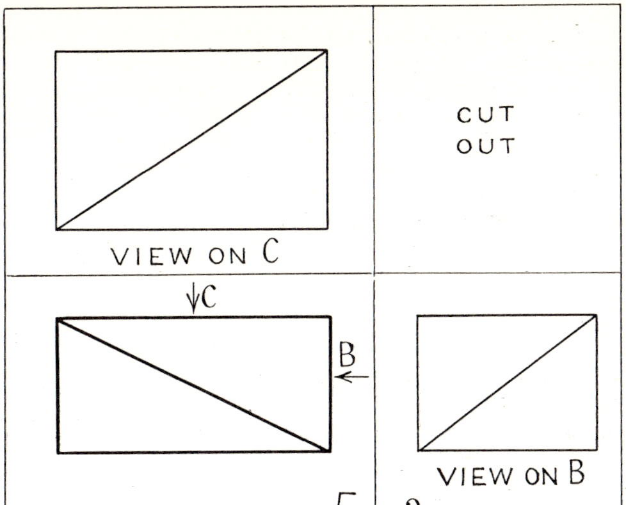

Fig 2

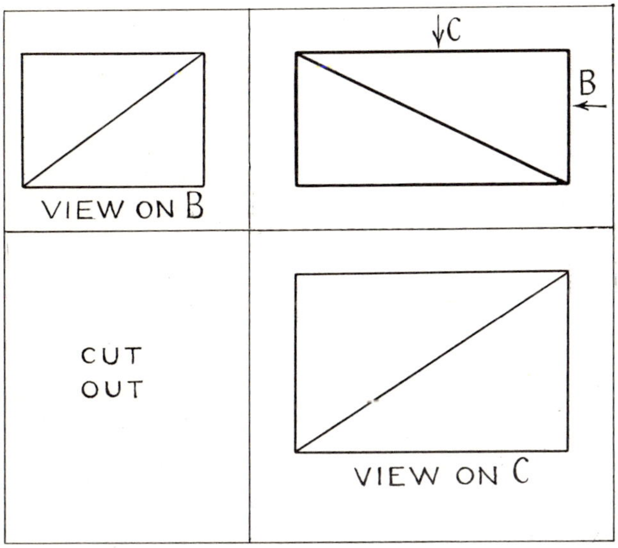

Fig 3

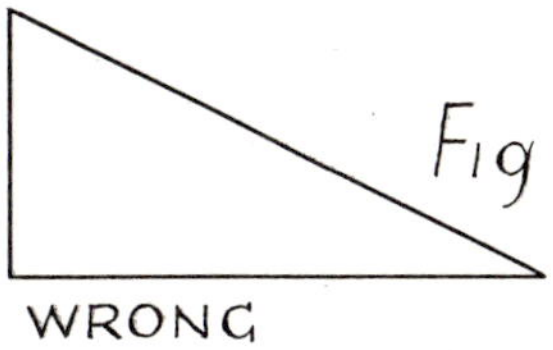

Fig 4

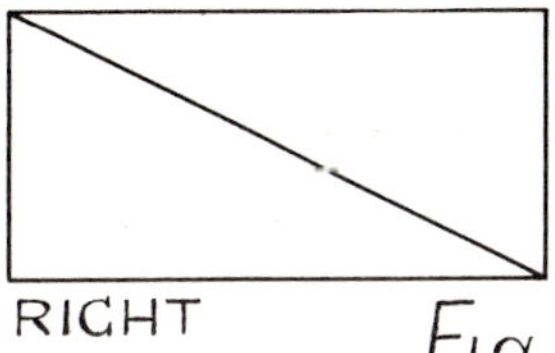

Fig 5

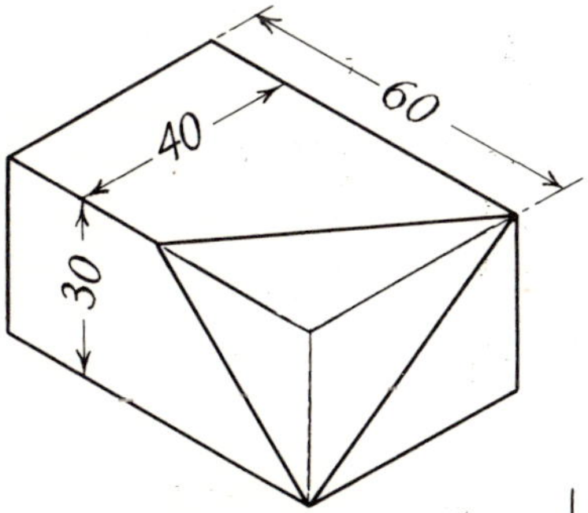

Fig 6

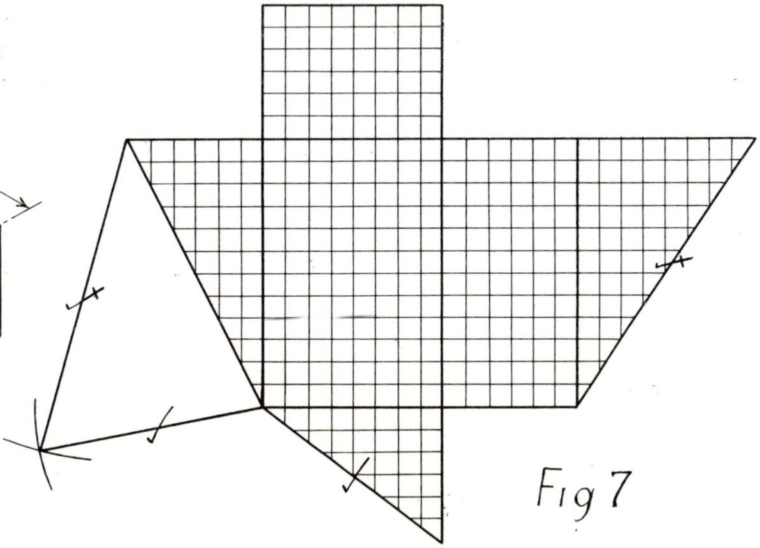

Fig 7

Hidden Lines

A hidden line is one which represents an edge or a surface which is invisible to the observer when viewing an object. If the object has yet to be made, then the draughtsman has to visualize which will be the edges or surfaces hidden from view. *If a hidden line is important to the understanding or interpretation of a drawing,* then the line should be shown dotted, the dashes of the dotted line being uniform in thickness and spacing (see p. 34).

The rectangular block in fig. 1 has a recessed area which can be described by using dotted lines in the ordinary views, as shown in fig. 2.

It is desirable that the student should always adopt a neat practice when showing dotted lines. The most important conventions are shown in fig. 3: a dotted line should begin with a dash; dashes should meet at corners; a dotted line should meet another at the centre of a dash. The practice of these simple rules should be persevered with until it is automatic.

Exercises

The articles shown in pictorial form (to different scales) in figs. 4 to 10 carry dimensions in millimetres.

Make *three* drawings of each object, full size, labelling the views and arranging them either as in fig. 2, or as in fig. 3, on page 35. Show dotted lines for hidden parts *only* where these lines are necessary for a complete understanding of the drawing.

Note.—In the absence of dimensions or notes, the student must make reasonable assumptions. For example, he must assume that the vee in fig. 6 is central, that the widtn of the edge in fig. 7 is 6 mm everywhere, and so on.

Dovetailed Packing Piece, fig. 9. Draw full size using 1st angle projection (see page 34)

(*a*) an elevation seen in direction A.
(*b*) a side elevation to the right of view (*a*).
(*c*) a side elevation to the left of view (*a*).
(*d*) a plan view below view (*a*).

Note.—Try to build up all views together, bit by bit.

Fulcrum Guide Bracelet, fig. 10. Draw full size in 3rd angle projection (see page 34)

(*a*) an elevation seen in direction of A.
(*b*) a side elevation to the right of view (*a*).
(*c*) a plan view above view (*a*).

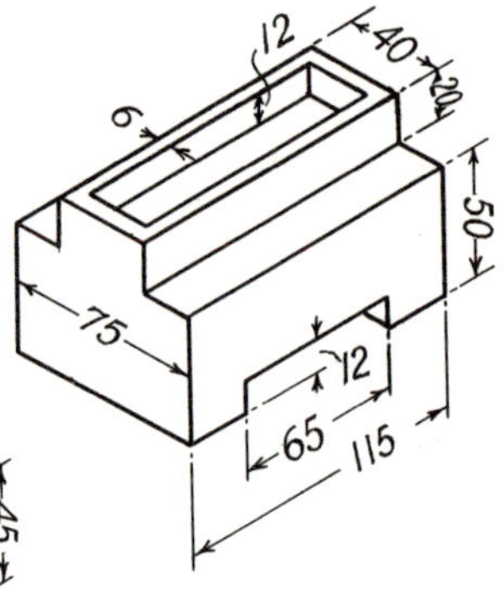

Fig 7

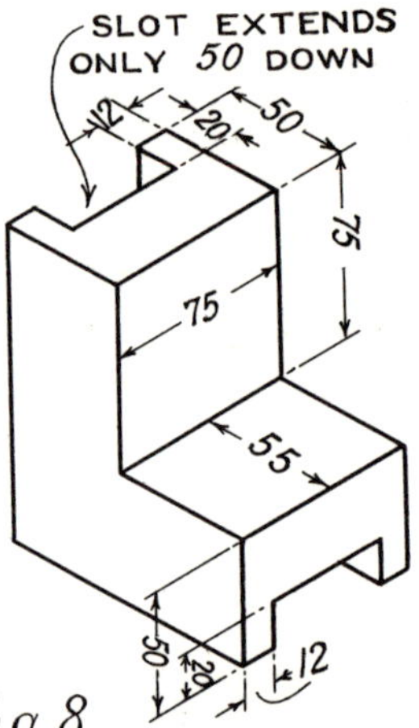

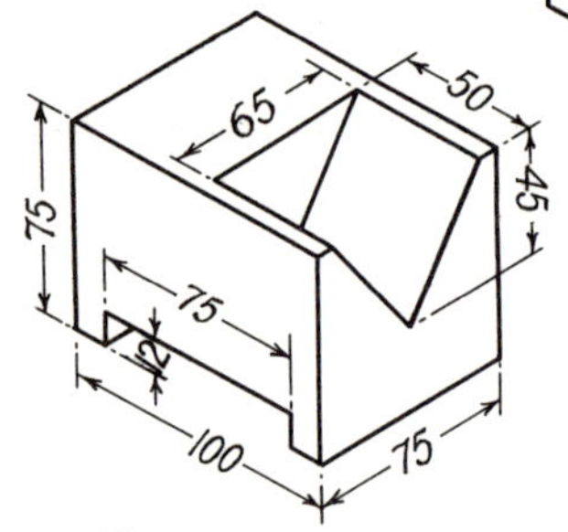

Fig 6

Fig 8

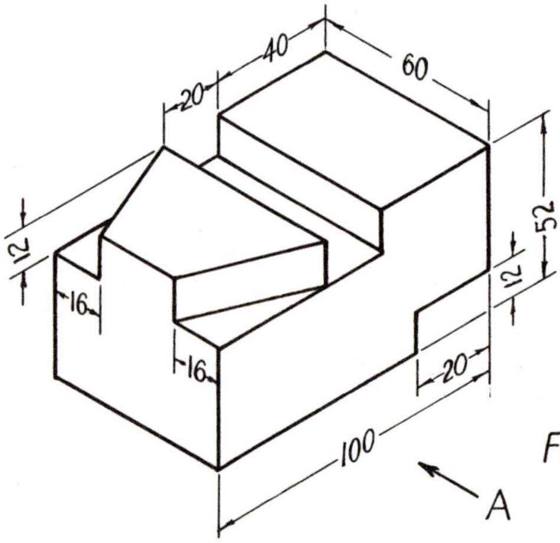

Fig 9

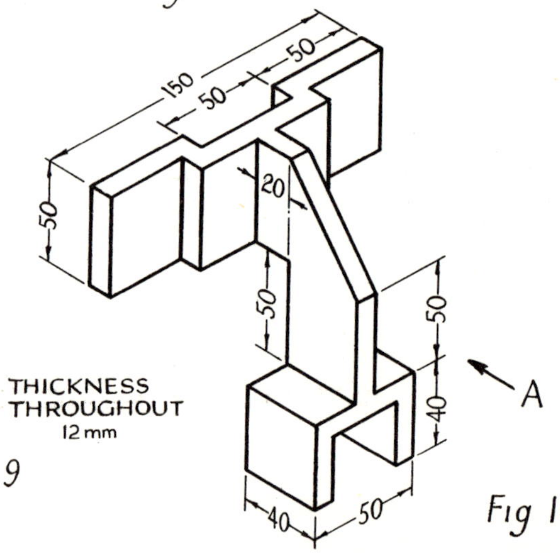

Fig 10

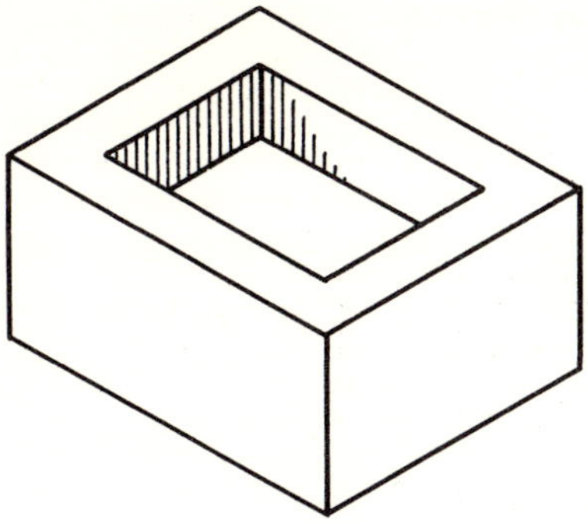

Fig 1

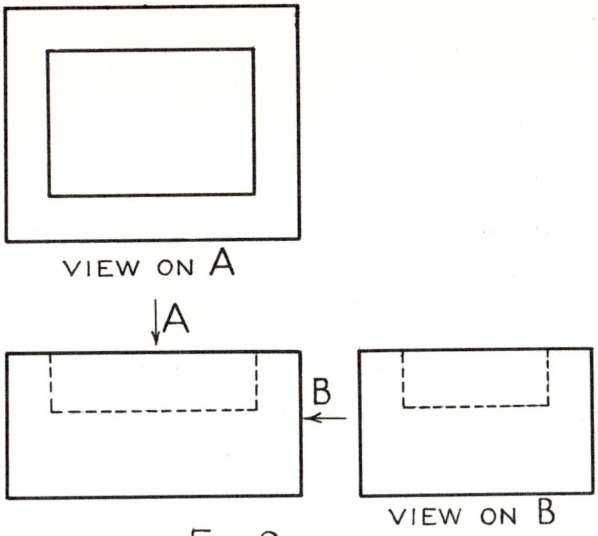

Fig 2

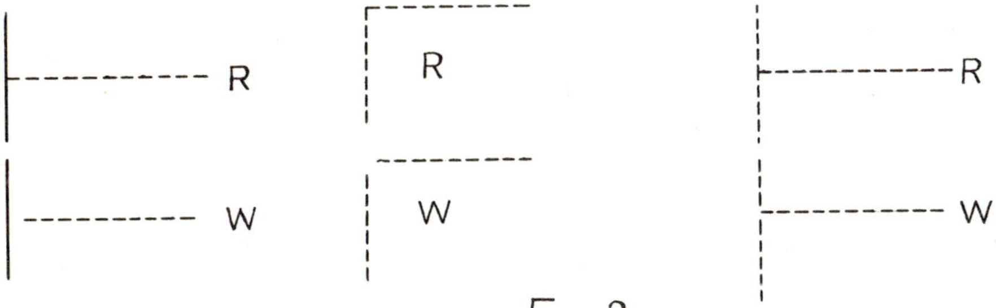

Fig 3

R RIGHT
W WRONG

Fig 4

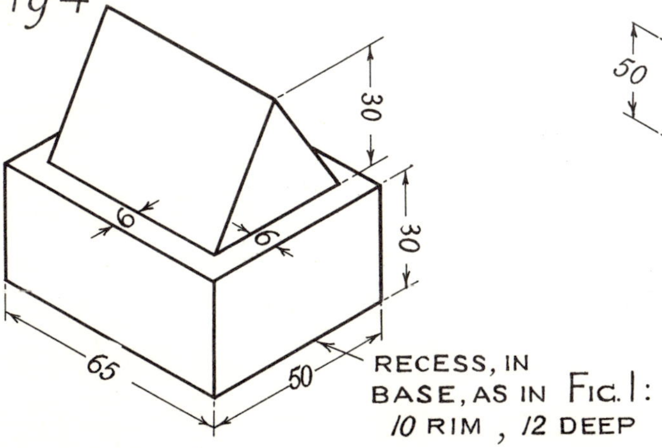

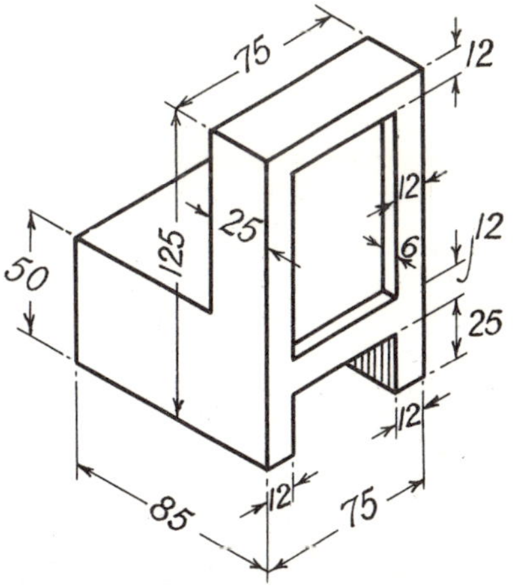

Fig 5

The Hexagonal Prism

This prism is exceedingly common in all applications of technology because it is the form of most nuts and bolt heads. As will be clear from fig. 1, the ends of the prism are regular hexagons, and the six sides are equal rectangles.

The drawing of a regular hexagon with its six equal sides is a very simple matter. Consider fig. 2. The student will know that the radius of a circle can be spaced off around its circumference exactly six times. *Lines joining these division points give a hexagon. The side of a hexagon is therefore equal to the radius of the circumscribing circle,* and this property helps us greatly in the construction of the hexagon.

If we draw lines to the centre of the circle from the six points on the circumference, as in fig. 3, we obtain six equilateral triangles: that is, triangles with three angles each of 60°. From this it will be clear that the hexagon can in fact be drawn, in two positions, by using the 30°–60° set square, according as the hexagon has a diagonal vertical, fig. 4, or horizontal, fig. 5. Normally, begin by drawing the circumscribing circle and a vertical or horizontal diameter; then start with points of intersection, such as P, and use the set square for inclined and vertical lines. Alternatively, the circle may be dispensed with and the hexagon drawn by using the set squares only, as indicated in fig. 6. *The numbers give the order in which the lines should be drawn.*

It should be noted that a side view, in the direction of the arrow A, fig. 1, can quickly be drawn without reference to the hexagon. The spacings of the four parallel lines are $\frac{1}{2}s$, s, $\frac{1}{2}s$, where s is the length of the side of the hexagon. If s is 20 mm, then the lines are spaced 10 mm, 20 mm and 10 mm apart. This is a valuable aid to the draughtsman. No such simple relationship helps us with the view in the direction of the arrow B; the spacing of the three parallel lines here should be taken from a drawing of the hexagon. Actually, the lines in this view are separated by the distance ·865s, as may be calculated from fig. 3.

The hexagonal prisms used for nuts and bolt heads are modified by the removal of the sharp corners, usually on one face only. This 'chamfering' process requires a small change in the drawing of the prism and is discussed later.

Exercises

1. Draw a hexagon, side 30 mm, in the two positions shown in figs. 4 and 5. First use the circumscribing circle. Then draw the hexagons without the use of the circle, as in fig. 6.
2. A hexagonal prism has a measurement across the corners of 100 mm and is 50 mm thick. Cut out a development, as in fig. 7, and fold to make a model of the prism. Draw three views of the prism, arranging them to the best advantage.
3. A rectangular hexagonal prism with a part cut away is shown in fig. 8. It is regular because all sides of the hexagon are equal and all angles are equal. Draw, full size, in 1st angle projection:
   (*a*) the given plan.
   (*b*) the given elevation.
   (*c*) a view as seen from the left of view (*b*).
4. An irregular hexagonal prism is shown in fig. 9. It is irregular because the sides and angles are not equal. Draw, full size, in 1st angle projection:
   (*a*) the given plan view.
   (*b*) the given elevation.
   (*c*) a view as seen from the left of view (*b*).

Note.—In both exercises the view seen from the left will be drawn to the right of view (*b*) since it is in 1st angle projection.

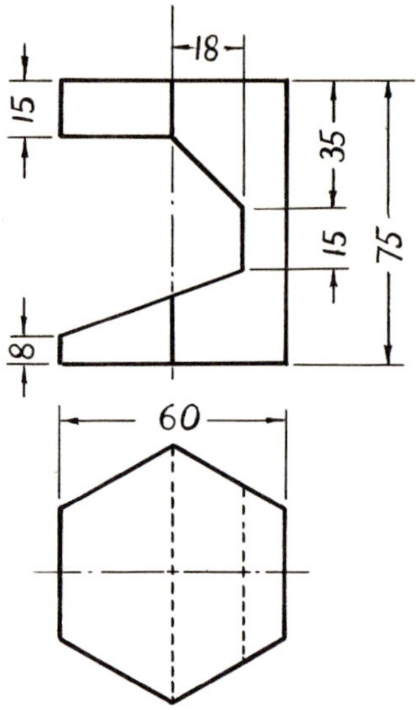

Fig 8

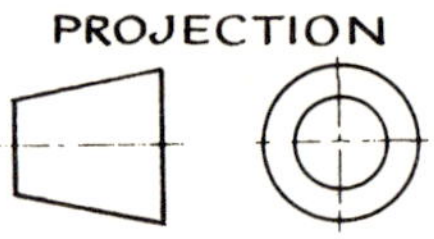

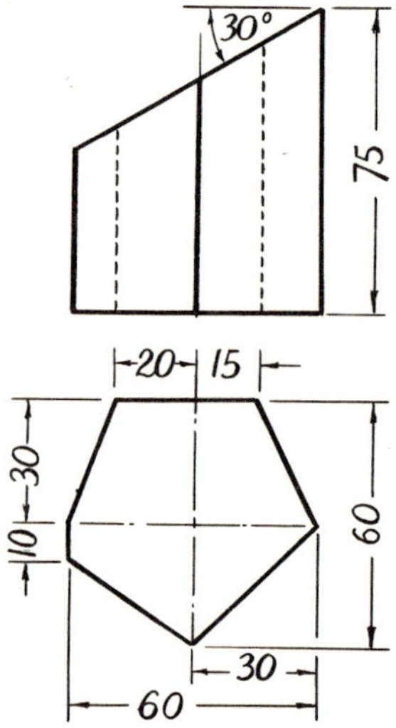

Fig 9

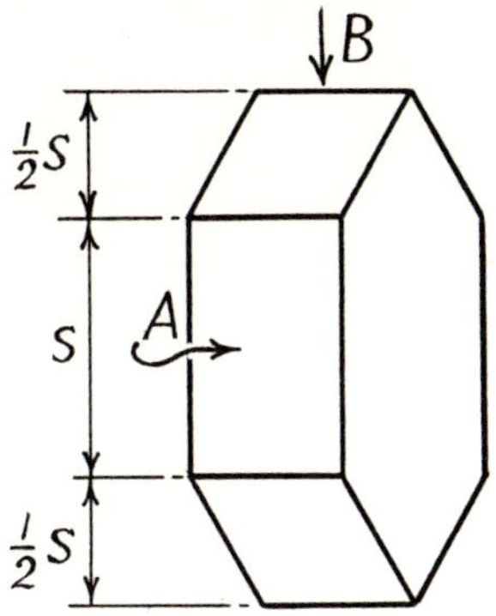

Fig 1

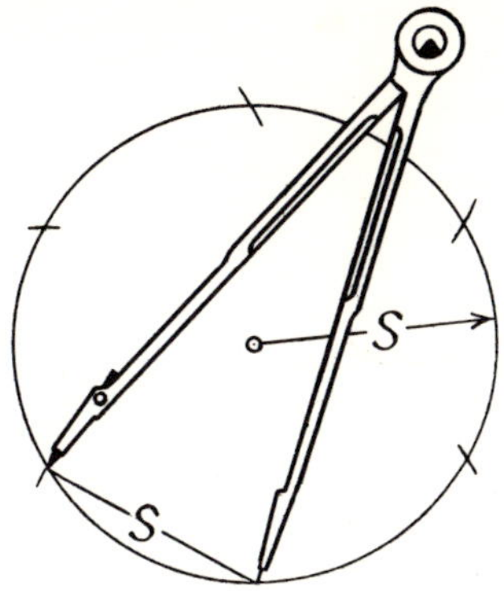

Fig 2

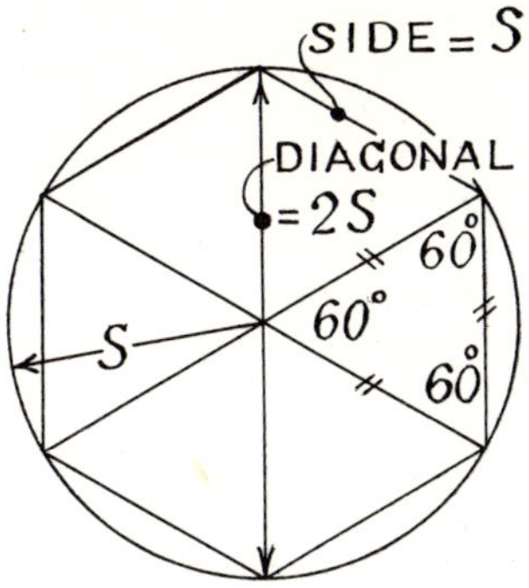

Fig 3

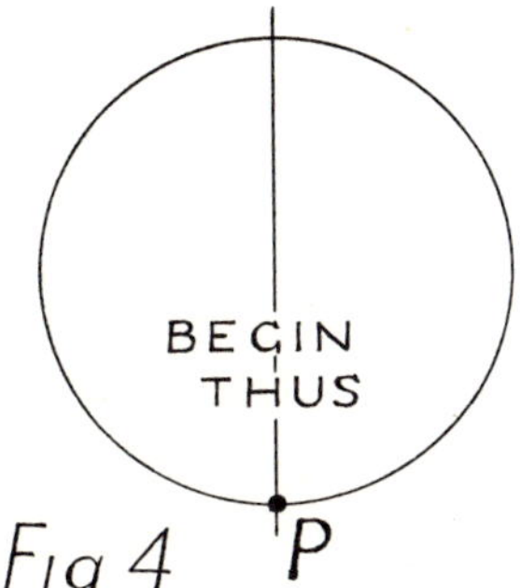

Fig 4

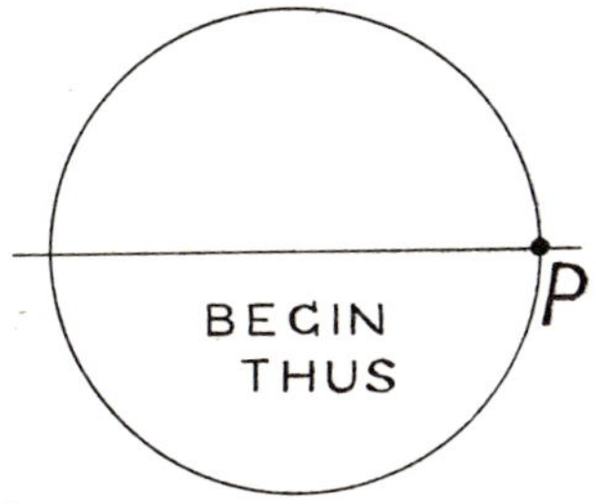

Fig 5

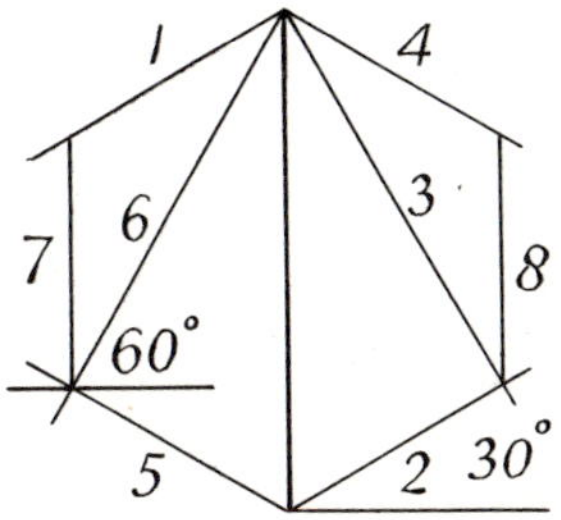

Fig 6

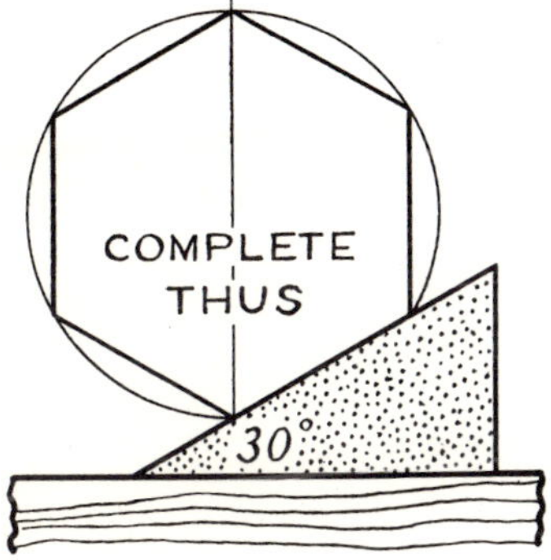

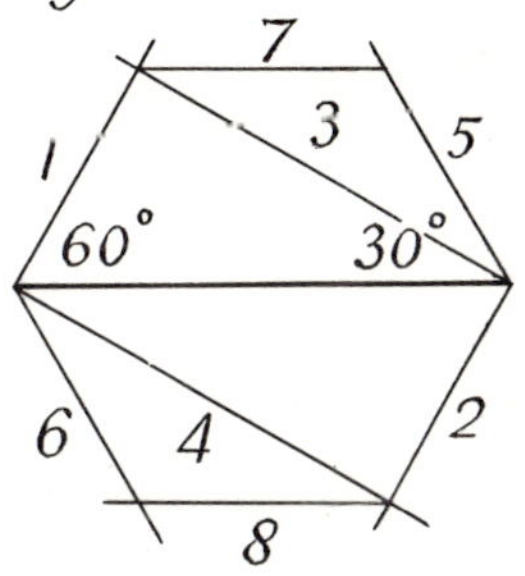

Fig 7

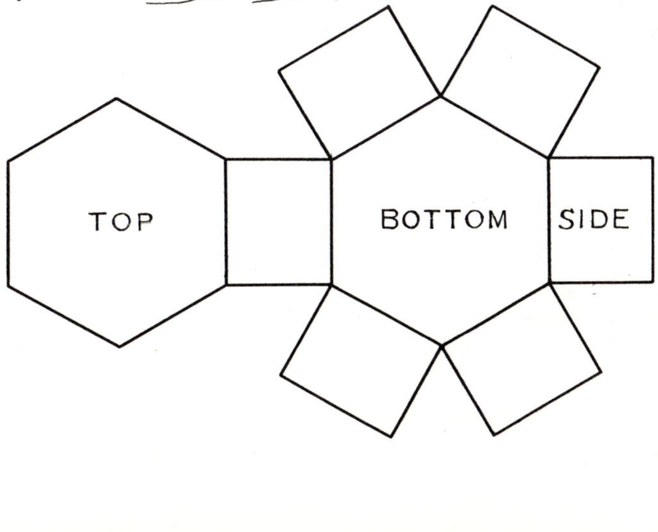

It will be clear that we can arrange an object in a number of positions and that there will be a corresponding number of drawings showing the views of the object. The student should refer to the many pictorial views of the same object on page 61. Any rectangular object will have six principal faces. We normally show three of these faces, in three separate views—from front side and top; and we can usually have freedom to select these views, to the best advantage. But if all six faces have features which must be shown, and if the use of dotted lines is undesirable, then we must add three more views—from the underside, back and other side. These six views can be arranged in many combinations, as will be clear from the discussion on page 60.

Let us consider the simple hexagonal prism. We may show it as in fig. 1, as in fig. 2, or in innumerable positions typical of that in fig. 3. In only one of these views do we see the true shape of a rectangular side face—that marked in fig. 1. All the other rectangles are shown narrower than they really are. We may infer that we shall see the true shape of a surface in a drawing only if this is what is seen when looking squarely upon the surface. If the surface is inclined to the line of vision, then the drawing representing that view will not show the true shape of the surface. This is a very loose way of expressing the matter and more accurate statements are given later.

Many objects are box-like in form or have parts which are so shaped; and one would normally arrange them so that each of the usual three views was taken squarely on a principal face and gave its true shape. But sometimes we require the true shapes of other faces. The object shown in fig. 4 has a sloping face, and none of the normal three views in fig. 6 will show the true shape of the sloping surface. Because the surface is rectangular, however, and because we are looking squarely on its length in one view and on its breadth in the other, we may readily construct its true shape. For example, (*a*) and (*b*) give us the length and breadth, and we could easily draw the rectangle upon the side *a* as shown in fig. 7.

We saw from the development in fig. 7 on page 37 that the true shape of the triangle could be obtained by using the true lengths of the three sides obtained from each of the views.

The determination of the true lengths of lines and the true shapes of surfaces are common operations in technical drawing and they will be dealt with comprehensively in the section on auxiliary projections in Part 2.

Exercises

1. **A coal scuttle is shown in fig. 5 with dimensions in millimetres. Using a scale of 1 : 10, make three views of the object, corresponding to those in fig. 6, with the lid raised to make an angle of 30° with the horizontal, as shown in fig. 5. What is the true size of the lid? (Answer: 800 mm by 673 mm.)**
2. **Fig. 8 shows a further example of an irregular hexagonal prism, this time the drawing being made in 3rd angle projection. The plan view is above the front elevation and is constructed to the following dimensions:**

   AB and ED = 45 mm   AF and CD = 20 mm
   EH and BG = 20 mm   FH and CG = 30 mm

   The vertical height of the prism = 80 mm. The cutting plane is at an angle of 45°.

   Draw, scale full size, in 3rd angle projection:
   (*a*) the given plan view.
   (*b*) the given front elevation.
   (*c*) the given side elevation.
   (*d*) a side elevation to the right of view (*b*).
   (*e*) the given surface development.

***Note.*—There are certain features shown in fig. 8 which will be referred to in later exercises in the book.**

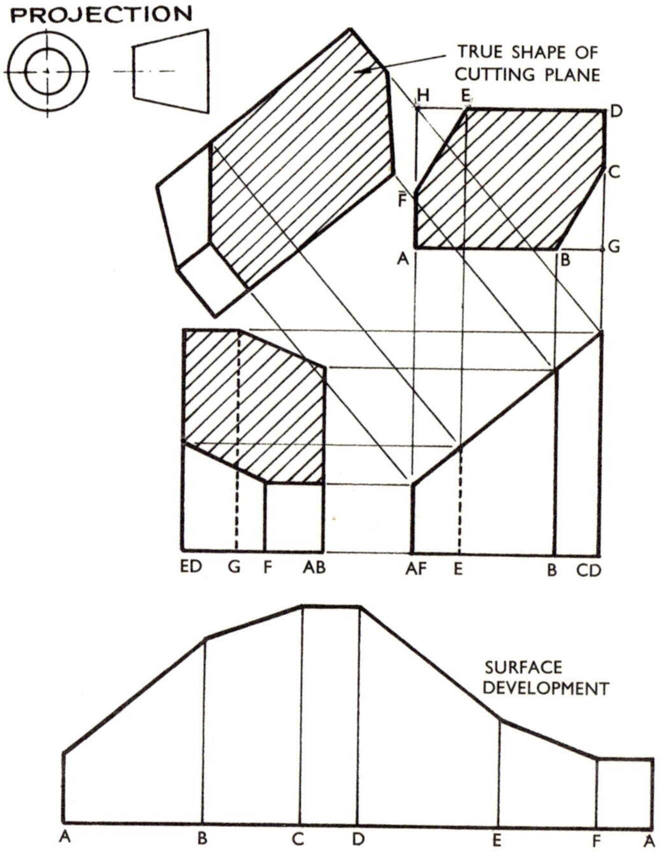

Fig 8

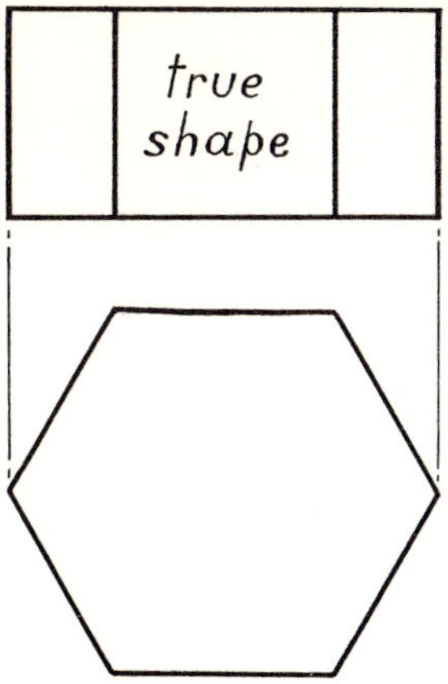

Fig 1

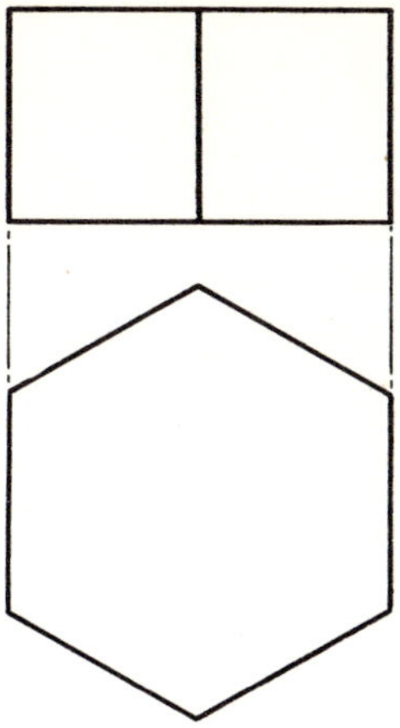

Fig 2

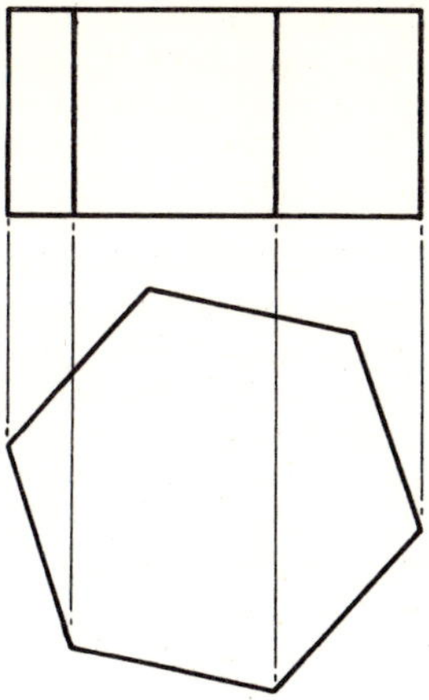

Fig 3

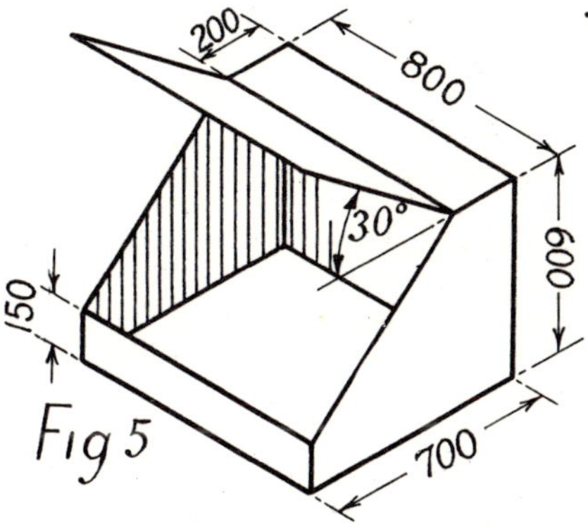

Fig 5

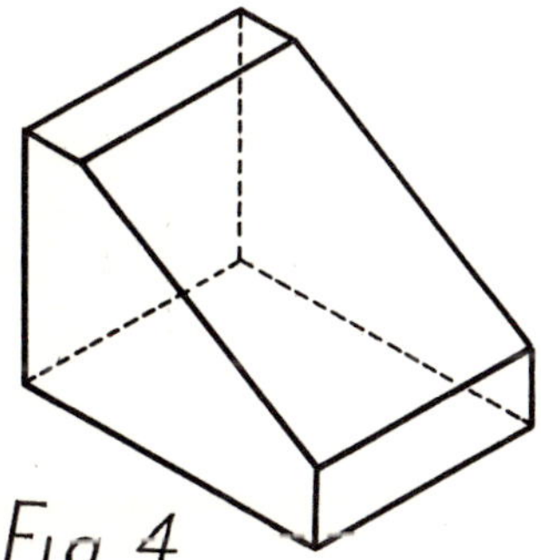

Fig 4

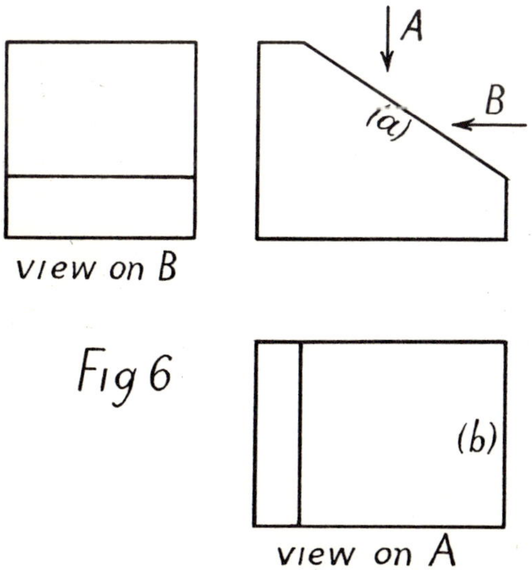

Fig 6

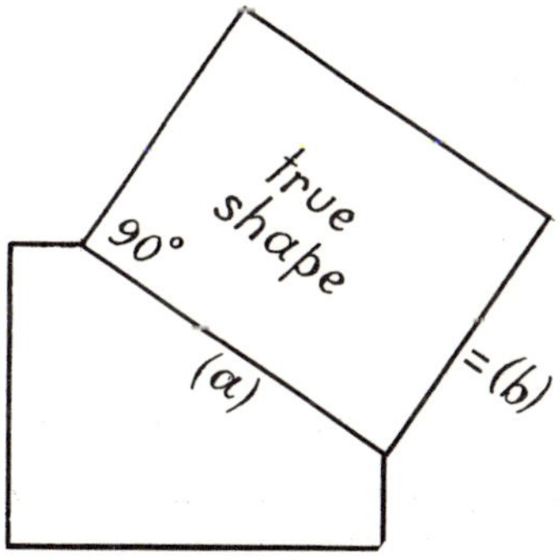

Fig 7

Pyramids

Pyramids are much less commonly used in engineering than prisms, cylinders and cones, and a study of this page may be deferred.

The hexagonal pyramid

The drawing in fig. 1 shows a hexagonal pyramid. The base is a hexagon and the sloping sides are equal triangles meeting at a point above the centre of the base. The three views of this pyramid shown in fig. 2 need hardly any explanation. The base hexagon of side *S* is constructed as on page 41, and diagonals are drawn to represent a top view of the sloping edges of the pyramid. The side views may readily be drawn if we know the height *H* of the pyramid. The triangles shaded are three views of the same sloping side.
Suppose however that we are given not the height *H* but the lengths of the sides of a triangular face; one side would be *S*, and the other two would be equal. We can construct fig. 3 which shows, on the flat, the base and two opposing triangular faces the sides of which we know. Let us cut out this figure. Now fold the triangles upward until they meet, as in fig. 4, which shows almost an edge view of fig. 3. We have now obtained the height of the pyramid. We can perform the same operation on the drawing by the construction shown by the projection lines on the right, swinging around the radius H_1 (on both sides) until the arcs meet at the apex. If we are given the height H_1 of a triangular face, then we can use this as a radius at once. But there is another and perhaps more obvious way of dealing with the problem. Consider the three views in fig. 2. Only two of the lines in these views give us *the true length* of a sloping edge. These two lines are marked with a tick. If therefore we are given, or if we find, the length H_2 of a sloping edge, we can use this as shown on the left and thus obtain the apex.
The development of the surface of the pyramid may be set out as shown in fig. 5. The triangles should be folded about their common sides. Small extensions are shown in fig. 5 to help in the final sticking together.
Note.—The base of a pyramid may have any number of sides, from three upwards. Some of these cases are considered later. If the number of sides is very large, then the pyramid becomes more like the cone.

Exercises

1. The side *S* of a hexagon is 40 mm and the sloping edge H_2 of a triangular side is 100 mm. Construct fig. 5 and make the pyramid.
2. Refer to the model you have made and draw views corresponding to those in fig. 2. Measure and record the distances *H* and H_1. (Answer: $H = 91{\cdot}7$ mm, $H_1 = 98{\cdot}0$ mm.)
3. Fig. 6 shows the elevation and plan view of a rectangular hexagonal pyramid. Notice that none of the slant edges in the elevation indicates a true length, nor will one be shown in the side elevation.
   Draw, scale full size, in 1st angle projection:
   (*a*) the given plan view.
   (*b*) the given elevation.
   (*c*) a side elevation drawn to the right of view (*a*).
   (*d*) a development of the surface area.

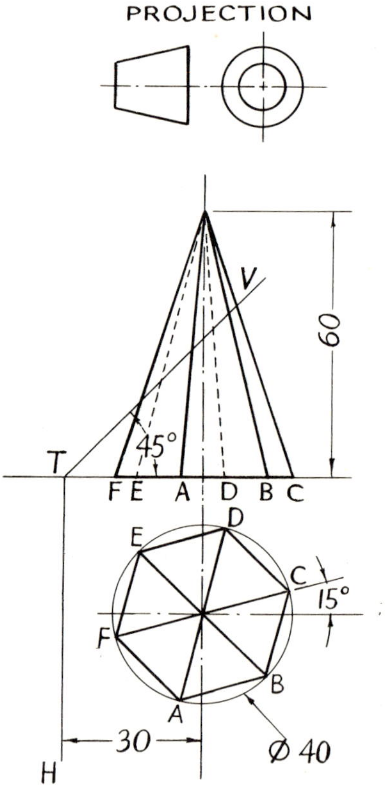

Fig 6

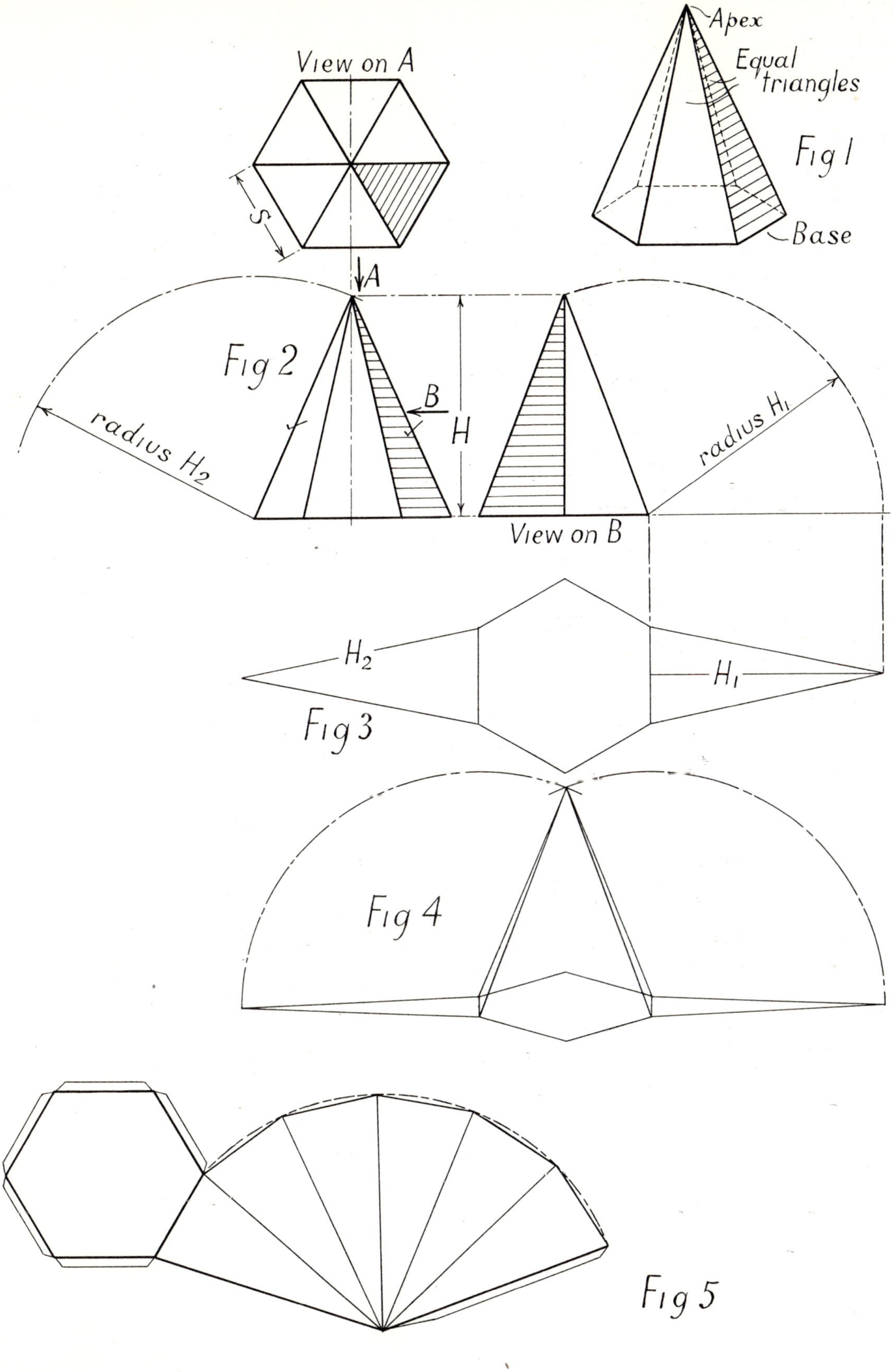
View on A
S
A
Apex
Equal triangles
Fig 1
Base
Fig 2
B
H
radius H_2
radius H_1
View on B
H_2
H_1
Fig 3
Fig 4
Fig 5

Cylinder and Cone

The cylinder and the cone are among the commonest engineering forms and we shall now consider the simpler drawings of them. They are two of the many *solids of revolution*, so called because of the way they are formed. A piece of material, fig. 1, revolves between centres and is cut with a tool. An imaginary line about which the object turns is called the *axis*. If the tool moves along a line inclined to the axis, a cone is formed.

It is useful to consider other ways of defining these solids. If the wire rectangle in fig. 2 is spun about one side, the opposite side traces out a cylindrical surface. Again, in fig. 3, if a triangle is spun about one side, another traces out a conical surface.

Material may be removed from a solid to form holes of cylindrical or conical shape. Cylindrical holes are as common as cylindrical bolts.

The simpler drawings of the cylinder and cone are shown in figs. 4 and 5. The end view of each solid is a circle. Side views give a rectangle for the cylinder and a triangle for the cone. The triangle has equal base angles, as marked.

At this stage we shall assume that each solid is described by a diameter D and a length L, as marked in figs. 4 and 5. Conical forms in engineering practice are usually truncated, or shortened, and are rarely complete; see fig.1, page 29. The part of the cone is called a *frustum*.

Centre lines

With the cylinder and the cone (as with all other solids of revolution) it is the usual practice to draw thin centre lines through the axis in all views, as shown opposite. To an engineer the drawing of a cylinder or a hole appears unfinished without centre lines. These lines should be broken, but they should be chain lines, not dotted lines, as shown opposite. See also page 62 which defines the various types of lines used in technical drawing.

The *development* of the surfaces of the cylinder and cone, while not difficult, is best dealt with later.

Exercises

All dimensions are in millimetres.

1. Make a freehand sketch, similar to that shown, of the object in fig. 6. It is a rectangular block carrying a cylinder, with a hole through the block. Both cylinder and hole are centrally arranged. Make three suitable full-size drawings of the object.
2. Make three full-size drawings fully describing the bracket shown in fig. 7. Only show dotted lines where they are necessary. The cylindrical boss and hole are centrally arranged. The recessed part in the underside of the base extends from front to back. Select a suitable radius for the junction of the vertical part with the base.

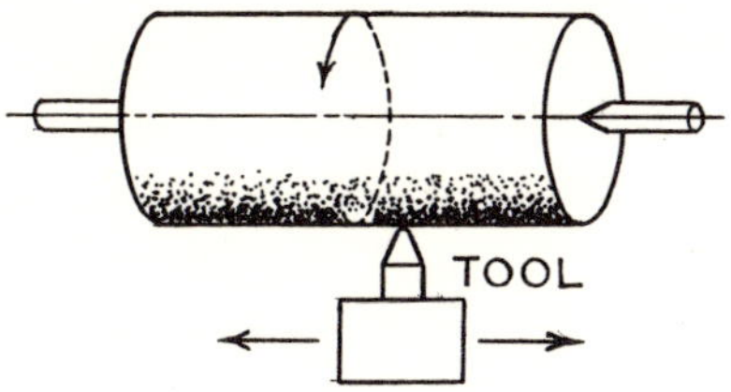

Fig 1

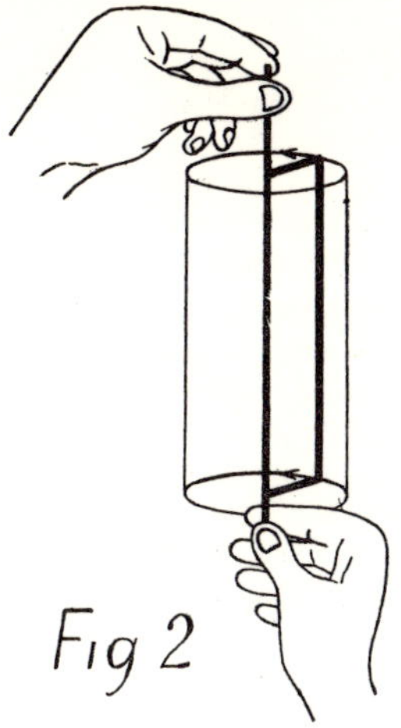

Fig 2

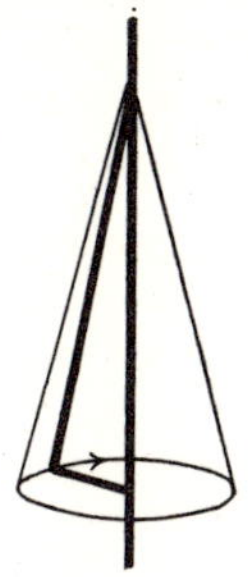

Fig 3

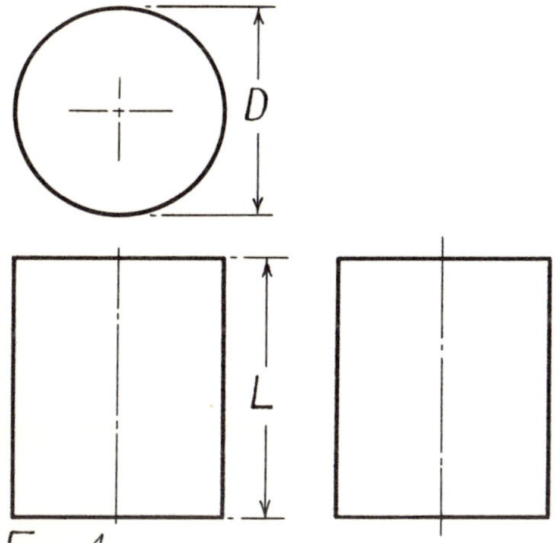

Fig 4

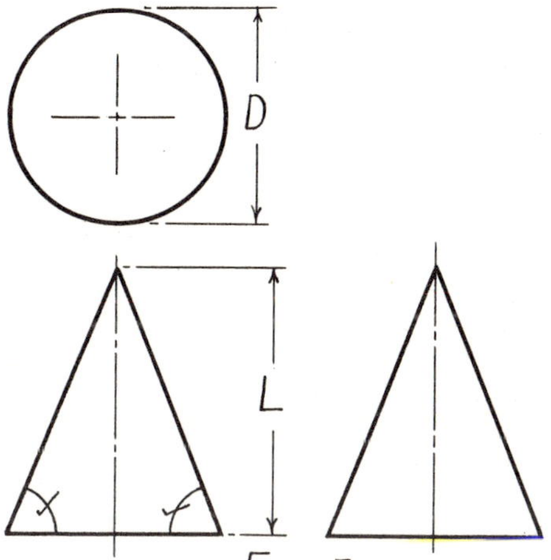

Fig 5

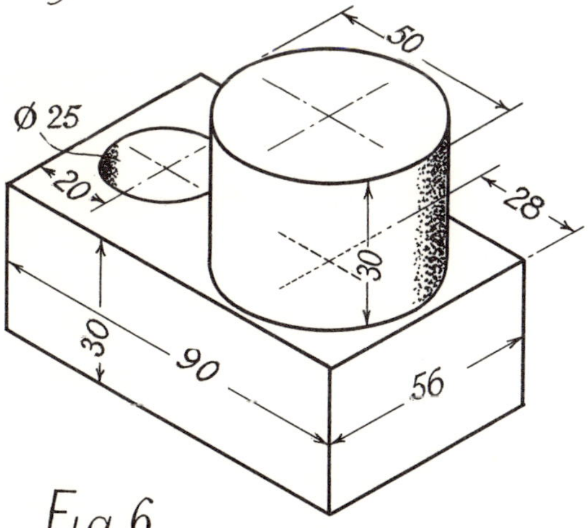

Fig 6

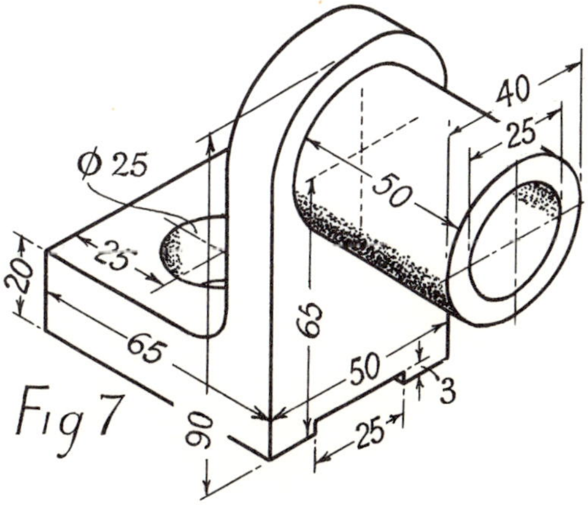

Fig 7

Simple Sectional Views

A sectional view is one which is given when a part of an object in front of a cutting plane is removed. The operation is imaginary, there being no cutting plane and no removing of the part. The view should show not only the section supposed to be cut but also those parts of the object which are visible beyond the section. Usually the purpose of a section is to show the shape of parts hidden from view which would otherwise have to be defined by dotted lines. Such an example is given in figs. 4 and 5. The rectangular object in fig. 1 is shown cut by a plane which may be imagined to be a sheet of glass. When we suppose the plane, and that part of the object in front of it, to be removed, we see the shape shown in fig. 2. To make a drawing of this we might begin with a view on the top of the object, marked A in fig. 3. The points *a* and *b* must be known, and they can therefore be marked. The line joining them represents an edge view of the cutting plane. We may now draw two other views, arranged as in fig. 3. It will be seen that the edges of the cut surface are given by 'projecting' from *a* and *b*, the 'projectors' being thin chain lines.

The area of the actual section in all views is usually filled in by thin lines crossing the surface at 45° and drawn carefully at regular intervals. This is sometimes called 'crosshatching'. The spacing of the lines depends on the area to be covered but, normally, they should be about 3 mm apart. They should be drawn by eye, uniformly distant and equally thin.

The example discussed above and shown in figs 1, 2 and 3 is an interesting exercise in geometry, but it is not the kind of example common in engineering practice. Usually the purpose of the section is to reveal an interior; further, the portion imagined removed is *not* omitted when the *other* views are drawn. An example is shown in figs. 4 and 5. Here a section plane is used to reveal an internal shape which would otherwise have to be shown entirely by dotted lines. Two normal views are first drawn, an underneath view and a side view, the latter having dotted lines to show the hidden interior. The cutting plane is marked by the line CC. The section reveals very clearly the shape of the hollow interior, and this sectional view is placed where a view on CC would normally be placed. Here it has been shown on the right, but the alternative position on the left could be used if preferred.

The line indicating a section plane should be a thin chain line with a thick portion at each end (see p. 34).

Exercises

See page 50.

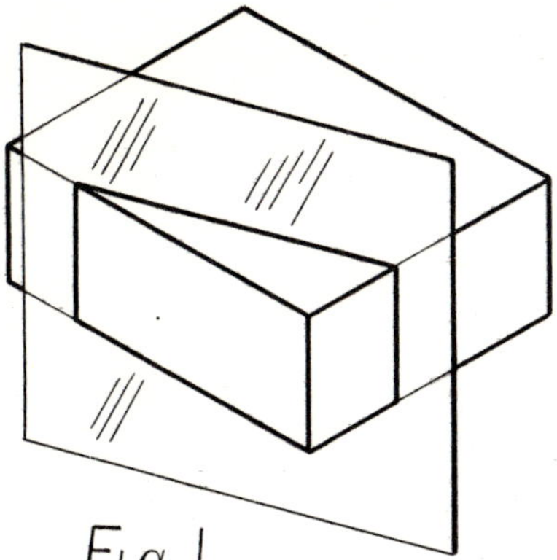

Fig 1

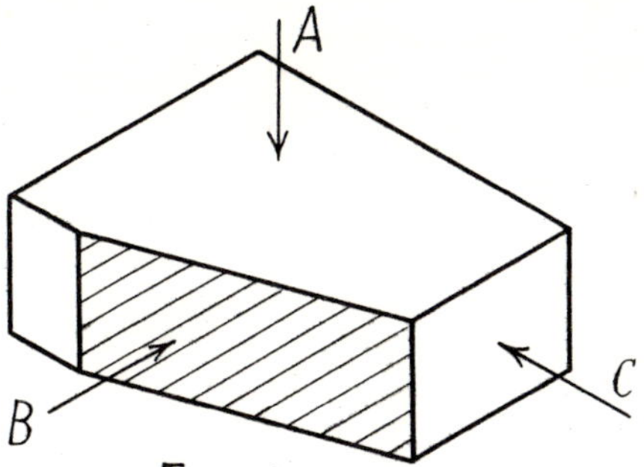

Fig 2

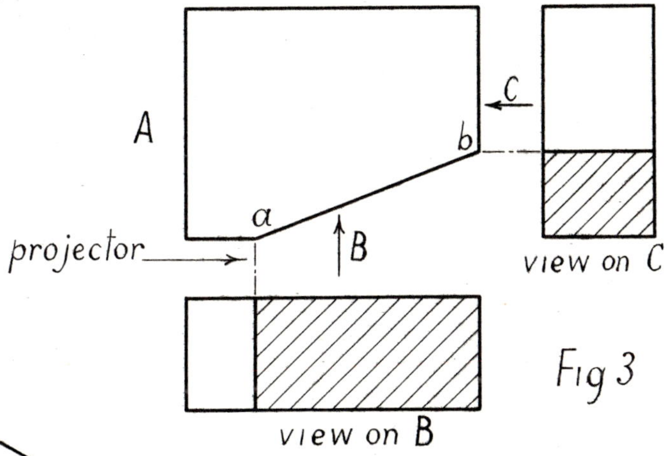

Fig 3

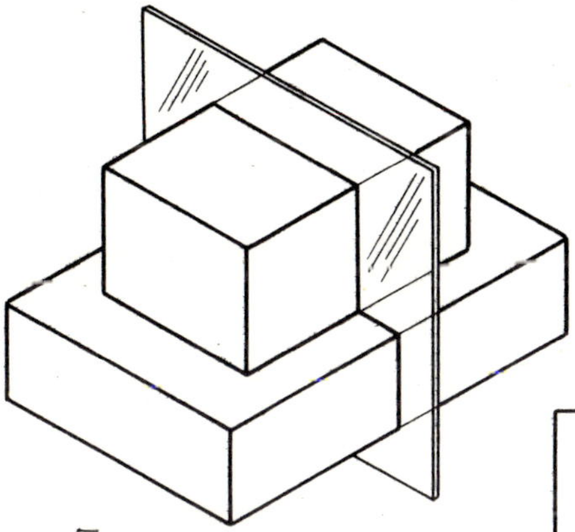

Fig 4

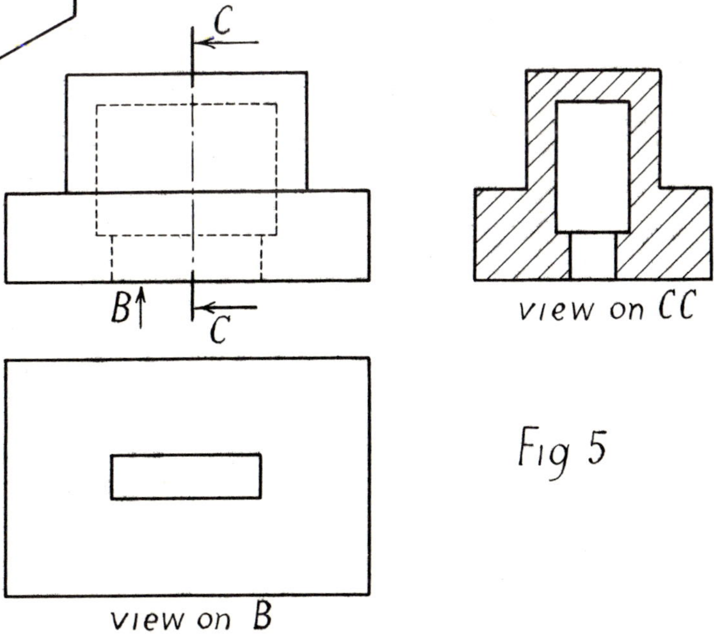

Fig 5

Half-sectional Views

Very often only a half section need be given for objects which are symmetrical about a centre line or axis. The article shown pictorially in fig. 1 is an assemblage of cylinders of varying outside and inside diameters, on a square base. The object could of course be described completely in a drawing using full lines for the outside and dotted lines for the inside; but, even so, a clearer and more satisfying impression would be conveyed if we showed a view *half in section*. A complete section could be produced by the cutting plane XX in fig. 2. This would create a view which would be the same on both sides of the centre line. In order to describe the object more fully, one half of this view could be used, with an outside view on the opposite side of the centre line. Thus two sets of details could be shown in the one elevation as in fig. 2. There is no actual cut made into the object, it is purely an imaginary view, so the centre line *ab* will remain in the elevation of fig. 2. The arrangement is adequately illustrated in the pictorial view of fig. 1, where two planes show a quarter of the object cut away. The cutting plane should be shown as XX completely through the plan of fig. 2, and the note "Half section on XX" added to the elevation.

Exercises

1. (*a*) Measure carefully the sizes of the parts of the object shown in fig. 2. Draw, twice the size given, two views corresponding to those shown.
   (*b*) More difficult exercise. Imagine the object turned 45° about its axis so that a diagonal of the square base is horizontal in the top view. Draw, twice the size given, two views corresponding to those shown *but with the section plane taken through a hole in the base.*
2. The object shown in fig. 4, which is dimensioned in millimetres, has a recessed top with a central hole, and a rectangular slot passing from one side to the other. Draw, full size, three views: one view on B, another on A, and a sectional view as given by the central plane CC.

3. The various parts of the object shown in fig. 5 are rectangular except the hollow rounded surface which is in fact half circular. A shaft of 80 mm diameter fits into it.
   Draw, full size, three views of the object: one view in the direction A, one in the direction B, and a sectional view given by the plane CC.
   All dimensions are in millimetres.

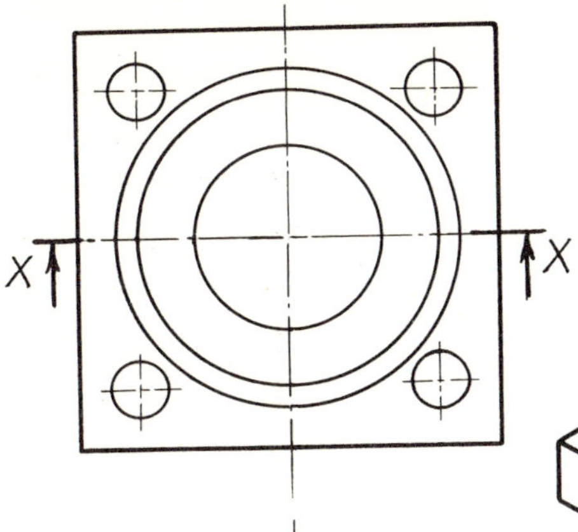

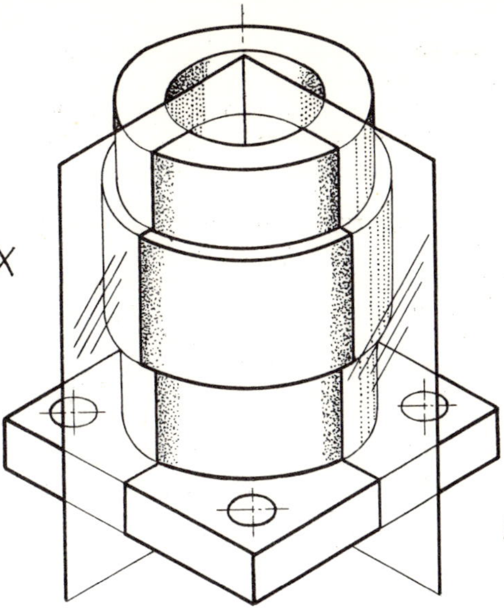

Fig 1

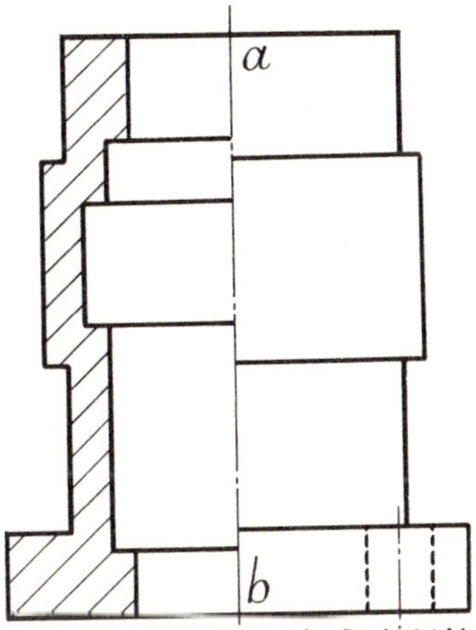

Fig 2

HALF SECTION ON XX

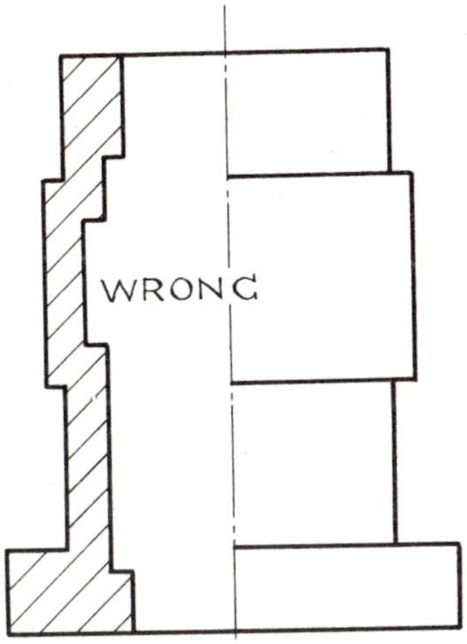

Fig 3

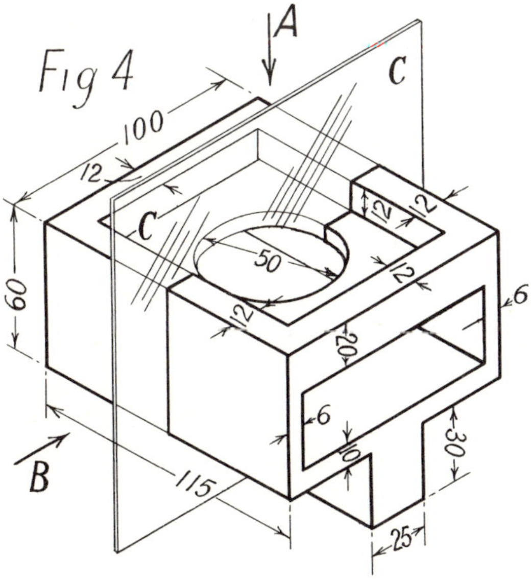

Fig 4

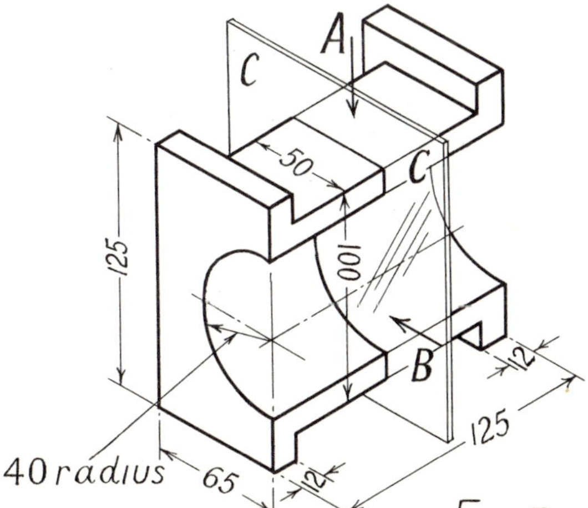

Fig 5

Pictorial Drawings

The student will have noticed that the pictorial drawings included in this book are much more easily understood than are the other more usual drawings—compare figs. 1 and 2 on page 37. The question will probably be asked: Why not always make drawings in this pictorial way? The reasons for not doing so are that drawings of this kind are often difficult and costly to prepare, usually look distorted, and are frequently difficult to dimension, as will be seen from some of the examples on earlier pages. The kind of article best suited for pictorial representation is one without smooth flowing junctures or curved parts; but these are just the features which many engineering articles must possess. Consider, for example, how well the sharp edges of the dovetailed joint lend themselves to the view given in fig. 1; but against this, the outline of the smoothly rounded link in fig. 2 is not easily drawn pictorially, its details being shown more effectively and drawn more quickly by ordinary views, with sections, as in fig. 3.

Nevertheless, the pictorial view has many advantages, and the student should master the methods of constructing these drawings. He will find much satisfaction in making them and in discovering for himself their limitations.

Isometric drawings

The system we shall use here is known as isometric projection, further discussed in Part 2. For the moment, we shall simply use it, without bothering about the whys and the wherefores.

In this system we shall make full use of the 30°–60° set square or triangle, using three base lines or axes, arranged as in fig. 4. Imagine them to represent the directions of the three edges at the corner of a rectangular prism. Suppose a box has edges 20, 30 and 40 millimetres long. Let us mark off these distances along the three axes, as in fig. 5. We may complete the view of the box by drawing through the extreme points we have marked other lines which are either vertical or inclined at 30° to the horizontal. If we quarter the intervals of fig. 5, as in fig. 6, we get the kind of drawing shown on page 31. This squared-paper drawing seems to exaggerate one defect in all isometric drawings, which is: that they do not look right. What we mean by this is that they do not look like photographs. We are so used to photographic views that we are quick to notice any differences between what is supposed to be a picture, and what the camera gives us. The isometric drawing disturbs us because, in fact, the sets of lines are parallel, whereas in a photograph they are not. The same box is shown in fig. 7 drawn with the groups of lines closing together —to give the converging effect which we notice when we look along railway lines. The view in fig. 7 is much more satisfying.

Exercises

See page 54.

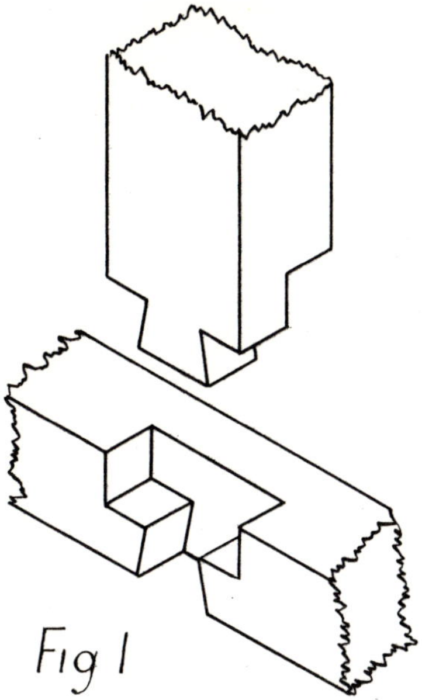

Fig 1

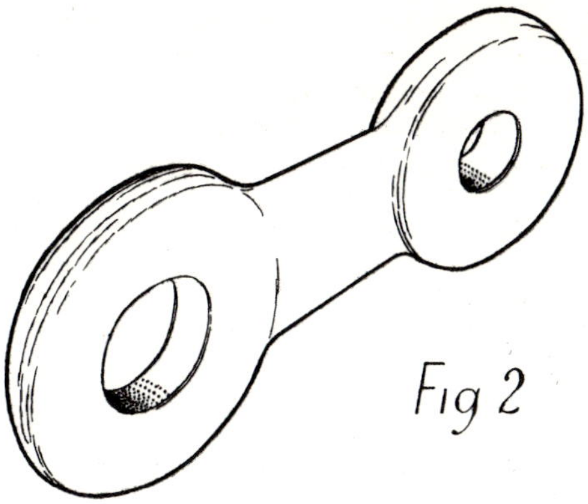

Fig 2

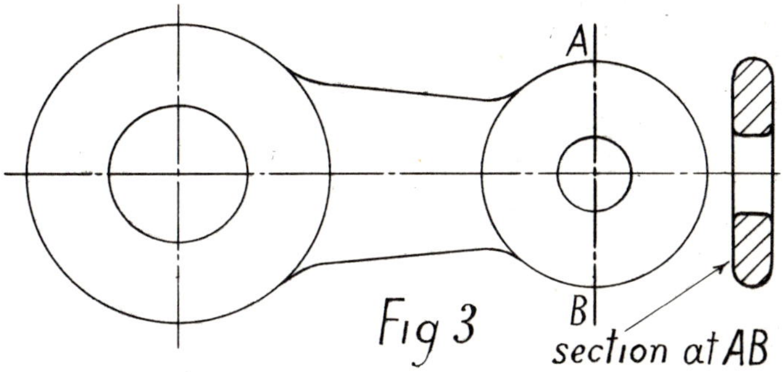

Fig 3

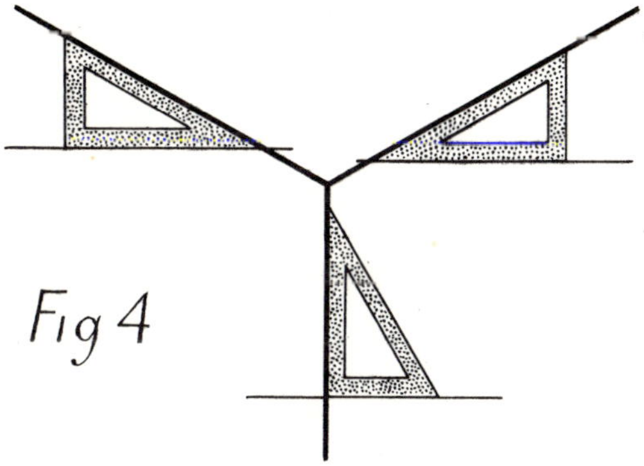

Fig 4

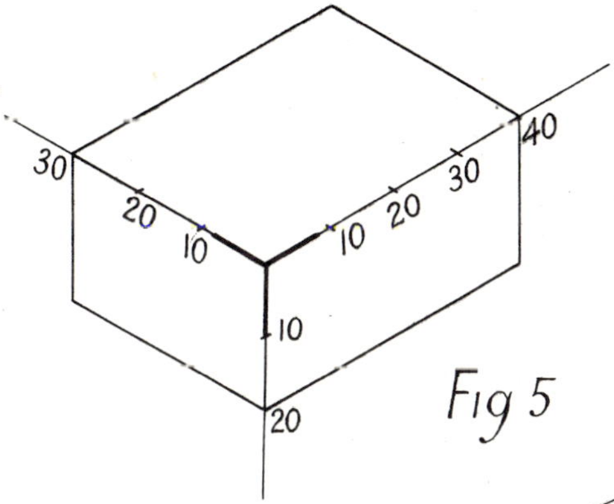

Fig 5

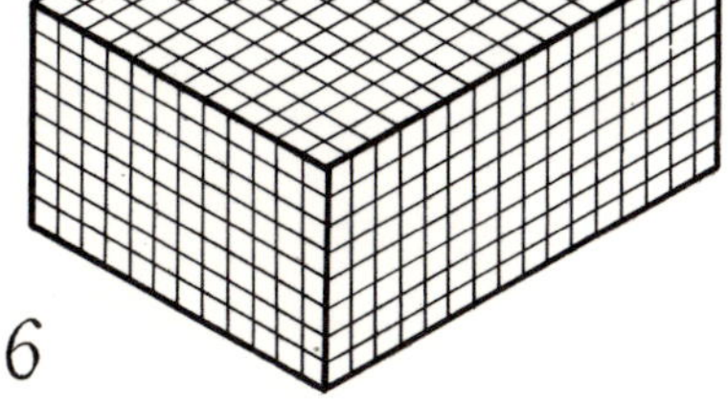

Fig 6

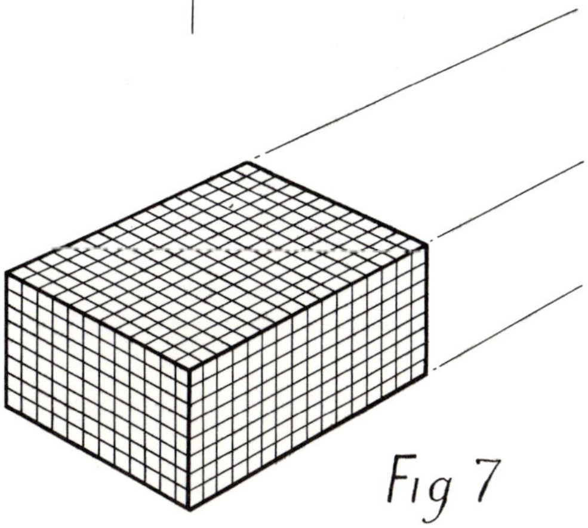

Fig 7

Isometric Drawings

Rectangular solids

We are now in a position to make isometric drawings of many of the objects illustrated in the previous pages.
Refer to page 31, showing the cube standing on the prism. The drawing of this composite solid involves us in first drawing the prism, as on the previous page, and then marking out one side of the cube in position on the top face of the prism, as shown clearly in fig. 1. We then complete the cube, as in fig. 2, using the same length for the vertical edges as for the sides of the square base. The drawing is completed by erasing the lines which will not be seen and thickening the others.
The drawing of the prism with the recessed top, fig. 3, calls for little comment. The depth of the recess is given by the length of the line *ab*, and this can be set off vertically from the corner *a*.

Non-rectangular solids
Reference frame

The shape of many articles may be defined by using an imaginary rectangular frame or box to contain them and then using the edges of the frame as reference lines. An example is shown in fig. 4, where a hexagonal prism is shown within a rectangular reference frame. We can draw the frame, with sides equal to the long and the short diagonals of the hexagon, and transfer distances, such as *a*, *b*, *c* and *d* from fig. 4 to fig. 5. The complete hexagonal prism is shown in fig. 6.
Further applications of this method are dealt with on the next page.

The scale of isometric drawings

The student may have already noticed that lines in the isometric views earlier in the book, are shorter than corresponding lines in the normal drawings. Refer to page 31; compare fig. 3 with figs. 5 and 6. This has been done deliberately; the construction of the scale used, known as the *Isometric Scale*, is discussed in Part 2. When the student draws the following exercises, without using a scale, he will not be surprised if the articles appear a little larger than they should.

Exercises

In the following, the full dimensions are to be used for distances along the isometric axes.

1. Make an isometric drawing of the cube and prism shown in fig. 2. Take the following sizes: Prism, 60 × 50 × 30 mm; Cube 40 mm.
2. The block in fig. 3 has dimensions 60 × 40 × 30 mm. The recess is 10 mm deep and the size of the recess, which is centrally arranged, is 50 × 30 mm. Make a drawing corresponding to fig. 3.
3. A hexagon has a side 50 mm long. It is the face of a hexagonal prism 50 mm thick. Make a drawing of the prism as in fig. 6.

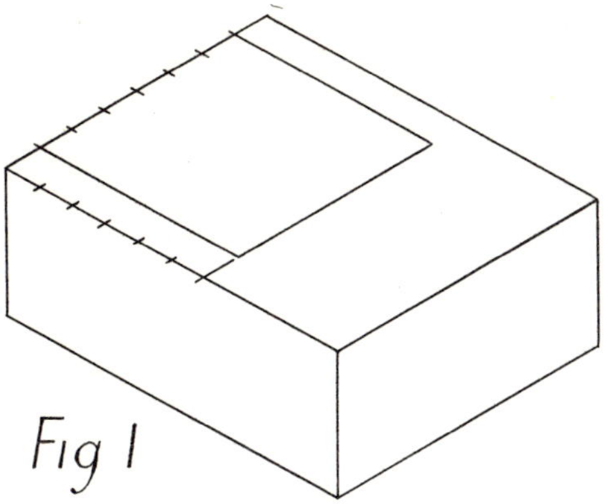

Fig 1

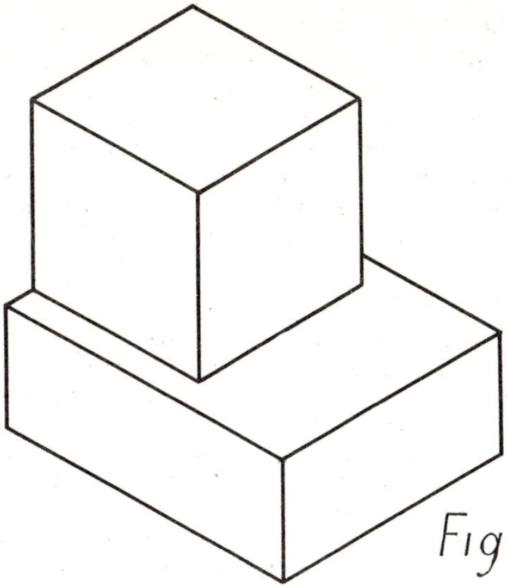

Fig 2

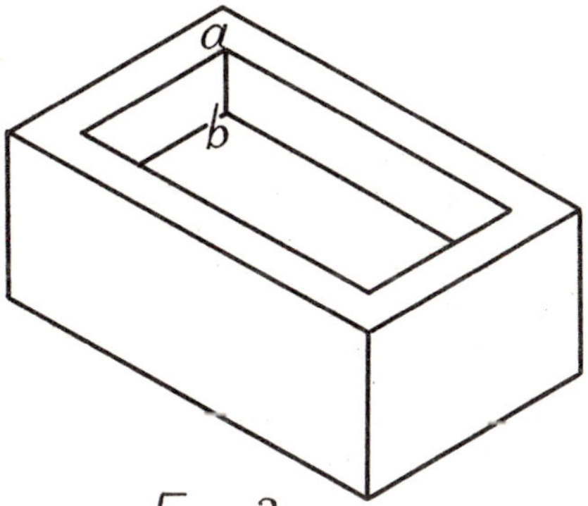

Fig 3

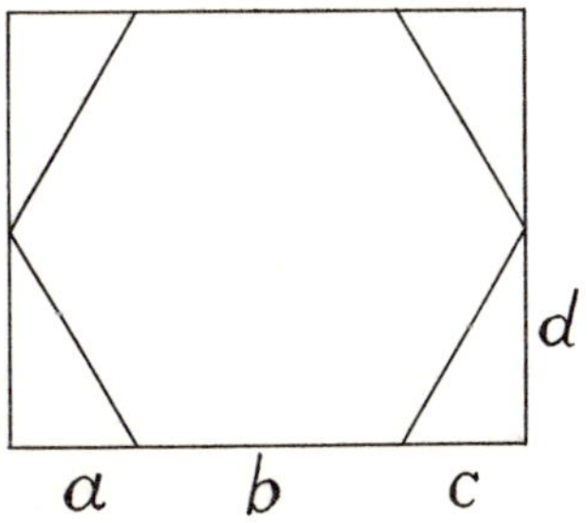

Fig 4

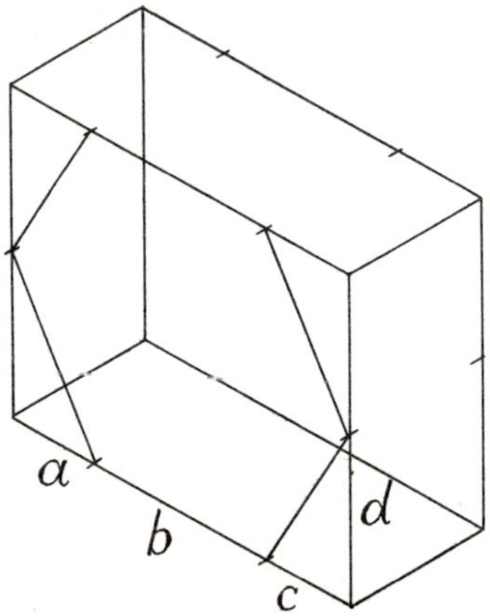

Fig 5

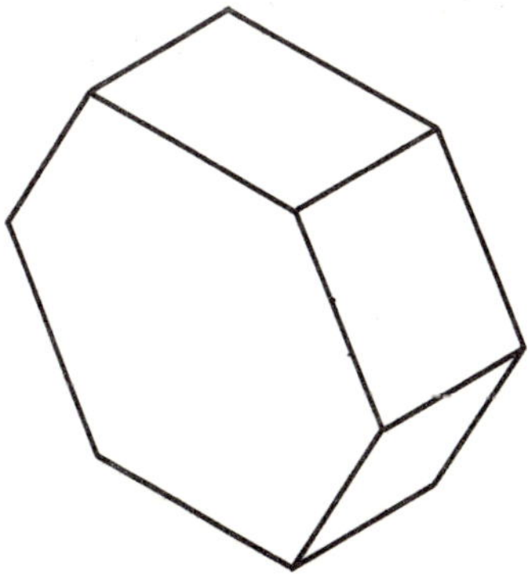

Fig 6

Isometric Drawings: Curves and Circles

The reference frame (continued)

Two further examples of the use of the frame are shown opposite.
(*a*) The solution for a *hexagonal pyramid* is shown in figs. 1, 2 and 3. The hexagon is set out on the base of the frame, the height of which is the vertical height of the pyramid.
(*b*) The two views of the *bracket* shown in fig 4 can be drawn on the sides of a reference frame as in fig. 5. The complete isometric view requires a selection of lines and is given in fig. 6.

Plane curve drawn isometrically

It will be clear that we may deal with any curve or outline on a plane surface by using some method of locating points on the outline. We can use the method of squares. Fig. 7 (left) shows the shaped handle of a plane. Let us enclose this with a rectangle of easy measurement and divide it into a suitable number of squares. We can reproduce this squared rectangle on the side of the isometric frame, as shown in fig. 7 (right). Corresponding points are marked on the squares and the isometric outline drawn through them. The thickness of the handle may be marked on a few selected lines such as *ab* and the second curve drawn parallel to the first.

Circles in isometric views

Although a quick way of drawing nearly true isometric circles is given on the next page, the method shown in fig. 8 is worth noting because of its accuracy and general application. The isometric square which will contain the isometric circle is first drawn, with its diagonals. On one edge of this square we draw a true square with its inscribed circle. A number of points on the true circle are chosen and these are transferred to the isometric view. Four points of obvious importance are those where the circle cuts the diagonals of the square, because they give the greatest and least diameters of the isometric figure; their positions can be obtained by drawing the dotted lines. Four other points are the mid-points of the sides of the isometric square. One other point, *a*, is shown to illustrate the method of transferring any point. We may plot as many other points as necessary, but in fact the eight key points are usually enough to enable us to draw a fair and accurate curve through them.
The resulting curve is an *ellipse*. The longest diameter of the ellipse is called its *major axis;* the shortest diameter is called its *minor axis*.

Exercises

1. **Refer to pages 38 and 39. Make isometric drawings like those in figs. 4, 5, 6, and 7, setting off the full distances along the isometric axes.**
2. **Refer to page 38. Imagine the object shown in fig. 8 turned around so that the slot at the back now comes to the front. Make an isometric view of the object in this position.**
3. **Refer to page 43. Draw, to a scale of 1:10, an isometric view of the object shown in fig. 5 but with the hinged front in a *horizontal* position.**

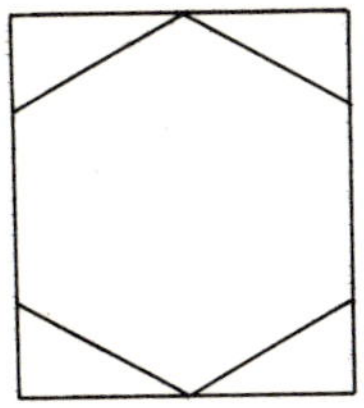

Fig 1

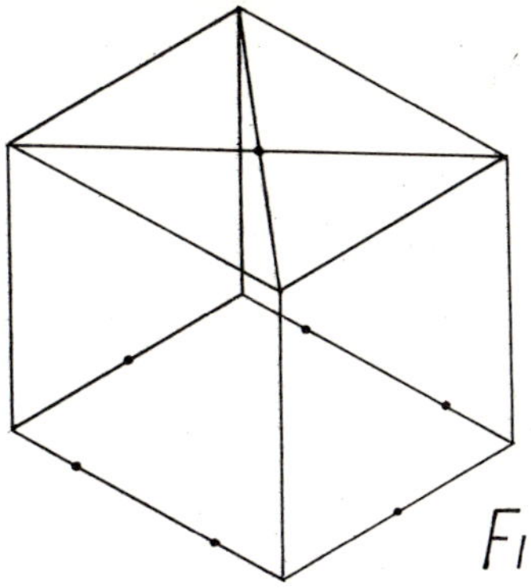

Fig 2

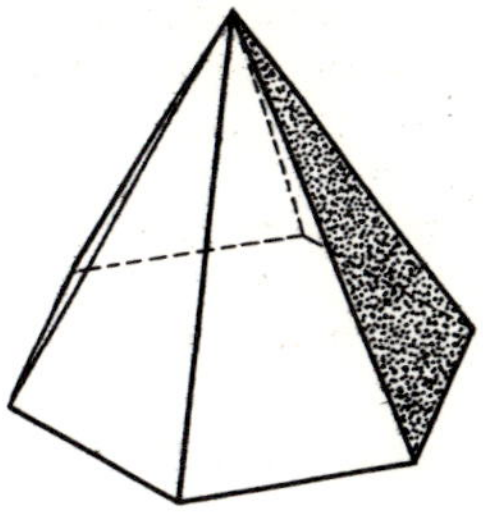

Fig 3

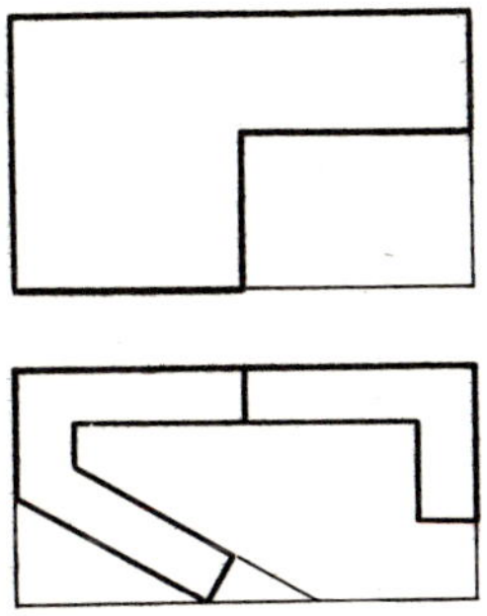

Fig 4

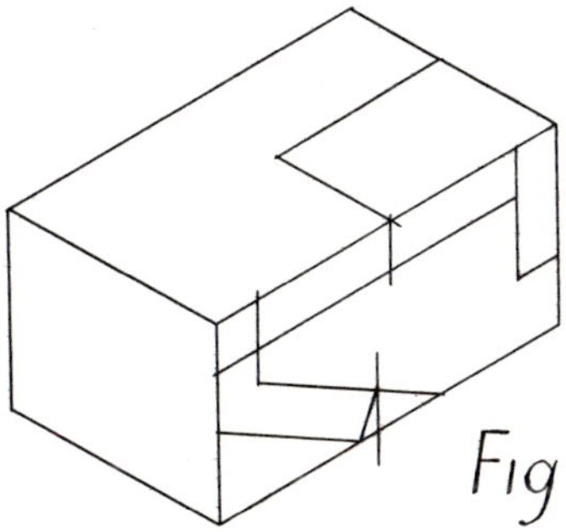

Fig 5

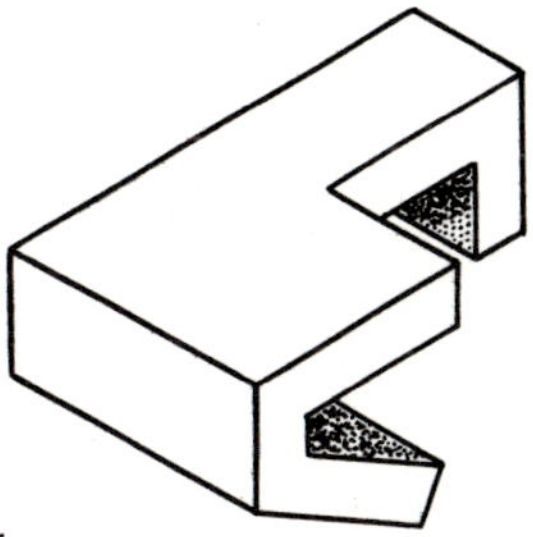

Fig 6

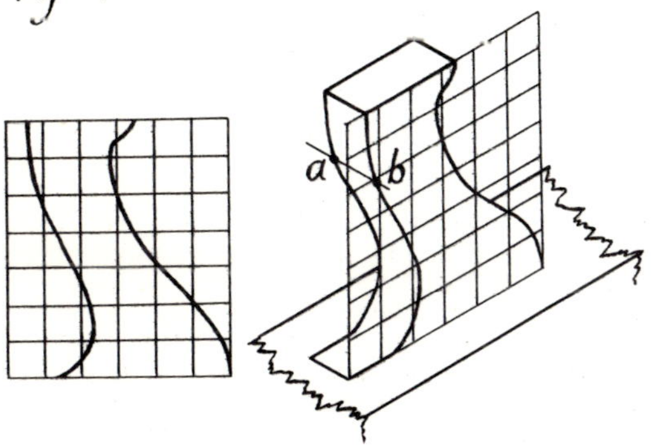

Fig 7

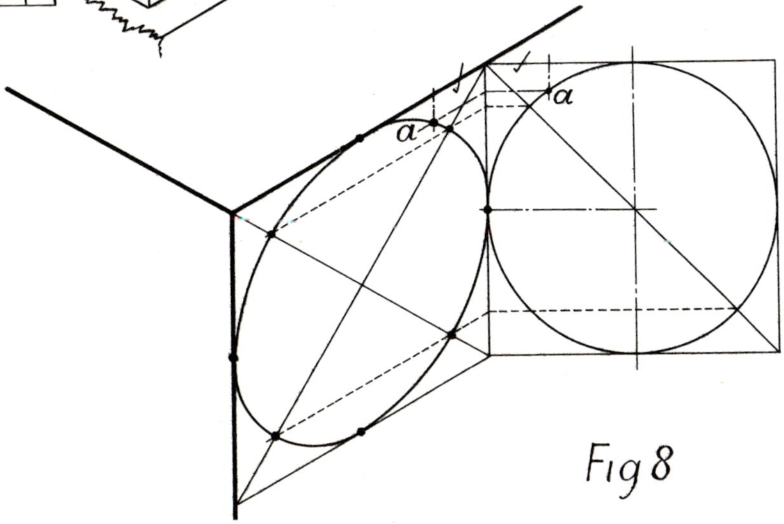

Fig 8

Isometric Drawings: Circles

A method for quickly constructing *approximate isometric circles*, using compasses and four points for circular arcs, will now be considered. The method is described pictorially in the step-by-step drawings opposite. In fig. 1, the isometric square has two centre lines drawn parallel to the sides. Four additional lines, thickened, are added in fig. 2. Four points on these lines are marked in fig. 3; these points are the centres of arcs, the radii of which are given by the two thickened lines. The drawing of the two pairs of arcs gives the curve, in fig. 4, which is nearly an ellipse. A shorter construction is given in fig. 5 and this should be adopted; the same centres and arcs are given by drawing three lines only—one, the diagonal, at 60°, and two horizontals. Using this method, near ellipses are drawn in the three isometric squares in fig. 6. *Note that the four centres and the two radii, on each face, are given by drawing horizontals, and lines at 60° to the horizontal, through the corners of the isometric squares.*

It should now be clear that circles lying in the faces of rectangular solids (or which are parallel to these faces) will normally appear as the ellipses shown in fig. 6. From this we may deduce two very important conclusions: firstly, that the major axis of an ellipse will be either horizontal, as at *a*, or inclined at 60° to the horizontal, as at *b* and *c*; and secondly, that the minor axis of an ellipse will be at right angles to the major axis, so that it will be either vertical, or inclined at 30° to the horizontal. These facts are of great help when we draw freehand isometric views of objects with circular parts.

The form of a circle in an isometric view is a true ellipse, which is narrower and longer than the approximate figures drawn with compasses. A comparison between the two curves is shown in fig. 7. The curve drawn with compasses is the less attractive, but it is quite good enough for use in most isometric drawings.

Omit at first reading

Another important factor concerns the major axis of the ellipse. This must always be equal to the true diameter of the circle. If the student holds a saucer in his hand and looks edgeways at the rim, he will see an ellipse, the major axis of which is the full diameter of the saucer. By the construction we have adopted, marking actual distances along the three isometric axes, the major axis of each ellipse in fig. 6 is greater than the diameter of the circle which it represents (which is equal to the side of the square). As may be checked by measurement, the increase is about 22 per cent. If this enlargement is undesirable, then a scale must be used which makes the major axis equal to the diameter of the circle. The construction of such a scale is dealt with in Part 2.

Exercises

1. Draw an isometric cube of 100 mm side and draw inscribed circles as in fig. 6, using the approximate method.
2. Draw one 100-mm isometric circle, by plotting a few points as in fig. 8 on page 57, thus obtaining a true ellipse. Compare the two curves.
3. More difficult exercises. Refer to pages 47 and 51. Draw isometric views like those in figs. 6 and 7, page 47, and figs. 4 and 5, page 51. Omit the section planes.

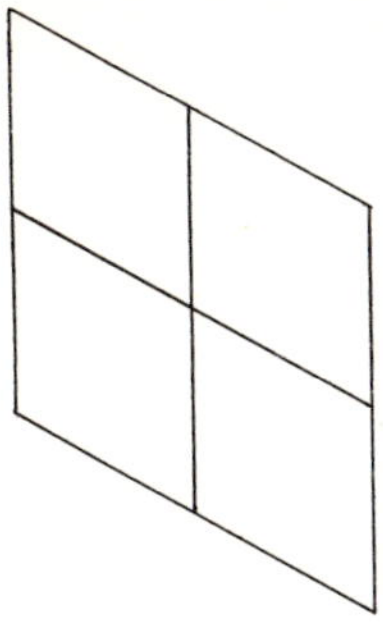

Fig 1

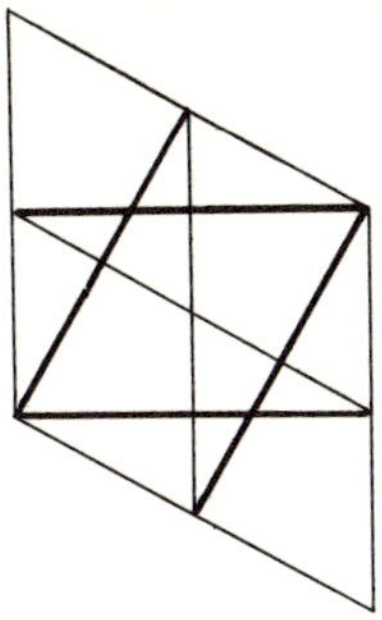

Fig 2

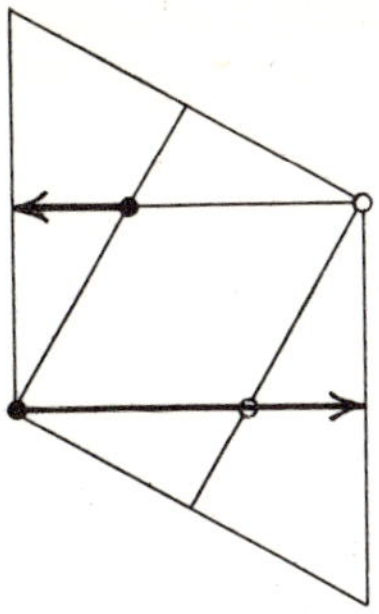

Fig 3

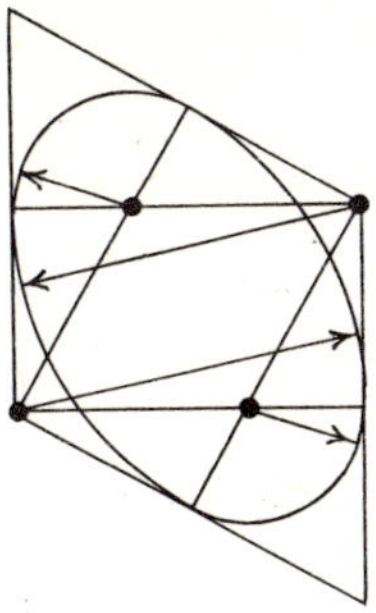

Fig 4

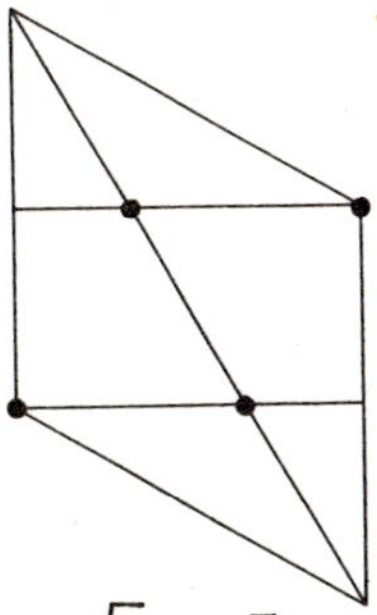

Fig 5

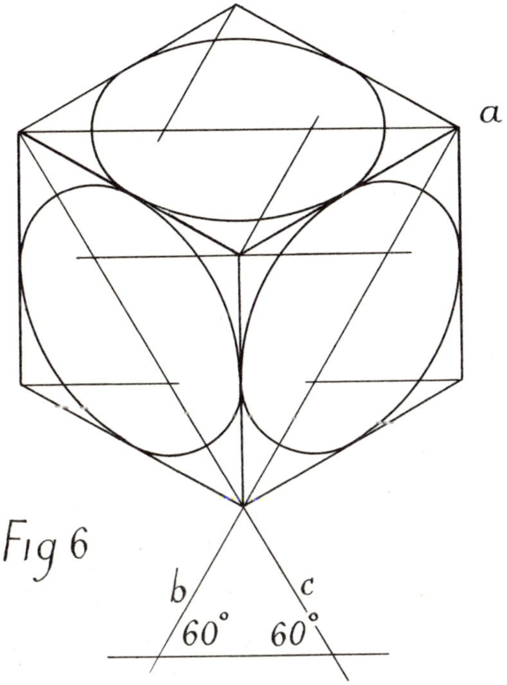

Fig 6

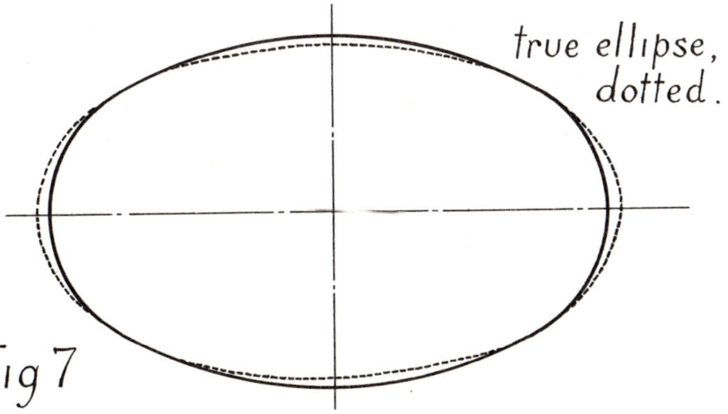

Fig 7

Isometric Drawings: Variety of Views

The student will have realized that it is possible to make many different isometric drawings of the same object. Eight such drawings are shown opposite.

Three faces of an object of rectangular form can be shown in one isometric drawing, but not more than three. As there are six faces, two drawings will show them all; but to do so effectively the views must be selected to the best advantage.

A rectangular block is shown with its edges in the directions of the isometric axes. The faces are lettered A B C D E and F. We shall apply this lettering to the main faces of the object shown by views 1 to 8, which all refer to the same object. In views 1 to 4, face A is uppermost; in views 5 to 8, this face is underneath. Similar sets of eight views can be prepared with face B, or face C, first uppermost and then underneath. These 16 additional drawings will differ from the eight shown, although they will have similarities. There are thus 24 views available, and we have usually to choose one, and sometimes two. The view or views selected must of course convey all the essential information.

A great deal may be learned from the construction and study of views such as those opposite. The student is advised actually to make the object shown, in wood, and then to arrange it before him so that all the various views can be seen and assessed. Suitable dimensions, in millimetres, are: Size of block before shaping, 60 x 40 x 30; recess, width 20, depth 15, length 50; parallel step, 20 wide, 10 deep; triangular step, 15 x 40, 10 deep. Dimensions have not been marked on a view opposite in order that the outlines may remain quite clear.

Avoidance of sharp corners and edges

It has been convenient for the purpose of illustration to show objects with sharp corners and edges. The student should not be left with the impression that this is common engineering practice; it is quite the contrary. Sharp corners give weak features in a design, and they are never accepted without question. The more usual practice is illustrated in fig. 1 on page 29.

Exercises

1. Using the dimensions given above, draw isometric views 1 and 2. By the help of these views, draw any two of the other 6 views shown, but without reference to the book.
2. Draw an isometric view of the object standing on face D, with faces A and B towards you.
3. Which two views together do you think give the clearest impression of the object? (The author suggests 1 and 5.) Draw these two views. Then put in the essential dimensions in such a way that the block could be made from your drawings.
4. Make four ordinary views of the object: on face A, on B, on C and on F.

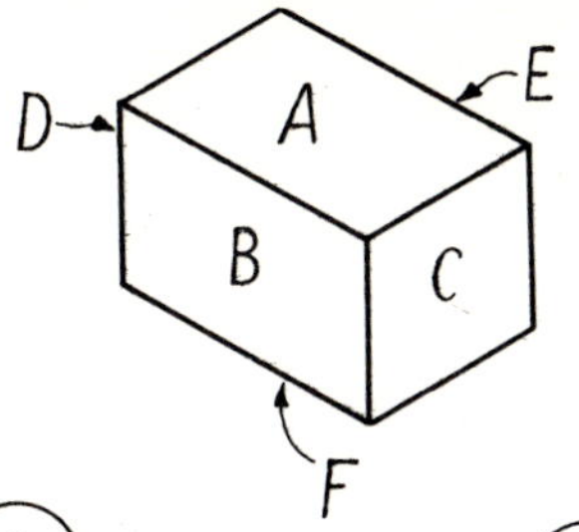

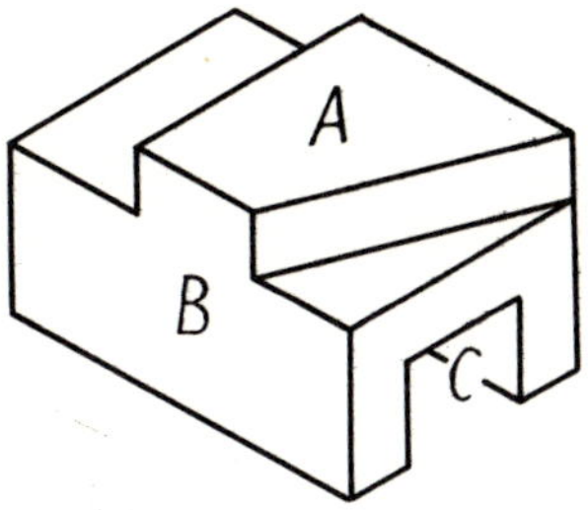

1

2

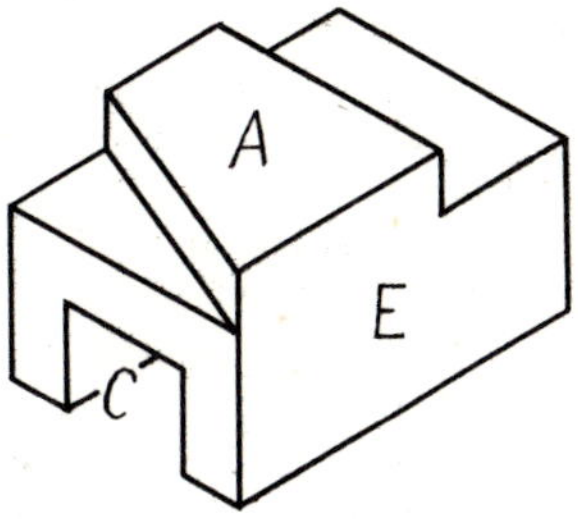

3

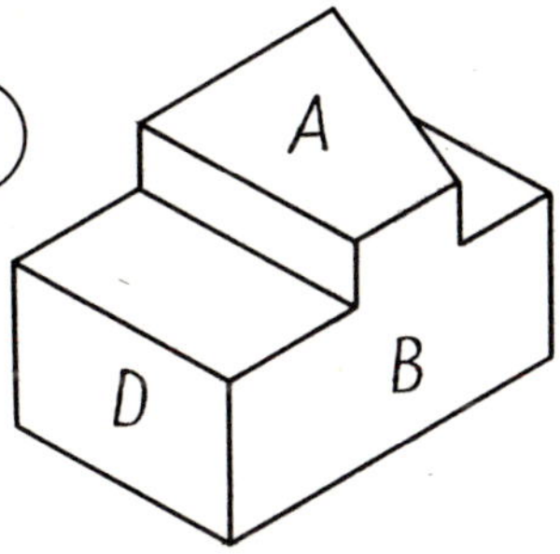

4

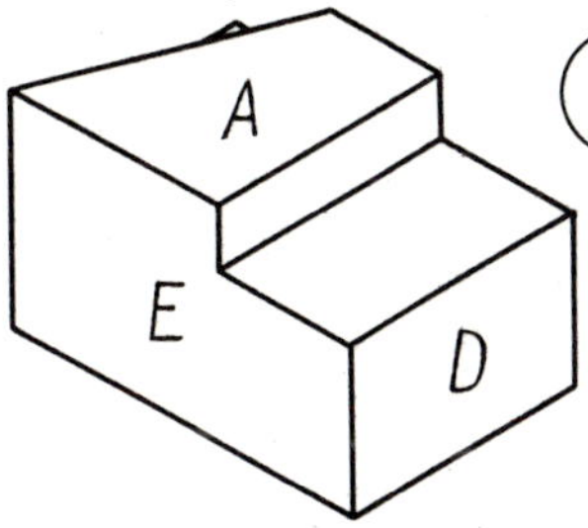

5

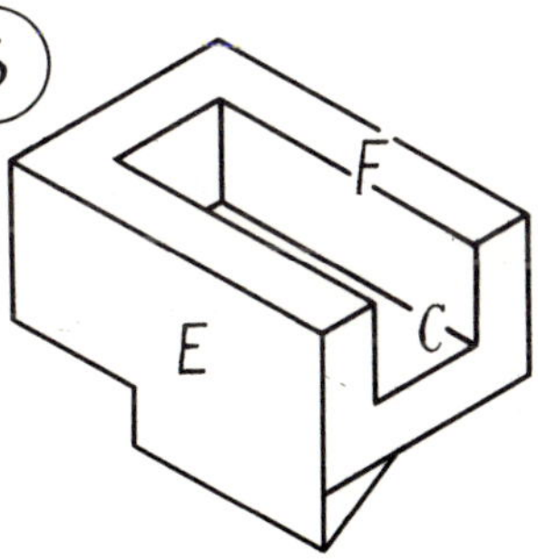

6

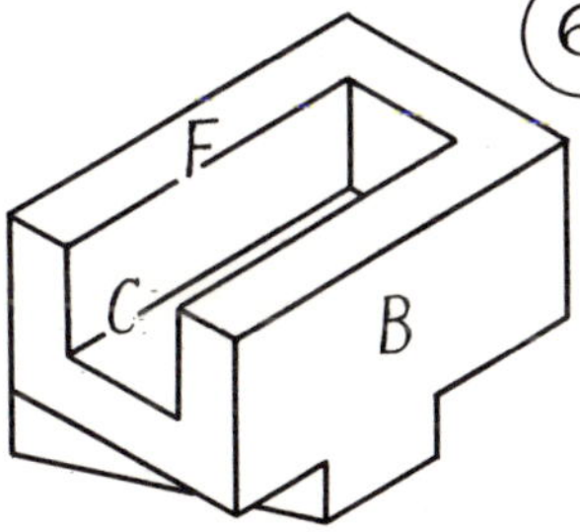

7

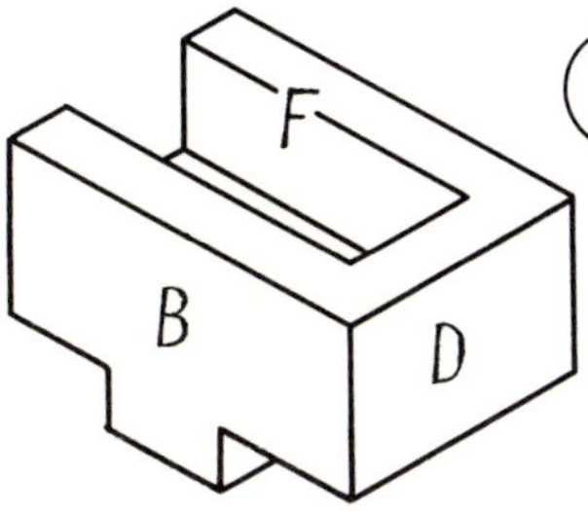

8

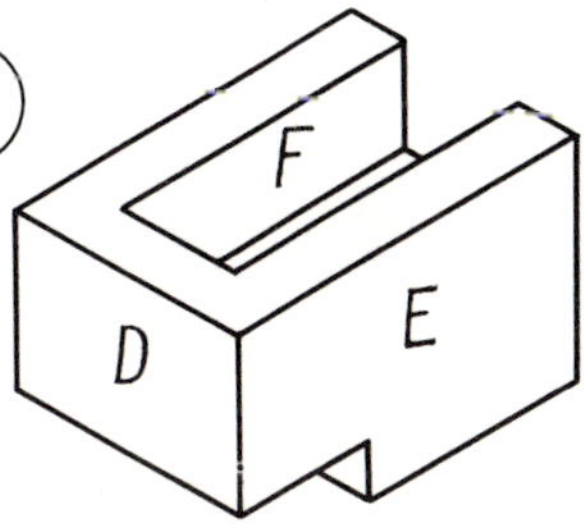

Dimensioning

The correct dimensioning of a technical drawing requires much thought and a knowledge of many rules. We shall therefore deal with the subject only briefly here, and give a few of the more important general rules.

Lettering and numerals

The types of printing to be adopted are dealt with later. It is sufficient to suggest here that the student uses the simple letter forms adopted in this book. Guide lines for lettering should be very thin, so that they are not noticeable when the lettering is completed. It should be remembered that the importance of lettering can be increased by broadening the letters without increasing their height. See the words DRAWING opposite.

Rules (see also page 154)

Extension lines should be about 1 mm clear of the outline and should extend about 2 mm beyond the dimension line. The dimension line may be broken for the dimension. Both extension and dimension lines should be thin and continuous. Figures should normally be at right angles to the dimension lines and should read either from the bottom or from the right-hand side, as in fig. 1.
Arrow heads should be sharp and either open, as elsewhere in this book, or blackened in; see opposite.
The principal view of an object should carry the main dimensions, which should not generally be repeated. Overall sizes should be given.
Centre lines should not be used as dimension lines. Holes and cylinders should be dimensioned from their centre lines, not from their boundary surfaces. Diameters are preferable to radii. Dimensions should not normally be taken from dotted lines.
In general, dimensions should be placed outside of the main outlines.

Line thicknesses

Outline: bold; *extension and dimension lines:* thin and continuous; *centre lines:* thin chain; *hidden parts:* bold, dotted; ***sections:*** thin chain dotted, thickened at the ends. See page 34.

Example

Two similar drawings of the object in fig. 3 are shown in figs. 2 and 4. Fig. 2 is correct. There are many deliberate faults in the dimensioning of fig. 4. The student should identify all the errors.

Exercises

See pages 64 and 66.

DRAWING DRAWING DRAWING

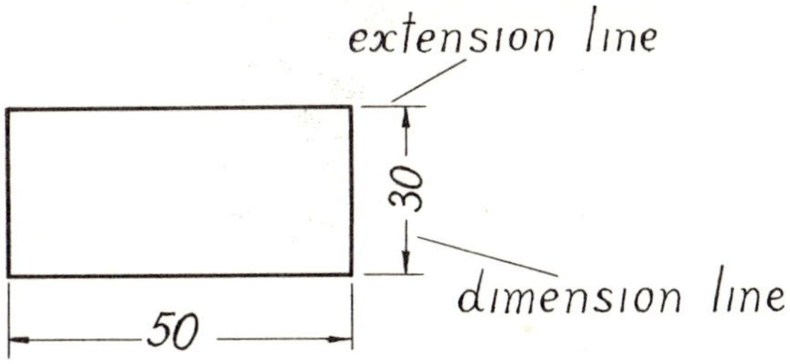

Fig 1

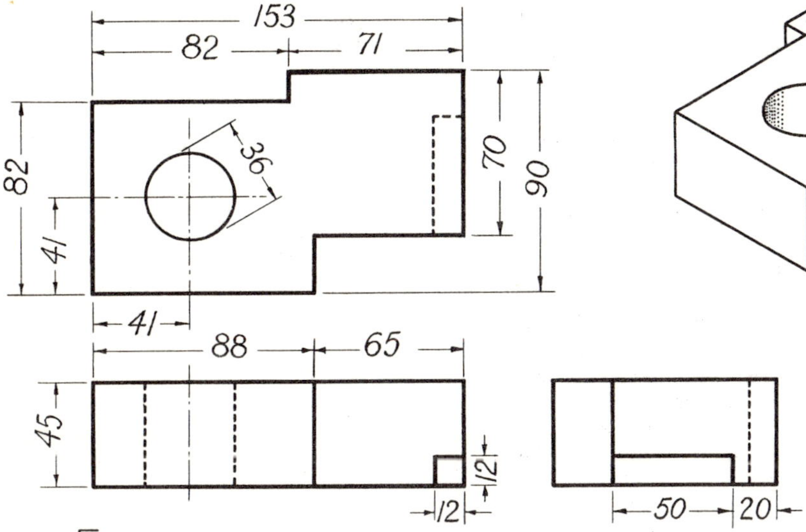

Fig 2 CORRECT

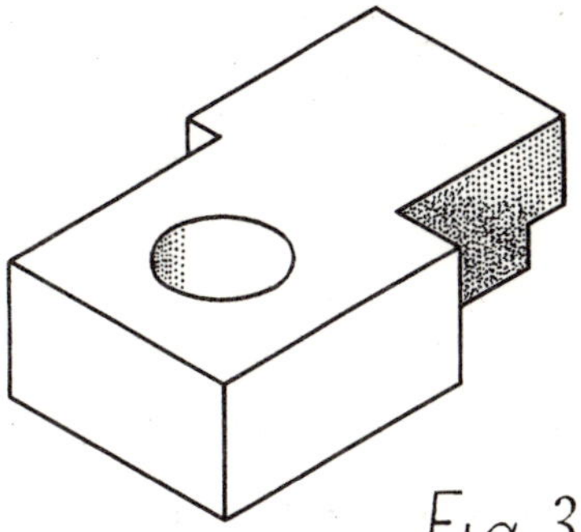

Fig 3

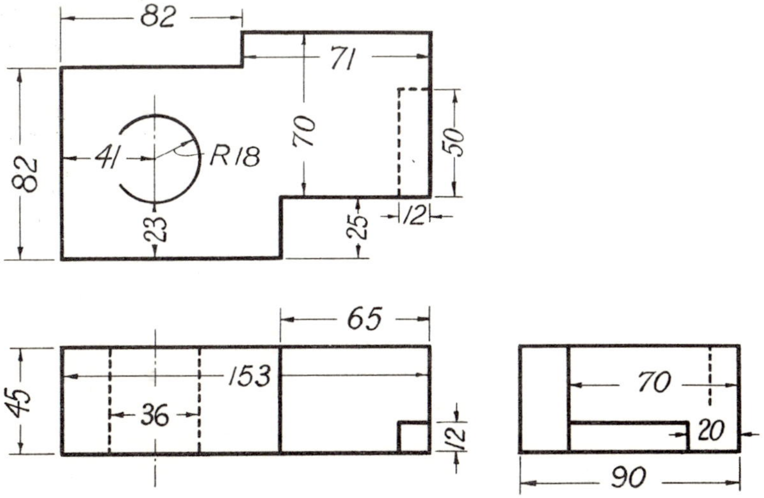

Fig 4 INCORRECT

Freehand Sketching

The ability to make freehand sketches is a valuable asset, acquired by practice. The student will find that sketching makes him observant and accurate. He should lose no opportunity of sketching articles in the workshop and the laboratory. Some suitable objects are illustrated on the following page. If these or others like them are not available in model form, then the student must fall back on the sketching of sketches—never very satisfactory.

Requirements

Any sketch which is made should satisfy the following requirements: (*a*) It should describe the shape of the object completely, showing the relative parts in fair proportion but not to any particular scale. (*b*) It should carry all essential dimensions. (*c*) It should have notes to specify, for example, the material of which the object is made and the method of manufacture.

Procedure

The rules for making sketches are few. Straight lines are best drawn lightly in pieces and then made continuous. Circles require a few marked diameters. The bare outlines of the views should be spaced out in thin lines, and centre lines should be inserted. See fig. 1.

The details can now be added, fig. 2. It may occur to the student that a sectioned view would make matters clearer,* also that a fourth view is desirable. Then follows the final thickening in and the addition of dimensions.

For pictorial views, the three isometric axes should be drawn, together with, if necessary, the rectangular frame which would hold the object. To draw the projecting cylinder we must concentrate on the hidden face and locate the position of the circle on that face, as in fig. 3. Note the directions of the major axes of the two ellipses. The final stage is shown in fig. 4.

A small part of the sketch of a foundry by Leonardo da Vinci, drawn about 450 years ago, is shown to indicate a high standard of execution. Another fine example is given in the frontispiece. The student should examine the sketches of this great draughtsman and artist.

Exercises

1. Visualize the block in fig. 4, and then sketch it unseen. Compare results. Dimension your drawing to the following particulars (mm): size of block, 100 × 80 × 55; recess, 50 × 30 × 25; hole, 40 diameter, central, 30 from end; pin, 25 diameter, 30 long, central, 20 from end.
2. Using your pictorial drawing only, make freehand views corresponding to those in fig. 2 and insert dimensions. Label your views and add notes stating that the block is to be of mild steel, machined all over.

* The section plane used here would remove the pin with the portion cut away. The pin may be shown by 'ghost' lines.

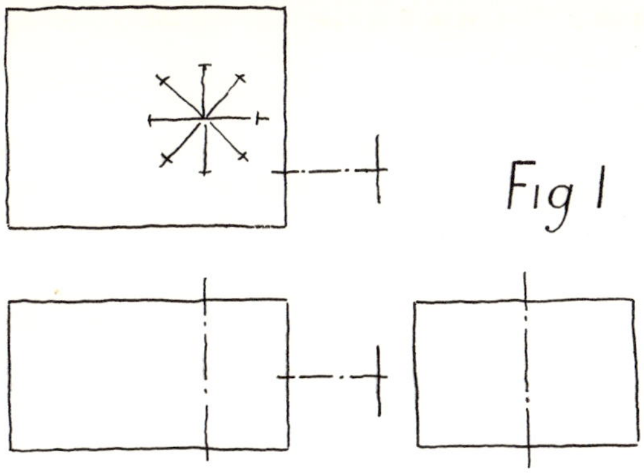

Fig 1

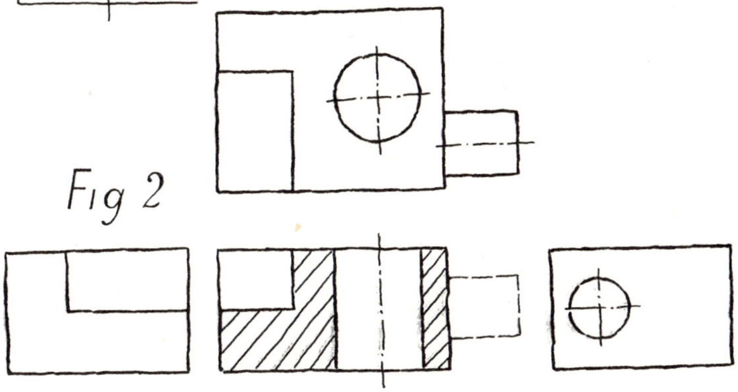

Fig 2

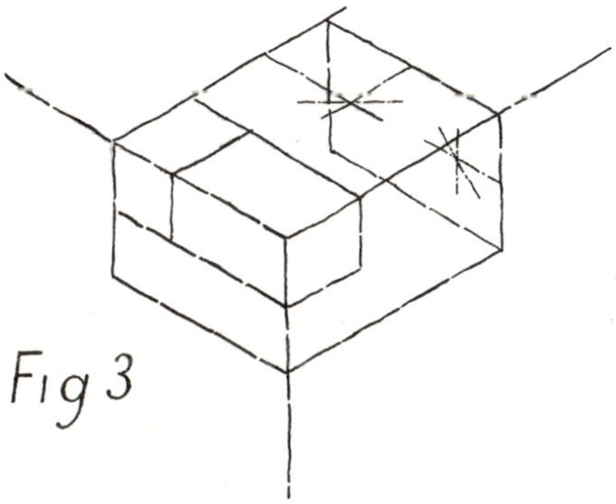

Fig 3

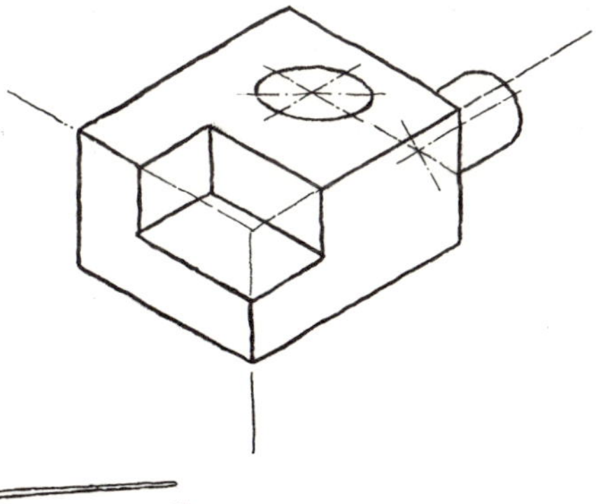

Fig 4

FRAGMENT
CANNON FOUNDRY
LEONARDO DA VINCI

Conversion to ordinary drawings
Exercises for freehand sketching and also scale drawing

On this page and that opposite are given isometric drawings of various objects to scale, graded in difficulty. The objects represent models, and if models can be used, so much the better.

The student should copy a pictorial drawing freehand. He should then measure the dimensions of the object shown and enter the sizes on his own drawing. Measurements must *always* be taken in the direction of the isometric axes, never in any other direction. All the objects are shown half size and measurements should therefore be made with a suitable scale. Alternatively, the drawings may be measured direct and the sizes multiplied by two.

From his pictorial sketch the student should make a two-, three-, or four-view drawing, full size, arranged in the best order, complete with dimensions.

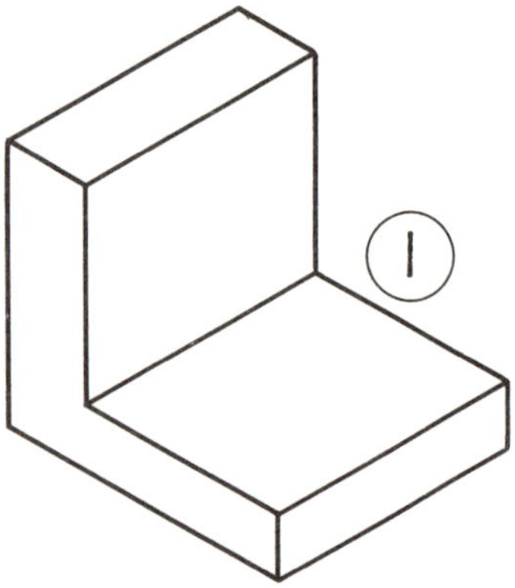

1

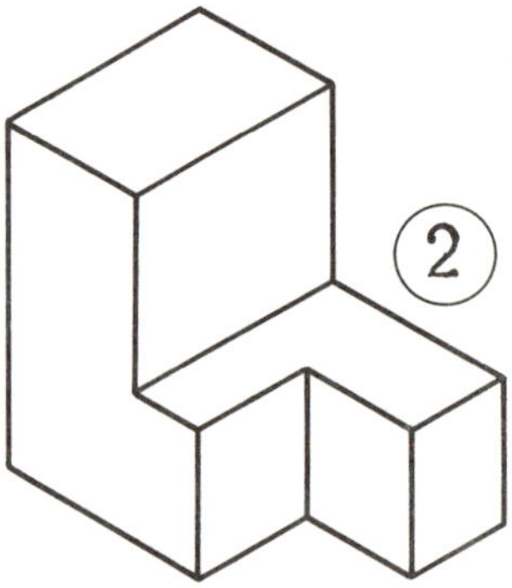

2

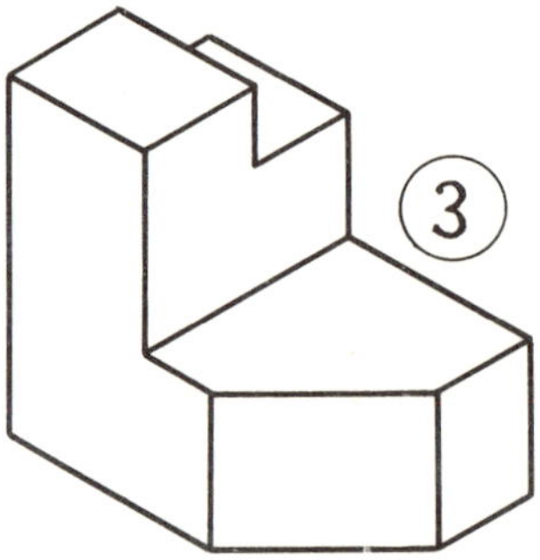

3

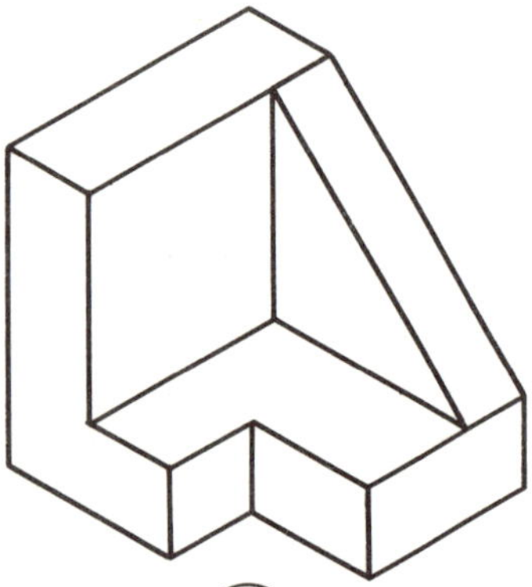

4

5

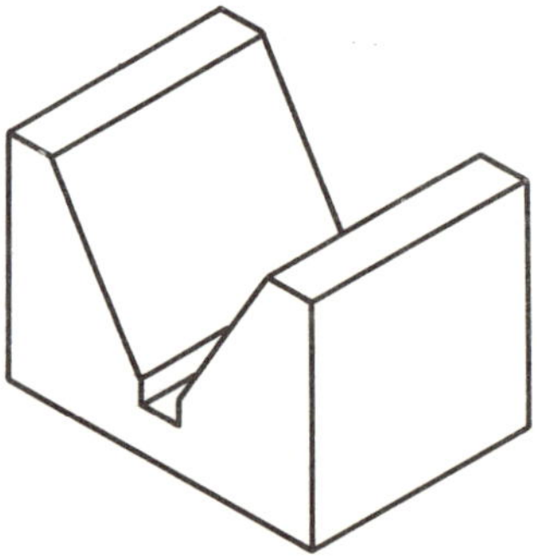

6

(7)

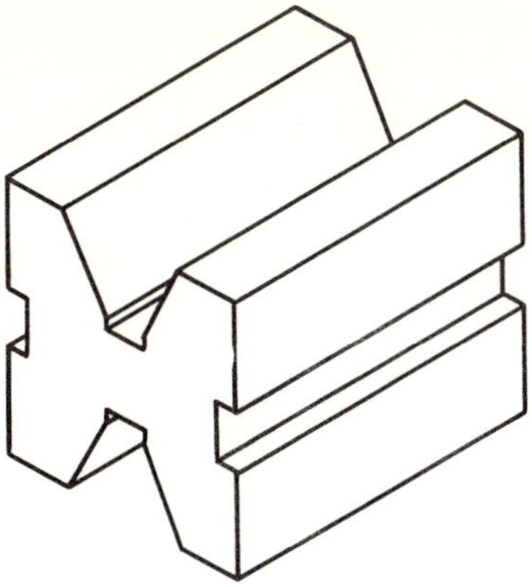

(8)

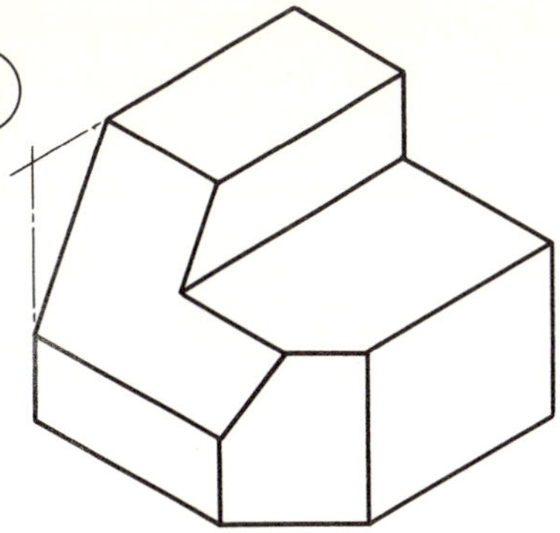

(9)

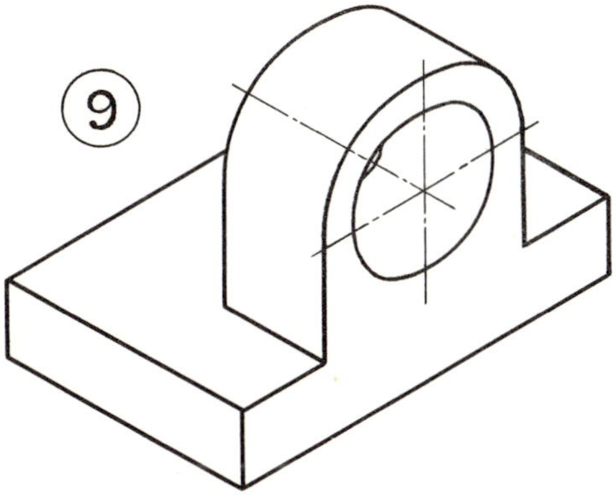

(10)

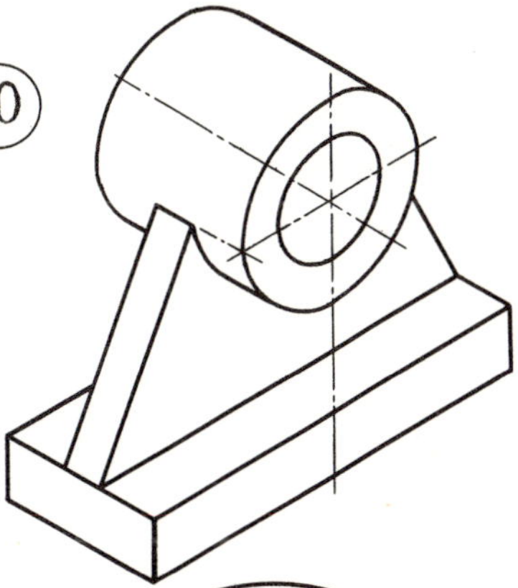

(11)

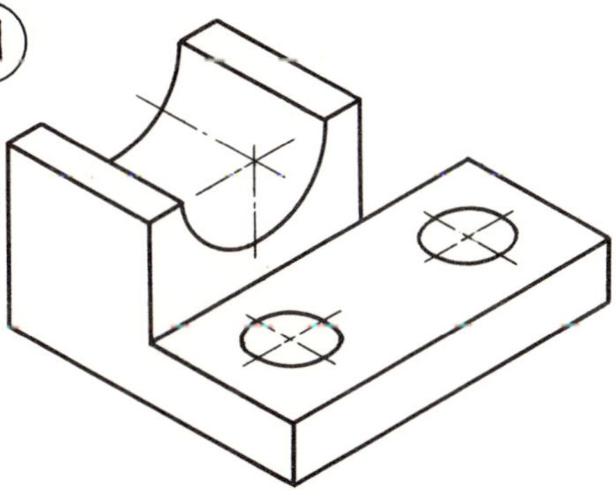

(12)

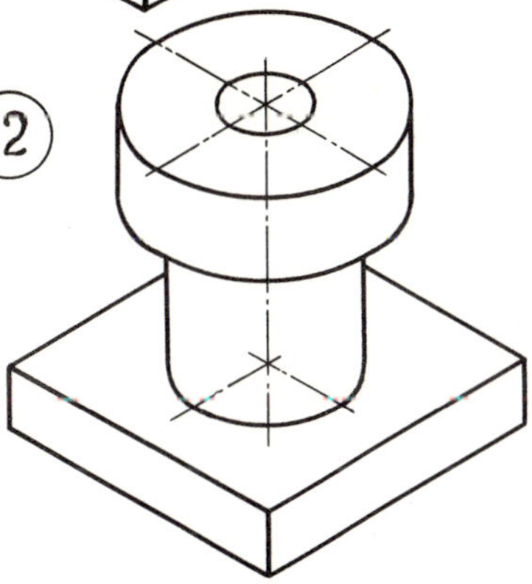

(13)

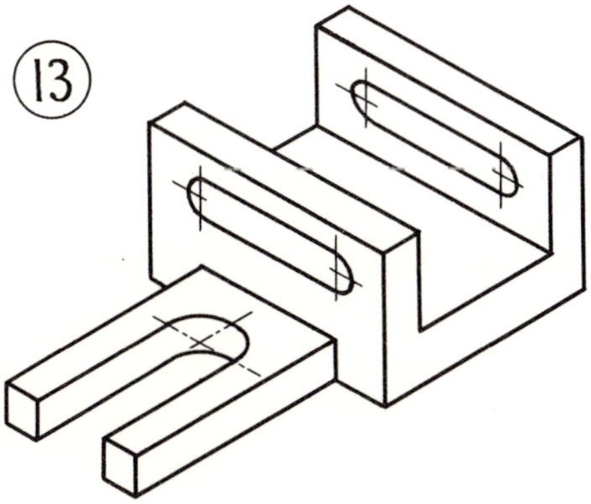

(14)

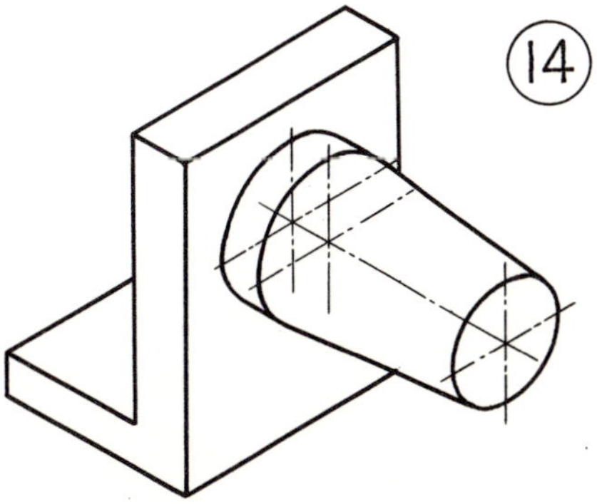

Interpretation of Drawings

Conversion to pictorial views (may be deferred)

It has already been said that the *interpretation* of a technical drawing may be more difficult than the *preparation* of the drawing; it is sometimes much more difficult. In other words, in this graphic language the reading is often more difficult than the writing.

When a draughtsman is preparing a drawing he has the shape of an object in his mind. Perhaps he has made a sketch such as those on pages 66 and 67. But others who will use his drawing have to build up this shape, mentally, from the lines he has used; and this is rarely easy.

It is always a great help to the understanding to convert the plain drawing into its pictorial counterpart. Some examples are given here for the student to try. He should first draw *freehand* the isometric box to include the largest dimensions. He should next sketch lightly on each face the lines in the corresponding view for that face. This has been done for examples (1) and (2) below. The rest is now a matter of visualization; and very little advice upon it can be given. It can be said however that once he has got the solution, the student will usually know that it is right. The solution for example (1) has been completed; lines such as *a*, *b* and *c* are often omitted in error. If scale drawings are made, distances should be marked only along the isometric axes.

The student should regard the group of exercises as a challenge. He should get as far as he can with them, and then refer to the solutions given on page 70.

The arrangement of views is the same for each example, the view being *nearest* the face it describes. The arrangement satisfies what is later described as *Third Angle Projection*.

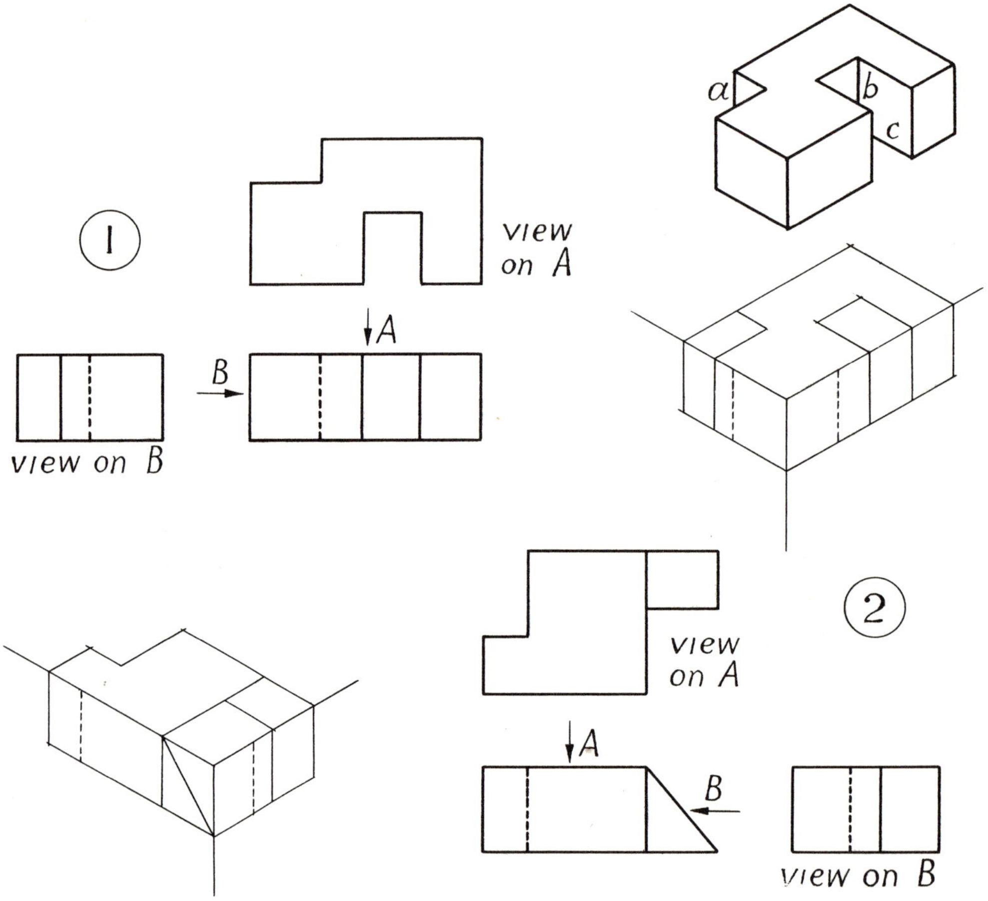

3

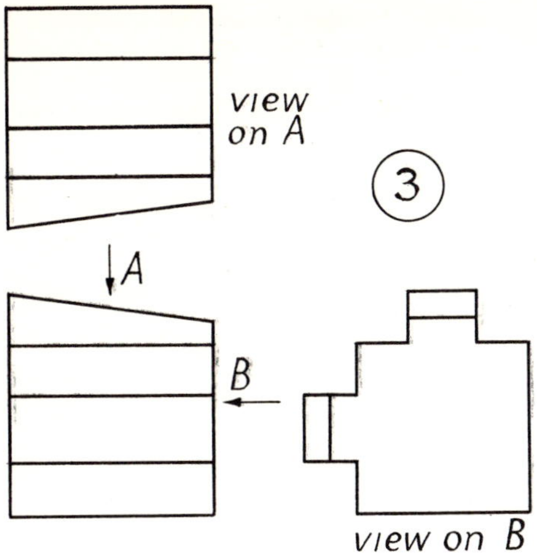
view on A
A
B
view on B

4

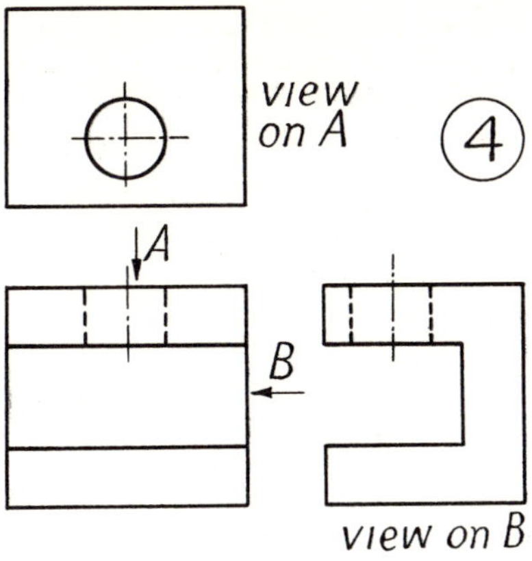
view on A
A
B
view on B

5

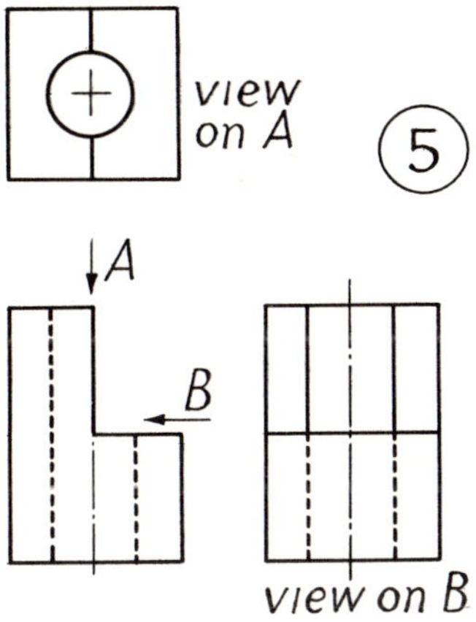
view on A
A
B
view on B

6

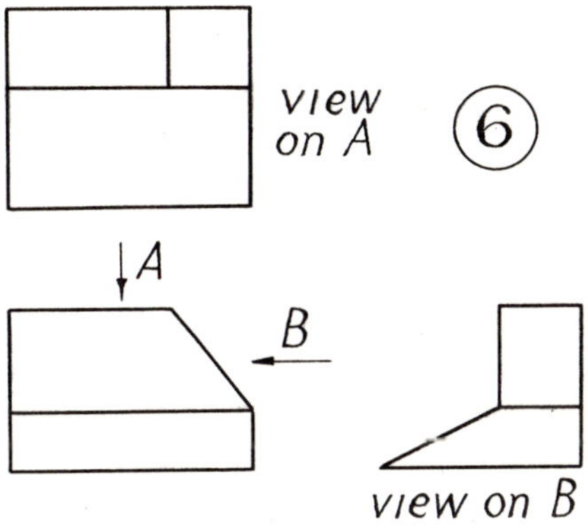
view on A
A
B
view on B

7

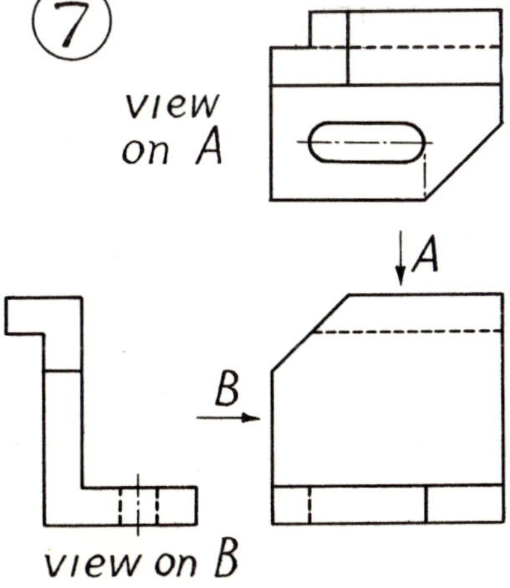
view on A
A
B
view on B

8

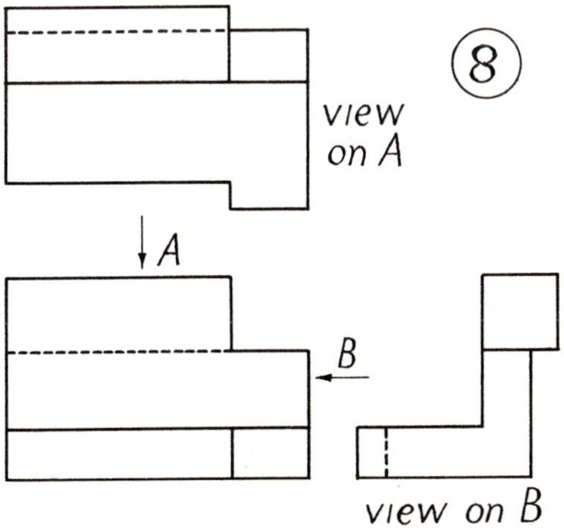
view on A
A
B
view on B

SOLUTIONS

(2)

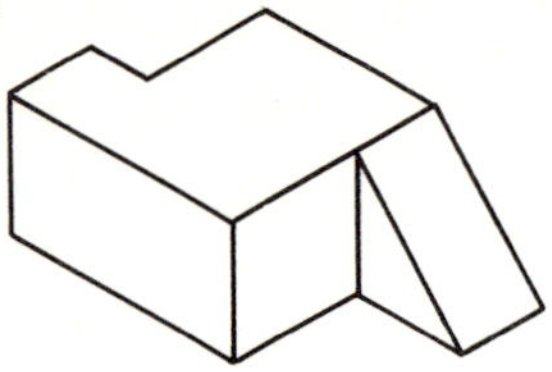

(3)

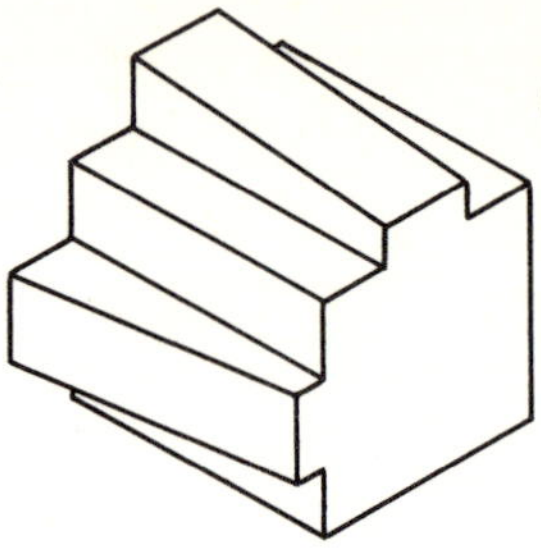

(4)

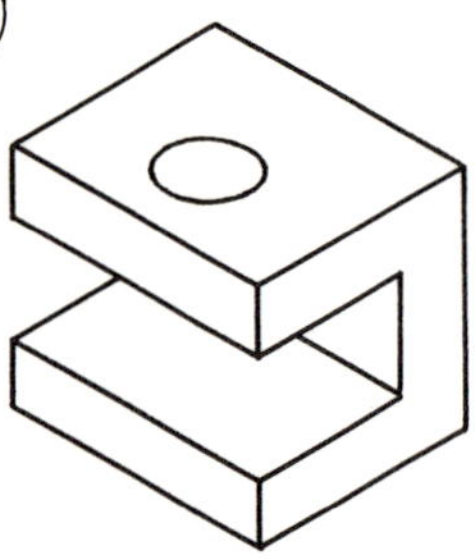

(5)

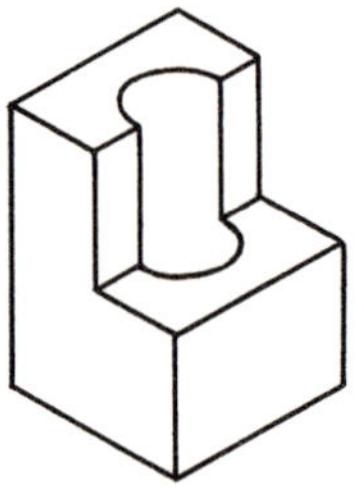

(6)

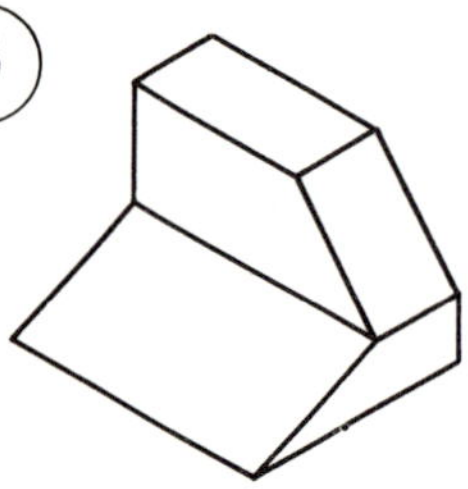

(7)

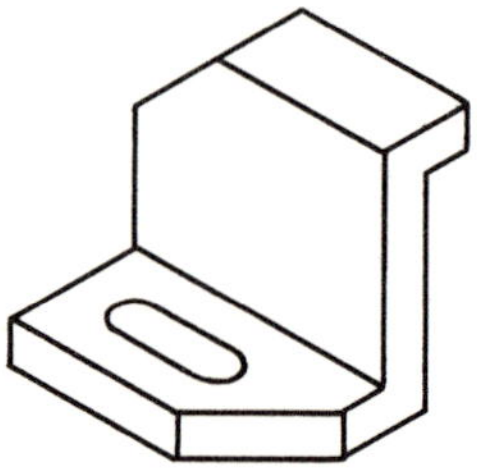

(8)

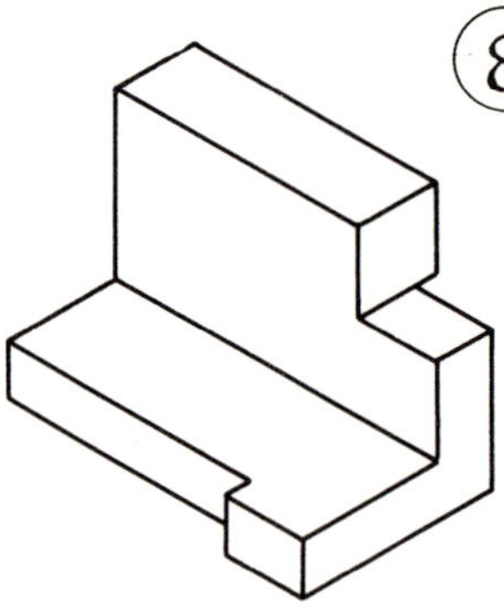

Part Two

Applied Geometry

Accuracy in Drawings

The use of the tee square and set squares has been discussed in Part I, and possible set-square errors were mentioned. We must now take this question of errors more seriously, for the effects of the combination of inaccuracies from equipment can be serious. The edge of the drawing board and the blade of the tee square are often not straight; wear occurs where some parts of these edges are used more than others. Set squares slightly change shape with time and use.

Checking a square

Let us test our equipment by using a square of, say, 150 mm side, drawn as in fig. 1. Using a fine hard pencil point to make the marks, apply a strip of paper to sides and diagonals, as in fig. 2, and compare the lengths. It will be surprising if obvious inaccuracies are not revealed. Small errors of this kind are enough to make it impossible to obtain the series of true rectangles on page 77, see ex. 1, page 76.

If it is not practicable for us to have the equipment corrected, then, when accurate results are important, we must use constructions which do not rely on the tee square and set square.

Euclid's constructions

In these we may use only a straight edge and compasses. We must check and correct the straight edge. The compasses must have a fine needle point and a hard sharp pencil point; the joints should not be slack.

Using only the compasses, we may set off a right angle at a point P in a line, as in fig 3. Continuing as in fig. 4 we obtain the four corners of a square. Errors may still creep in, however; it is surprisingly easy to miss the *exact* point of intersection of the arcs at *a*, *b* or *c*, fig. 3.

Drawing-board solutions

The graphical solution of problems appeals to engineers and others, not only because the steps are usually clear and the work speedy, but because the results are as accurate as the measurements justify. Nevertheless avoidable errors must not arise in the working, as is so often the case. The accuracy of the drawing equipment must never be taken for granted.

Exercises

1. Using tee square and set squares, draw a rectangle of sides 150 and 100 mm. Check the accuracy of your work. Then repeat the construction on tracing (or transparent) paper using a good rule and compasses. Place the tracing over the first drawing and compare the two rectangles.
2. Draw any square *abcd*, of side 50 mm, fig. 5. Bisect *ab* and *cd* to give *fe*. With centre *e* and radius *eb* draw an arc to intersect *dc* produced in *g*. Join *fg* and draw *fh* perpendicular to *fg* to intersect *gd* produced in *h*. Measure *gh*. [Ans. 100·625; *gh* is taken as 100 (i.e. 2 × side of square) in a system of proportions for architects. Refer to the Golden Rule, page 76.]

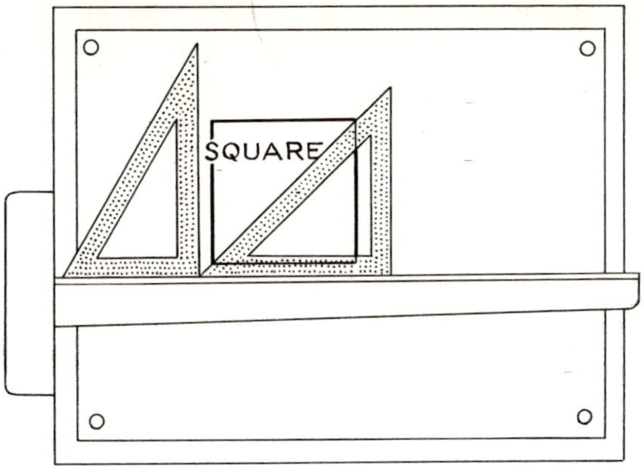

Fig 1

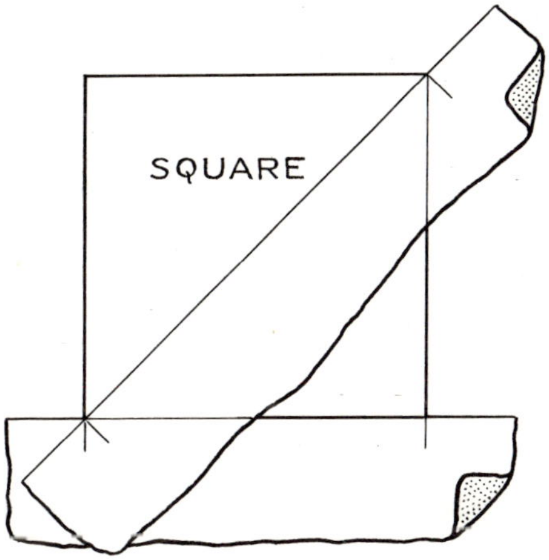

Fig 2

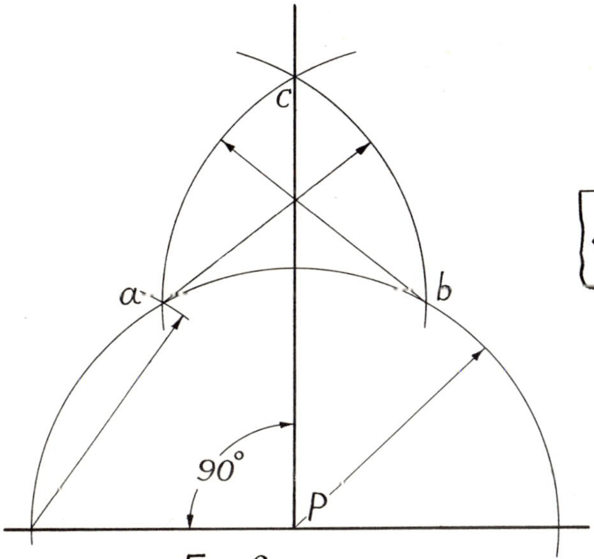

Fig 3

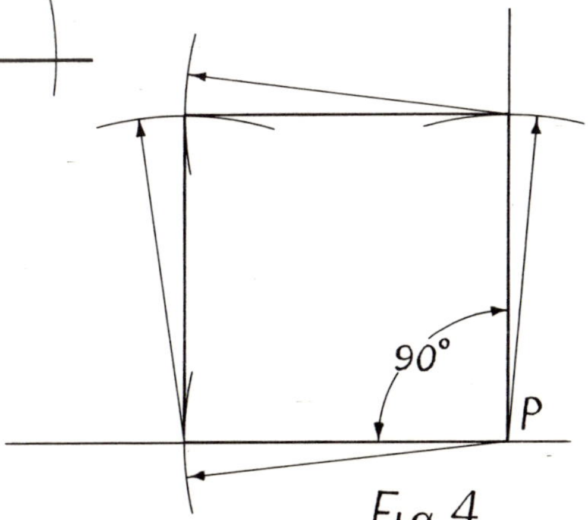

Fig 4

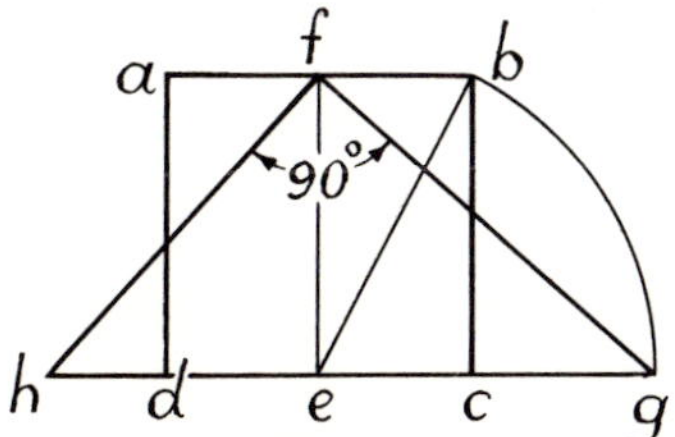

Fig 5

Important Ratios

Mean proportional, or geometric mean

A construction of importance in graphics is that giving the Mean Proportional C between two quantities A and B; that is, to find C so that $A/C = C/B$.

The quantities A and B are represented by the lines *oa* and *ob* in fig. 1, set off in opposite directions from *o*.
Describe a semicircle on *ab* and draw *oc* perpendicular to *ab*. The length of *oc*, C, is the mean proportional between A and B. That this is so follows from the similarity of the triangles *oac* and *ocb*, giving $oa/oc = oc/ob$.

Square root

It will be seen that if the length of B is 1, then $A/C = C/1$, i.e. $A = C^2$. Hence $C = \sqrt{A}$.
If A represents a number and B is unity, then C represents the square root of A.

Extreme and mean ratio

A line AB is divided by a point C in extreme and mean ratio if AB/AC = AC/CB.
The line AB is given in fig. 2. To find C we proceed as follows. Draw AD perpendicular to AB and equal to $\frac{1}{2}$AB. Produce DA, and mark off DF = DB. Mark off AC along AB equal to AF. Then AB/AC = AC/CB.
If we take AB = 2, then AD = 1 and DB = $\sqrt{5}$.

Then AC = AF = $\sqrt{5} - 1$; and CB = 2 — AC = $2 - (\sqrt{5} - 1)$.

The ratios are $\dfrac{2}{\sqrt{5}-1}$ and $\dfrac{\sqrt{5}-1}{2-(\sqrt{5}-1)}$; and cross multiplying shows that these are equal. The value of each ratio is 1·618.
A complementary construction, of considerable importance—dealt with fully on page 76, is shown in fig. 3. We are given a line AC and we wish to find a point B in AC produced such that AB/AC = AC/CB. Proceed as follows. Bisect AC in D. Draw CE perpendicular to CA and equal to CA. Join DE and mark off DB equal to DE along AC produced. Then AB/AC = AC/CB. The student should satisfy himself that the construction is correct—it is proved on page 76.
The ratios are expressed by the similar rectangles in fig. 4 where AB: AC :: AC : CB.

Exercises

Scale the lines P, Q and R, fig. 5. Then:

1. Find a line S such that $P/Q = R/S$. (Hint: set off any angle and mark $op = P$ along one line and $oq = Q$ along the other. Join *pq*. Mark $or = R$ along the *op* line and draw a line through *r* parallel to *pq*. Measure the intercept on the other line.) Ans. 80.
2. Find the mean proportional between (*a*) P and Q; (*b*) P and R. Ans. (*a*) 44; (*b*) 49.
3. Divide R in extreme and mean ratio. Measure the parts. Ans. 38; 24.
4. Draw a line 125 mm long and divide it into three parts having the ratio $P : Q : R$. Measure the parts. Ans. 32·2; 41·5; 51·3.

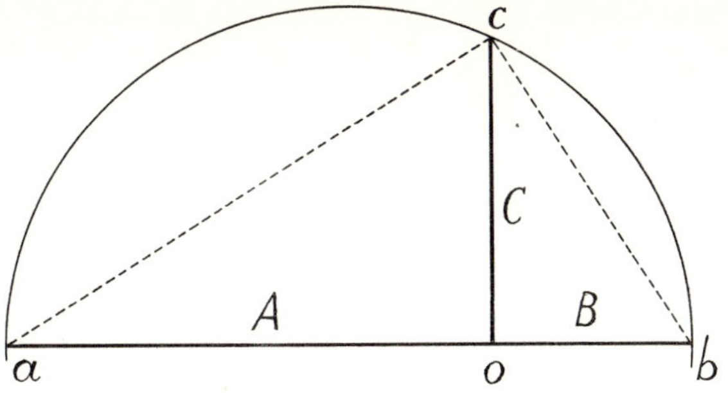

Fig 1

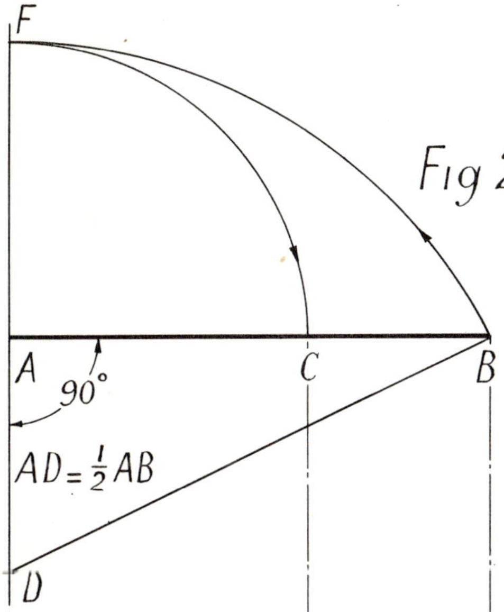

Fig 2

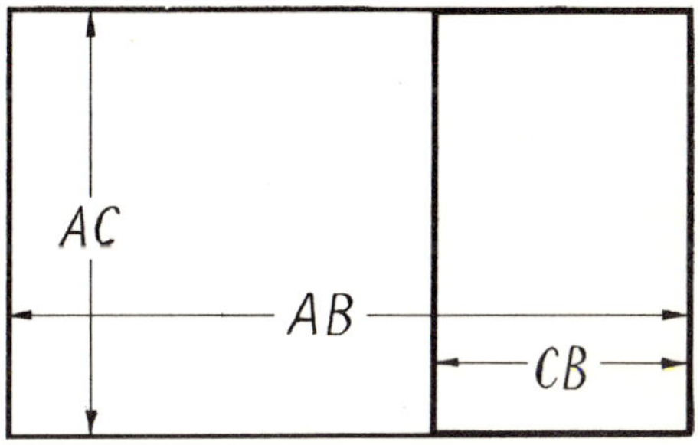

Fig 4

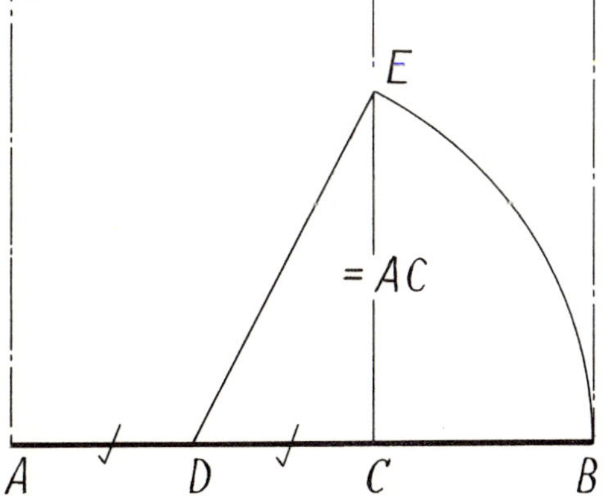

Fig 3

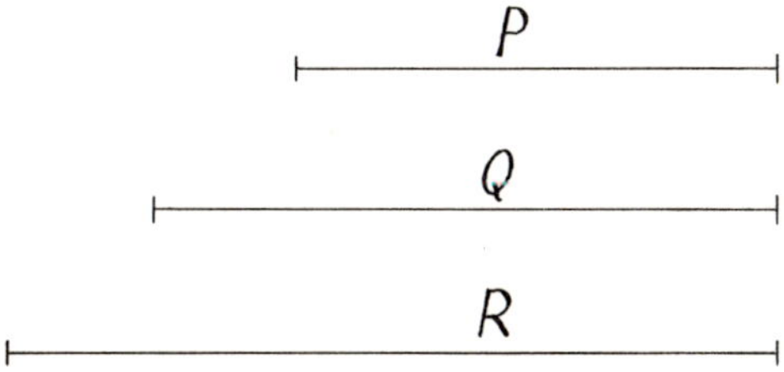

Fig 5

The Rectangle

A rectangle, fig. 1, has all its angles right angles and its opposite sides equal and parallel. If the opposite sides are equal and parallel but the angles are not right angles, fig. 2, the figure is a parallelogram.

The rectangle is the most common shape that we see: doors, walls, pictures, desk, cupboard, paper, envelopes, books. The question arises, is there a rectangular form of special merit which we might use when we can? That shown in fig. 3 is too nearly a square, that in fig. 4 is too elongated, to be artistically attractive. From the days of the ancient Greeks a form has been used which is peculiarly satisfying to the eye. It is shown in fig. 5. Its sides have the ratio 1 to 1·618, the ratio dealt with on page 74.

The golden rule

It has been said that the architects who designed the Parthenon, in Athens, one of the most satisfying buildings in the world, used the above relationship, called the Golden Rule, obtaining the proportions in the following simple way. Beginning with a square (*abcd*, fig. 5) they bisected a side of it (giving *e*). They then swung the length *eb* about *e* obtaining *g* on *dc* produced. The rectangle *adgf* thus obtained *without calculation* satisfies the Golden Rule and has some most interesting properties. Let us first find its proportions.

If $ad\ (=bc)=2$ then $de=ec=1$, and $eb=\sqrt{5}=2{\cdot}236=eg$.

Therefore $dg = 1+\sqrt{5} = 3{\cdot}236$, when $ad = 2$; and hence when $ad = 1$, $dg = 1{\cdot}618$. This is the Golden Rule proportion. It has been given various names and greatly used in artistic creations. The smaller rectangle *bcgf*, repeated in fig. 6, has the same proportion; for $cg = \sqrt{5} - 1 = 1{\cdot}236$ when $gf = 2$. Hence, when $cg = 1$ then $gf = 2 \div 1{\cdot}236 = 1{\cdot}618$.

Series of golden rule rectangles

If we begin with *adgf*, shown separately and larger in fig. 7, we may draw an unending series of similar rectangles by simply marking the length of the short side on the long. All these rectangles will have sides in the ratio 1 to 1·618. It will be found that *extreme accuracy* is required if the smaller rectangles are to be in true proportion. Ex. 1 will reveal this to the student.

Exercises

1. Obtain a Golden Rule rectangle from a square of 125 mm side, as in fig. 5. Then construct a sequence of 5 further rectangles as in fig. 7. Measure the sides of the smallest rectangle; they should be 18·3 mm × 11·3 mm.
2. Measure the sizes of the most satisfying rectangles you can see around you and find how they compare with the Golden Rule.
3. Write down the series 1 2 3 5 8 13 21 34 55 . . . formed by adding the last two numbers to give the next. Show that the further you go the closer the ratio of pairs to the Golden Rule: e.g. 5/3 = 1·666, 13/8 = 1·625, 55/34 = 1·619. . . . Note also that, taking any three consecutive numbers, the middle one becomes more and more close to the mean proportional of the other two: e.g. taking 8, 13, 21; $8 \times 21 = 168$, $13^2 = 169$; and so on. Can you explain these?

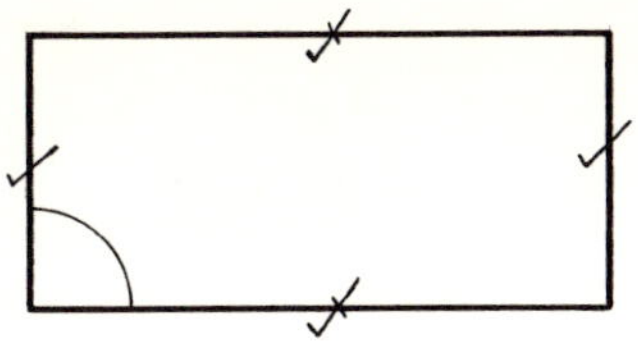

Fig 1

Fig 2

Fig 3

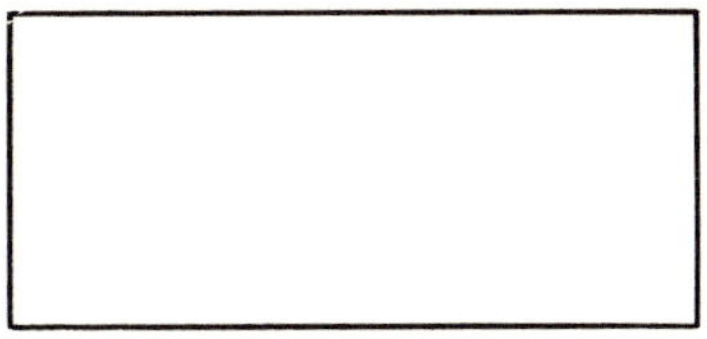

Fig 4

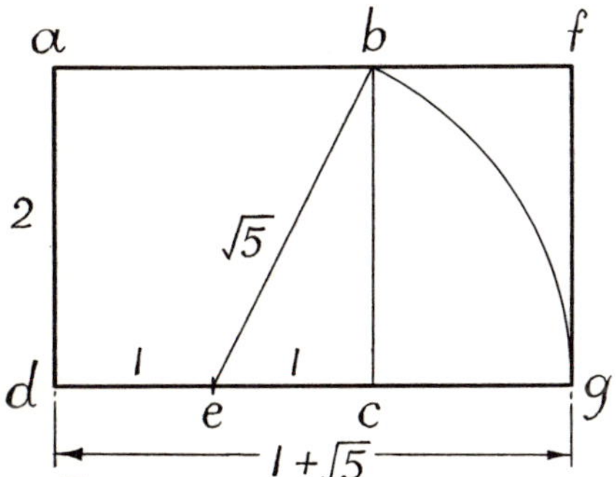

Fig 5

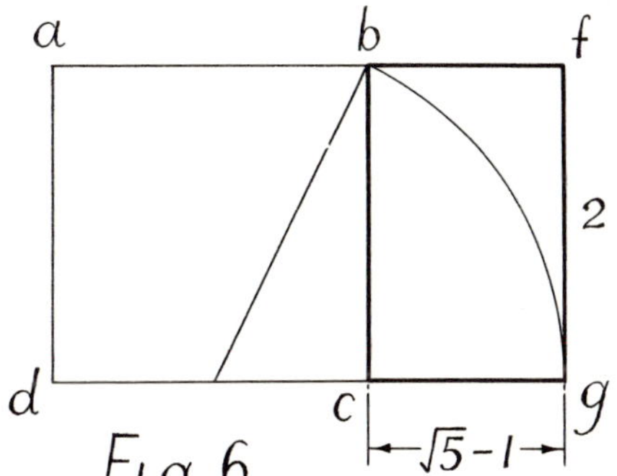

Fig 6

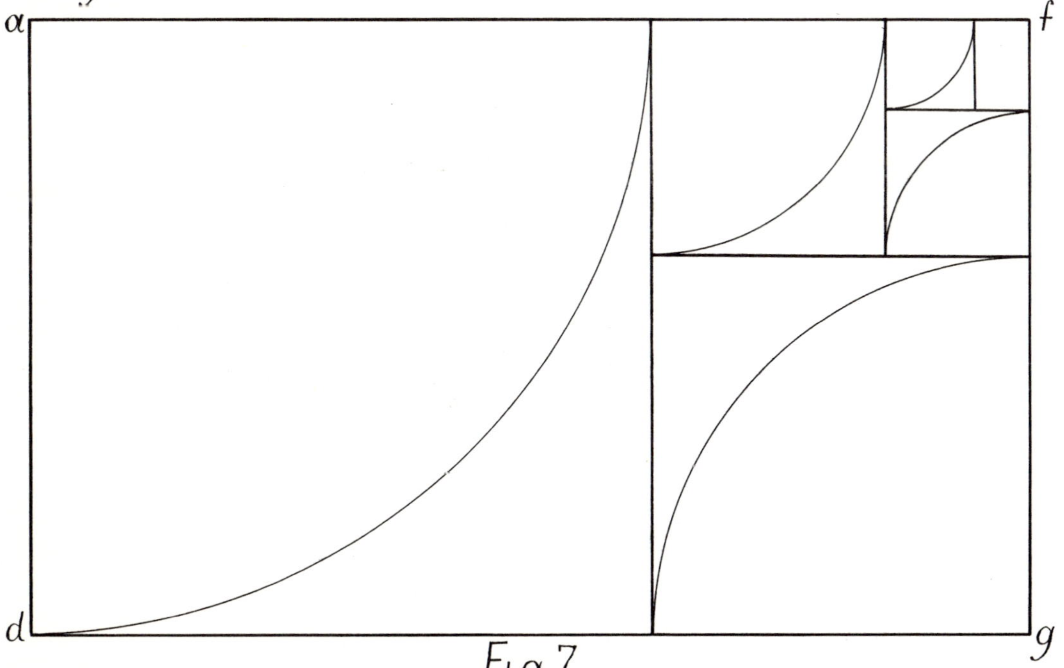

Fig 7

Triangles and Polygons

Triangles

A triangle may readily be drawn if we have the values either of its three sides or of two sides and the included angle. Some other ways of defining triangles will now be dealt with. They not only provide exercises in drawing but they illustrate the application of geometrical theorems.

Given: base AB, altitude *L* and vertical angle *Q*

Fig. 1

Draw AO inclined at 90°—*Q*, to AB to intersect in O a line bisecting AB at right angles. With centre O and radius OA describe the segment of a circle. Any angle in this segment (e.g. ADB) has the value *Q*. Draw a line parallel to AB, distant *L*, to intersect the segment in C. Join CA CB. ABC is the required triangle. There is a second position for C and a second triangle.

Alternative, fig. 2. Draw the given angle *Q* on tracing paper and apply it over the base AB and a parallel distant *L* from AB. Adjust the tracing paper until the lines of the angle pass through A and B while their intersection lies in the parallel.

Given: perimeter *P*, base AB and one base angle *Q*. Fig. 3

Mark off BD equal to *P* — AB. Join DA and bisect it at E. Draw EC at right angles to DA to intersect DB in C. Join CA. Then ABC is the required triangle.

Given: perimeter *P*, altitude *L* and vertical angle *Q*. Fig. 4

Draw lines CF, CG enclosing an angle *Q*. Mark off CD = CE = $\frac{1}{2}P$. From D and E draw perpendiculars to CD and CE, intersecting in O. With centre O and radius OD describe an arc; with centre C and radius *L* describe a second arc. Draw AB, a common tangent to these arcs. ABC is the required triangle. There is a second common tangent and a second triangle.

This is an elegant solution to a problem which is not very common. The proof will be clear when it is realized that AH = AD and BH = BE.

Polygons

To construct any regular polygon, given the length of a side

Suppose the polygon has 7 sides, one of which is AB, fig. 5. Produce AB, and with B as centre and BA as radius describe a semicircle. Divide the semicircle by trial into 7 equal parts. Join B to the *second* mark (always to the second mark whatever the number of sides). Bisect each side at right angles and obtain O, the centre of the circumscribing circle. The completion will now be obvious. When the polygon is a pentagon (5 sides) a special construction is that given in fig. 6—which is left for the student to reason out.

Exercises (see also foot of page 80)

Construct triangles to the following information. Measure the sides.

1. **Base 80 mm, vertical angle 30°, altitude 110 mm. Ans. 113; 153.**
2. **Base 80 mm, one base angle 45°, perimeter 280 mm. Ans. 83; 117.**

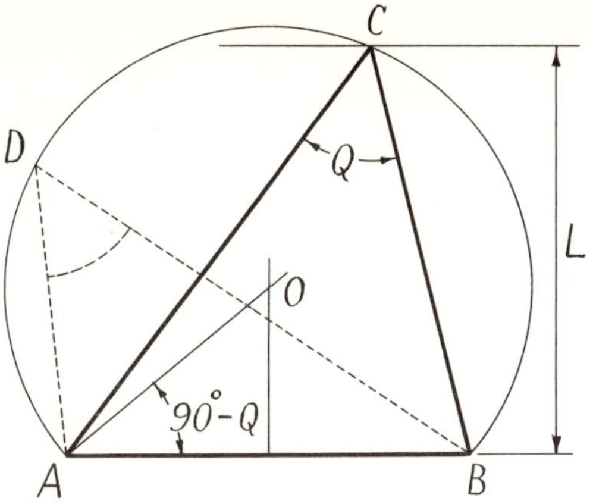

Fig 1

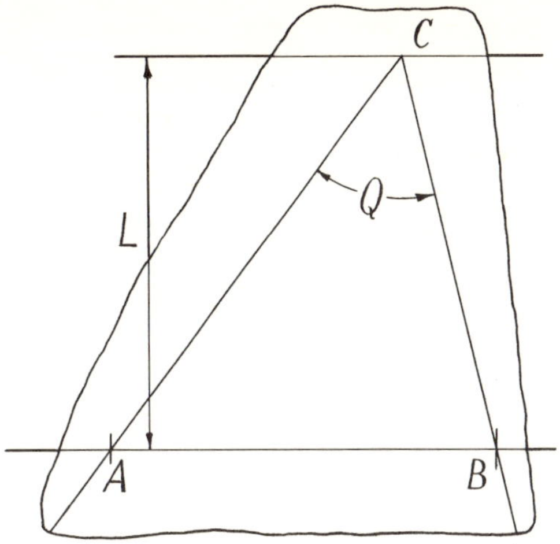

Fig 2

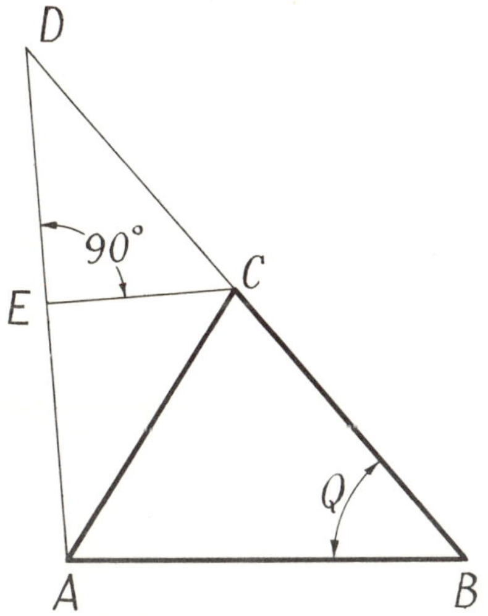

Fig 3

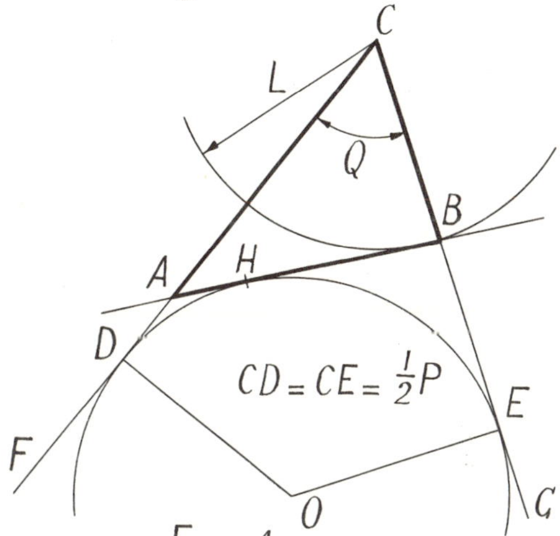

Fig 4

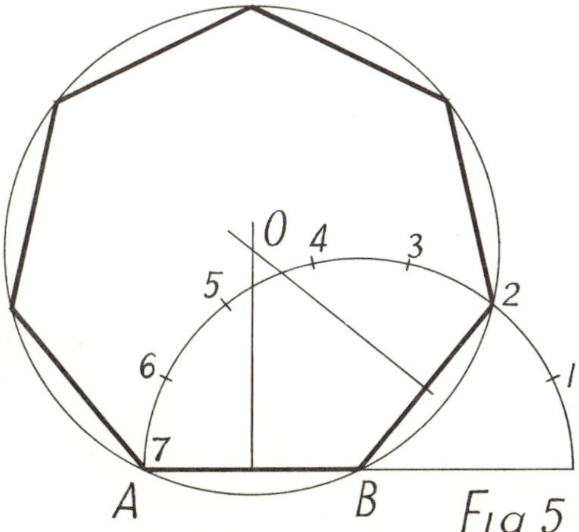

Fig 5

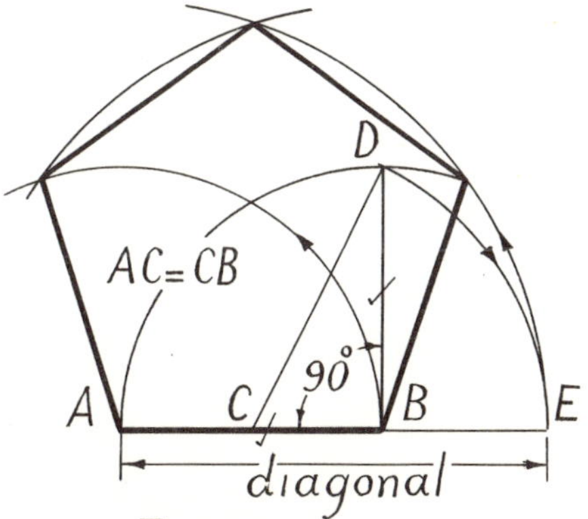

Fig 6

Areas of similar polygons

Lines are drawn from the corners of a polygon ABCDE, fig. 1 to any point O. The smaller polygon *abcde* has sides parallel to those of the larger polygon, bounded by the radial lines to O. So constructed, *abcde* is similar to ABCDE. The point O may be taken coincident with A (or B, C, D or E), to give a more compact construction. In fig. 2, O coincides with D. The same construction applies. A *larger* similar polygon may be drawn by continuing the radial lines—as in fig. 5.

The ratio of the areas of the polygons is given by $AB^2:ab^2$ (or $BC^2:bc^2$, $CD^2:cd^2$, etc.). If ab is made $\frac{1}{2}$ of AB, then the smaller polygon is $\frac{1}{4}$ the area of the larger. The same relationship is given by the lengths of the radial lines from O. For example, if $AO = 3$ (aO) then $AB = 3(ab)$ and $AB^2 = 9(ab)^2$. Hence the larger polygon would be 9 times the area of the smaller.

Given the ratio of areas: To draw the similar polygon

This is the usual problem. It may be very easily solved graphically. Suppose the ratio of areas to be 7 to 3.

Consider the side ED, drawn separately in fig. 3. Produce it and mark off three and seven equal divisions. Draw semicircles on E3 and E7, and erect a perpendicular at D to give P and *p*. Join PE and draw *pe* parallel to it. Then *e*D is the length of that side of the smaller polygon. This construction is incorporated with that in fig. 2 and shown in fig. 4.

If a larger polygon is to be constructed, say of the reverse ratio 3 : 7, then the slight amendment to the construction is shown in fig. 5.

Proof.—The proof of the construction in fig. 4 is as follows:

$DP^2 = ED \times 7$; $Dp^2 = ED \times 3$ (see page 74)

$$\frac{\text{Area ABCDE}}{\text{Area } abc\text{De}} = \frac{ED^2}{eD^2} = \frac{DP^2}{Dp^2} = \frac{7ED}{3ED} = \frac{7}{3}$$

A similar proof applies to fig. 5.

Note.—If we know the lengths of the sides of the *squares* having areas equal to the polygons, then we have *the direct ratio* between the sides of the similar polygons. This relationship is used on the next page.

Exercises

1. Draw a polygon ABCDEF making AB = 30, BC = 38, CD = 51, DE = 30, EF = 23, FA = 36, AC = 58, AD = 71, AE = 50 mm. Construct a polygon similar to ABCDEF but having an area larger in the ratio 5 : 3.

Further exercises for page 78.
Construct triangles to the following information. Measure the sides.

3. Perimeter 200 mm, altitude 50 mm, vertical angle 45°. Ans. 63; 87; 50.
4. Construct the following regular polygons: pentagon, side 80 mm; heptagon (7), side 60 mm; nonagon (9), side 40 mm. Measure the diameters of the circumscribing circles. Ans. 137; 139; 119.

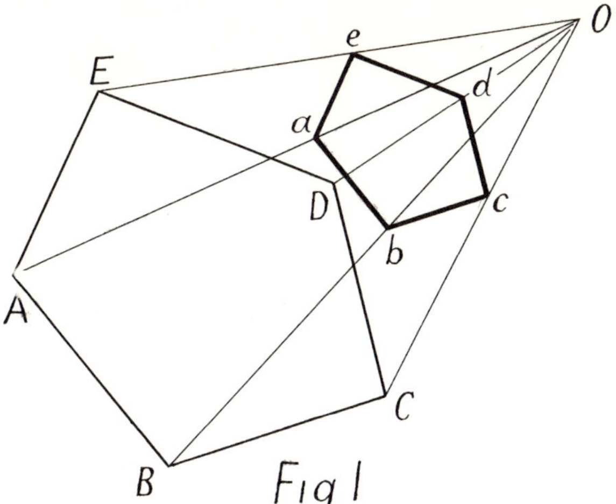

Fig 1

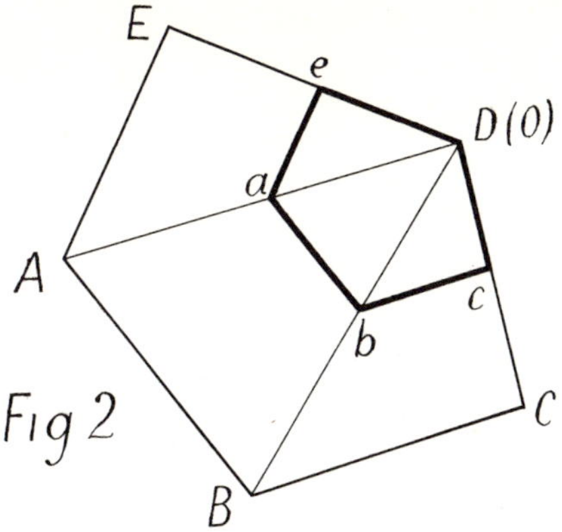

Fig 2

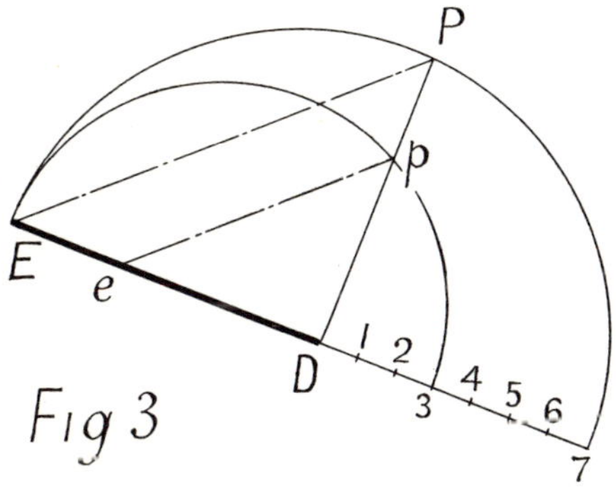

Fig 3

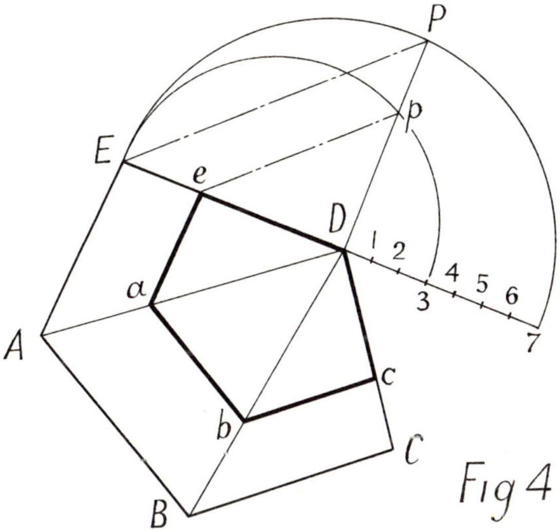

Fig 4

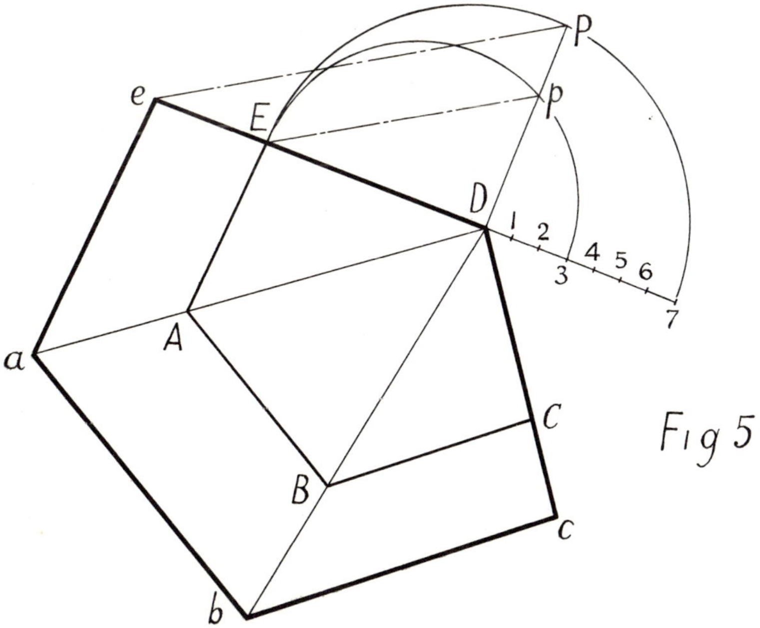

Fig 5

Polygons

Two constructions will now be dealt with concerning the areas of polygons and triangles. They are useful either alone or in combination.

To reduce a given polygon to a triangle of equal area

Consider the four-sided figure ACDE, fig. 1. Join AD, and through E draw EB parallel to AD to meet CD produced in B. Join AB. Then ABC is one triangle having an area equal to that of ACDE.

Any polygon can be divided up and dealt with in this way, as indicated in fig. 1*a*. Here the polygon is *abcdefg*, and the resulting equivalent triangle is *asr*.

To reduce a triangle to a square of equal area

The triangle ABC of fig. 1 has been used again in fig. 2. A *rectangle* of equal area is given by BCED, having an altitude half that of the triangle. To obtain a *square* of equal area, produce BC and make CF equal to CE. On BF draw a semicircle. Produce EC to intersect the semicircle in G. Then CG is the side of the required square CGPM.

To find the *area* of the square *by measurement*, set off (dotted) a distance MN equal to 100 mm say. Join NC and draw CO perpendicular to NC to intersect NMP produced in O. Then MO, measured in the unit MN, gives the area of the square—and hence of the polygon ACDE.

Proof—In fig. 1 the triangles EDA and BDA lie between parallels and have the same base DA; hence they are equal in area. The addition of ADC to each gives the equal areas ACDE and ABC.

Again, area ABC, fig. 2, equals (BC × half the altitude) which is the area of BCED. Then, from the discussion on page 74, BC × CE = CG^2 and OM × MN = CM^2.

Accuracy attainable

By measurement it will be found that BC = 281, CE = 65, CM = 135, MN = 100. Area of triangle = 281 × 65 = 18,265; area of square = 135 × 135 = 18,225. The length of MO = 18,225/100 = 182.

To construct a polygon A similar to a polygon B and equal in area to another polygon C

(no figure)

The solution comes directly from the constructions already dealt with. We must first reduce polygons B and C to squares of equal area; suppose their sides are *b* and *c*. Take any side of polygon B and mark a length which represents *c* when the side represents *b*. Then proceed as in fig. 2, page 81.

Exercises

1. **Find the area of the four-sided figure in fig. 3, which is dimensioned in millimetres. Reduce the figure to (*a*) a triangle, (*b*) a rectangle, (*c*) a square. Find the area of the square by one measurement. Ans. 3280.**
2. **Measure ACDE, fig. 1. Find a similar figure but equal in area to fig. 3.**

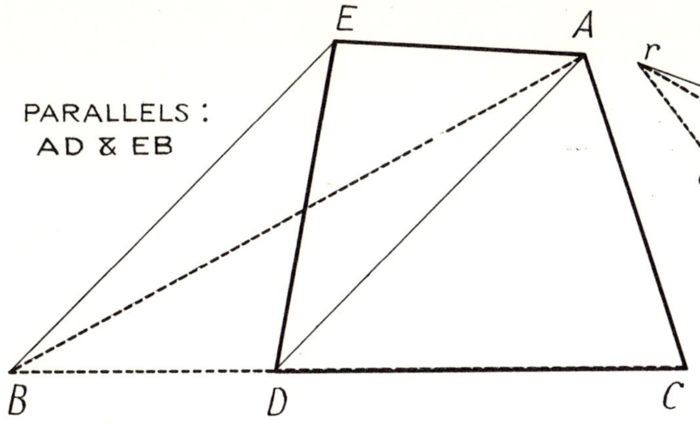

Fig 1

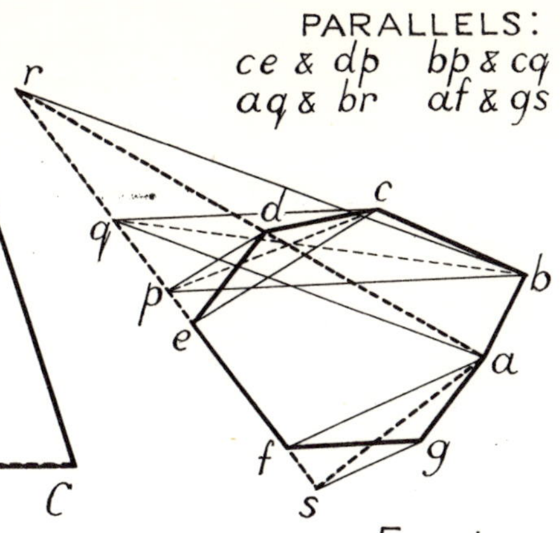

Fig 1a

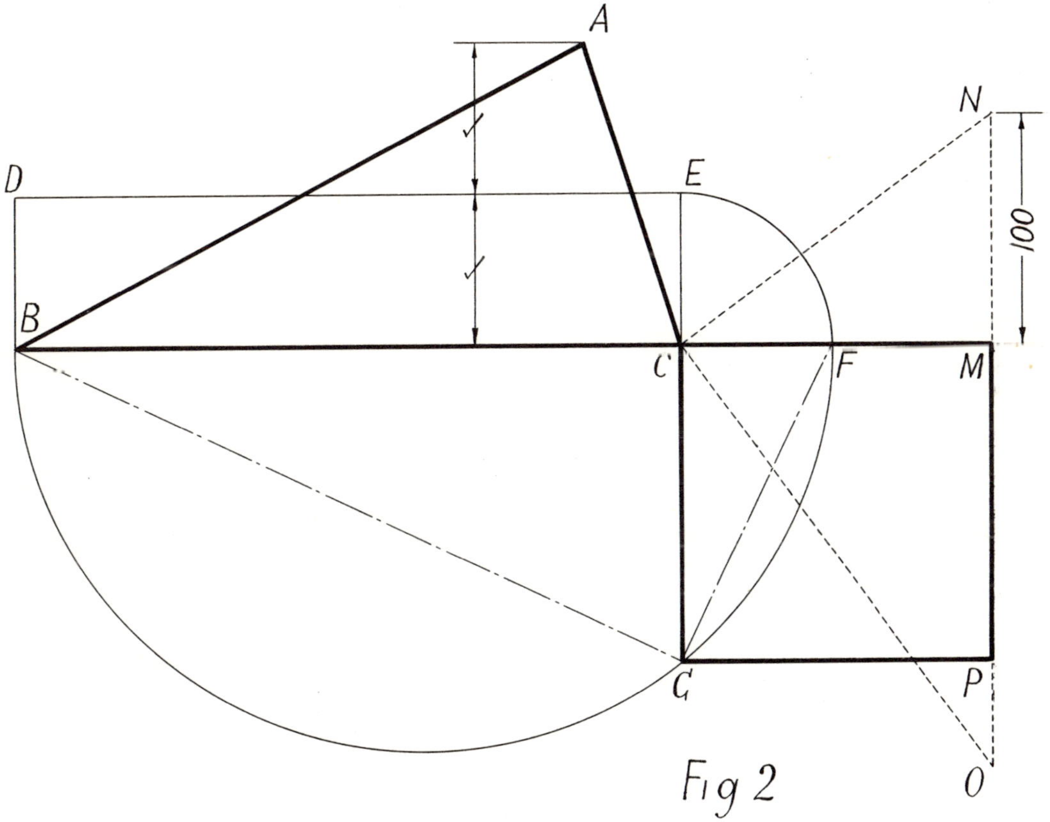

Fig 2

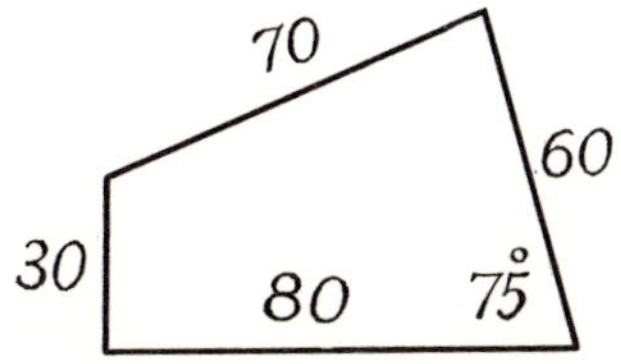

Fig 3

Circles

The drawing of arcs or circles to touch given lines or to pass through given points, frequently arises when technical drawings are made. The method of finding the centre of the arc or circle is usually a simple matter, but sometimes less obvious constructions are called for. A few examples are dealt with here. The drawings almost explain themselves and the exercises provide good tests for accurate work.

1. CIRCLE THROUGH THREE POINTS, P, Q and R (no figure)

Perpendiculars bisecting lines PQ and QR intersect in O, the centre of the required circle.

2. CIRCLE THROUGH ONE POINT P TO TOUCH LINES AB AND BC—Fig. 1

The centre lies on the bisector of the angle ABC. Draw *any* circle centre O_1 touching AB and BC. Draw PB and obtain R. A line PO parallel to RO_1 gives the required centre O.

3. CIRCLE THROUGH TWO POINTS P AND Q, TOUCHING A LINE CD—Fig. 2

Join PQ and produce to give R. RS is made equal to the mean proportional between RQ and RP (page 74), and a perpendicular to CD at S passes through O, the centre of the required circle.

4. CIRCLE THROUGH ONE POINT P, TOUCHING A LINE CD, AND HAVING ITS CENTRE ON A LINE AB—Fig. 3

We may convert this to problem (3) by obtaining a second point Q on a perpendicular to AB, making QR = PR.

5. CIRCLE TO TOUCH A CIRCLE, CENTRE R, RADIUS *r*, AND TWO LINES AB AND CD—Fig. 4.

If we draw parallels to AB and CD, distant *r* from them, we may use problem (2) to draw a circle centre O to touch these lines (dotted) and to pass through the centre R. The required circle is concentric to it.

6. CIRCLE THROUGH TWO POINTS P AND Q TO TOUCH A CIRCLE CENTRE C—Fig. 5

Join PQ, produce it, and bisect it at right angles. Draw *any* circle to pass through P and Q and to cut circle C in A and B. Join BA and produce it to meet PQ in D. Draw the tangent from D and at the point of contact draw a perpendicular to intersect the perpendicular to PQ in the required point O.

Exercises *

The grid shown in fig. 6 is marked in 3-mm squares and is to be used for the following. Measure the diameter of the resulting circle in all cases.

1. Draw a circle through P, Q and R. (Ans. 50.) 2. Draw a circle through Q and R to touch line PT. (Ans. 44.) 3. Draw a circle through P with centre on line UT, to touch line RS. (Ans. 39). 4. Draw a circle through P to touch lines QR, RS. (Ans. 34). 5. A circle centre U has radius 12 mm: (*a*) draw a circle to touch this circle and also lines QR and RV; (Ans. 42): (*b*) draw a circle through P and W to touch circle U. (Ans. 28.)

* (a) There is often more than one solution to a problem.
(b) The point of intersection of converging lines is sometimes inaccessible and other solutions have to be found.

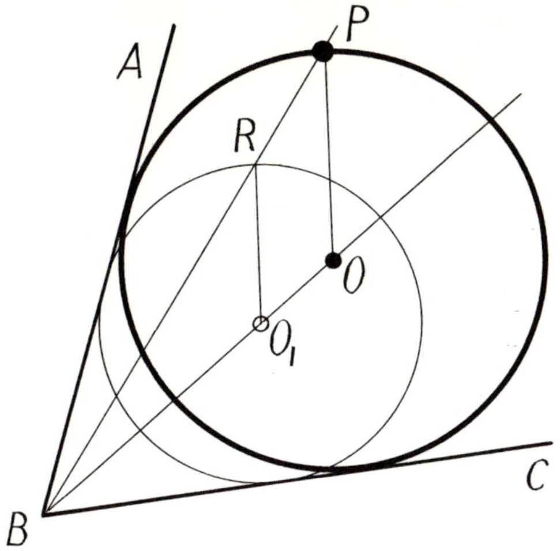

Fig 1

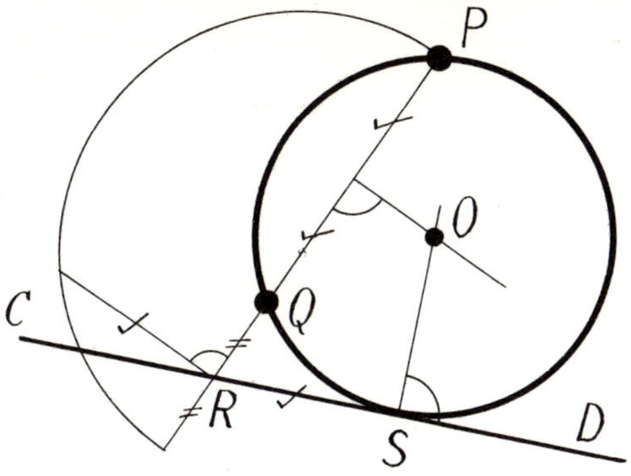

Fig 2

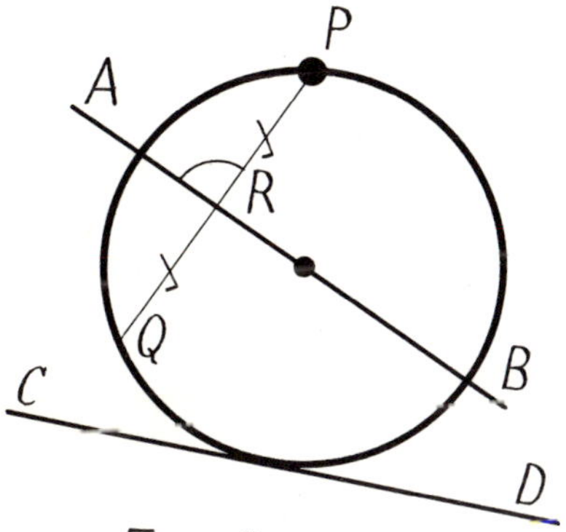

Fig 3

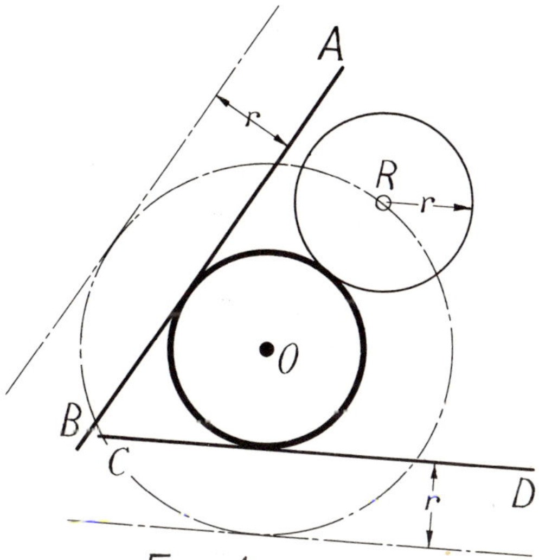

Fig 4

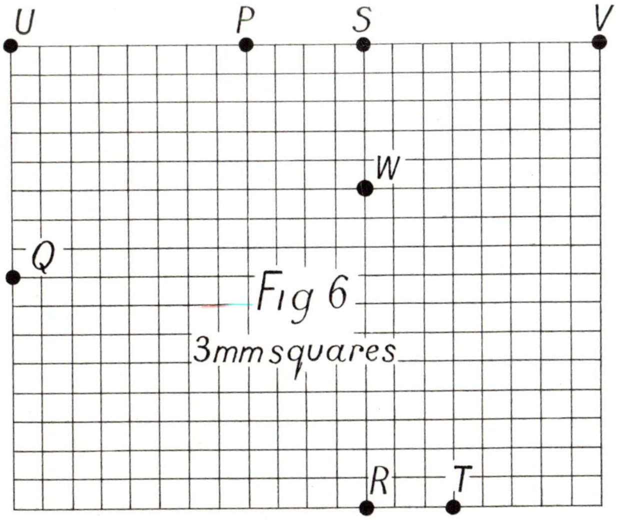

Fig 6
3mm squares

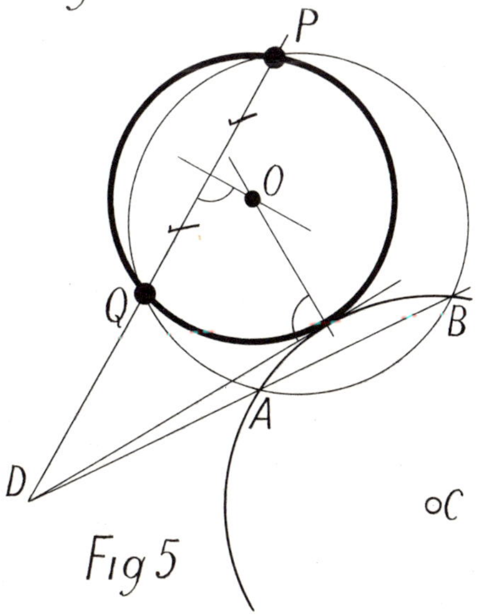

Fig 5

Parabola, ellipse and hyperbola

This important group of curves is shown in fig. 1. We have a fixed point F and a fixed straight line DD. If a point P moves so that its distance from F always equals its distance from DD, i.e. PF = PD, then the path of P is a *parabola*. If PF is *less* than the distance from DD, always in a fixed ratio, then the path of P is an *ellipse*. If PF is greater than the distance from DD, always in a fixed ratio, then the path of P is a *hyperbola*.

The three curves may also be obtained by taking sections of the cone (see page 121) and they are known therefore as *conic sections*. It will be seen that the ellipse is a closed curve which could be drawn by using another fixed point F_1, and another fixed line.

The fixed point F is called the *focus*; the fixed straight line DD is called the *directrix*; the ratio PF/PD is known as the *eccentricity*. A line through F perpendicular to DD is called the *axis*, and the point V in which the curve cuts the axis is called the *vertex*.

A *ratio scale* which may be used for marking positions of P, for the ellipse, is shown on the right. A similar scale may be made for the hyperbola.

The ellipse

The greatest and least diameters of the ellipse are known as the *major* and *minor axes* (VV_1 and BB_1 respectively, fig. 2). Using these, points on the ellipse can be marked by means of paper strips called *trammels*. We mark off, along an edge, distances equal to half the axes, as shown by $PQ = \frac{1}{2}BB_1$ and $PR = \frac{1}{2}VV_1$, fig. 2. The distances may be added or subtracted, giving the long or the short trammel. We move the trammel so that Q and R (or Q_1 and R_1) lie on the axes, and mark the positions of P to give the ellipse.

Tangent and normal at a point T on an ellipse

Fig. 2

In any ellipse, the sum of the distances of a point T, on the ellipse, from the foci F and F_1, is a constant and is equal to the length of the major axis VV_1. Hence if we take $\frac{1}{2}VV_1$ as radius, and the ends of the minor axis as centres, we may obtain F and F_1 by drawing arcs to cut VV_1.

If we join T to F and F_1, then a line bisecting the angle gives the *normal* to the ellipse at T. A line at right angles to the normal gives the *tangent* at T.

Tangent and normal to a parabola (no figure)

The tangent at P (fig. 1) bisects the angle DPF; and the normal is at right angles to the tangent.

Exercises

1. Taking a point F 50 mm from a straight line DD construct (*a*) a parabola (*b*) an ellipse, eccentricity $\left(\frac{PF}{PD}\right)\frac{2}{3}$ (*c*) two hyperbolas, eccentricities $\left(\frac{PF}{PD}\right)\frac{3}{2}$ and $\frac{9}{8}$
2. An ellipse has a major axis of 130 mm and a minor axis of 80 mm. Draw the curve.
3. A disc is 100 mm diameter and 25 mm thick. It is inclined to a plane at 45°. Draw the projection on the plane.

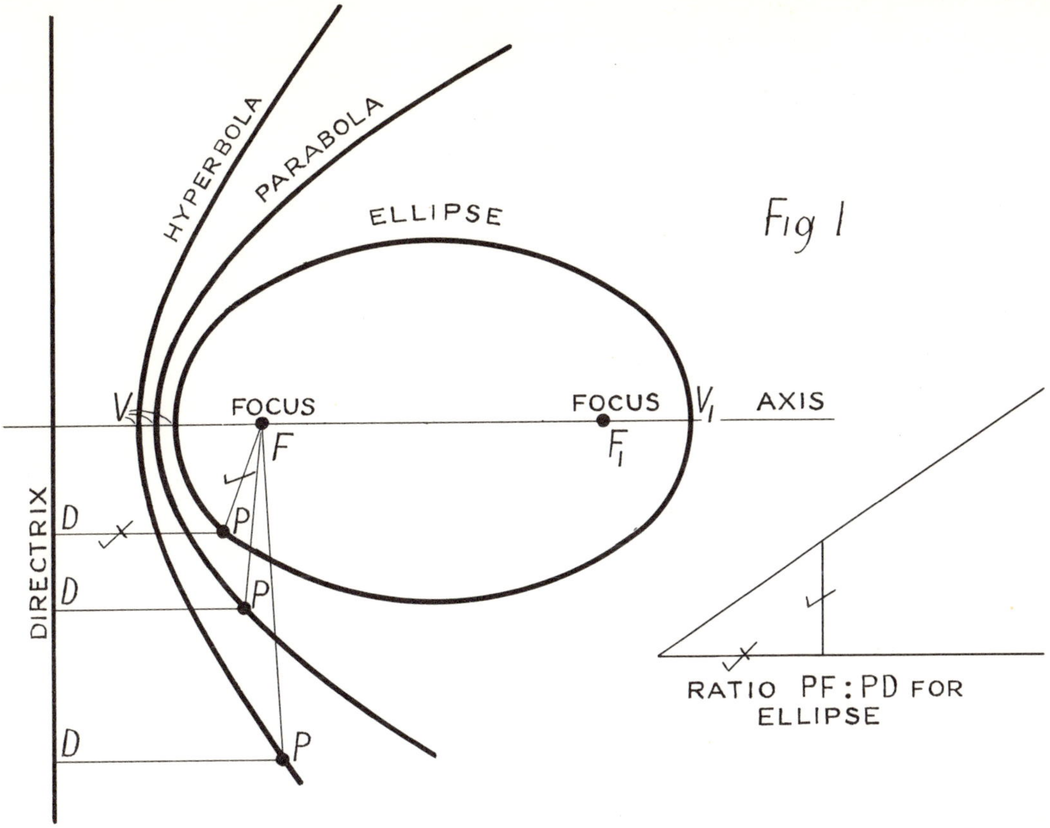
HYPERBOLA
PARABOLA
ELLIPSE
FOCUS
F
FOCUS
F_1
V
V_1
AXIS
DIRECTRIX
D
P
D
P
D
P
RATIO PF : PD FOR ELLIPSE

Fig 1

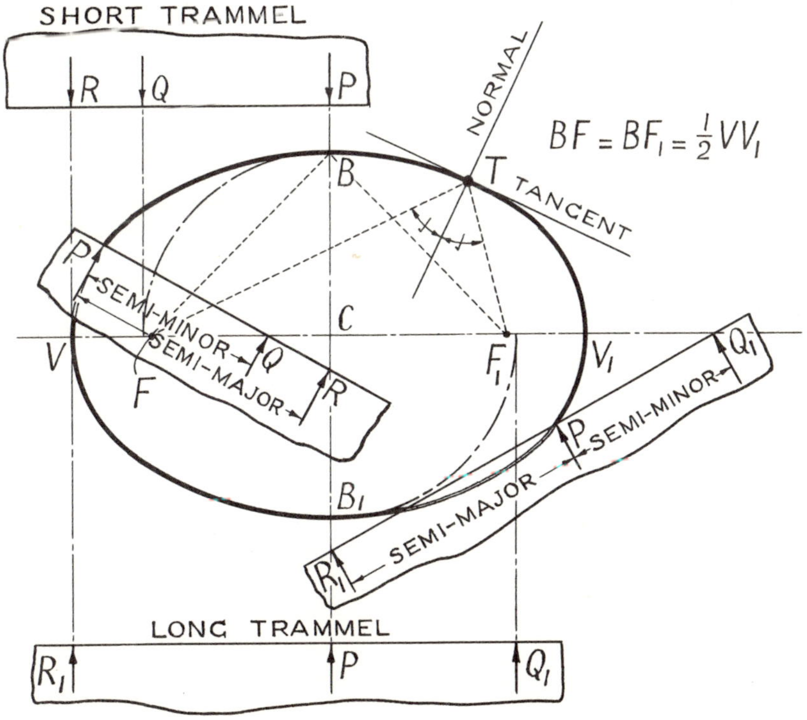
SHORT TRAMMEL
R
Q
P
NORMAL
$BF = BF_1 = \frac{1}{2}VV_1$
B
T
TANGENT
P
SEMI-MINOR
SEMI-MAJOR
Q
R
C
V
F
F_1
V_1
Q_1
P
SEMI-MINOR
SEMI-MAJOR
B_1
R_1
LONG TRAMMEL
R_1
P
Q_1

Fig 2

The Ellipse (continued)

In view of the importance of the ellipse some other constructions will now be considered.

(*a*) Given the major axis VV_1 and the foci F and F_1.

Because the sum of the distances of a point P on the ellipse from F and F_1 is constant and equal to VV_1, the construction in fig. 1 applies. Take any point A on VV_1; with centres F and F_1 and radii VA and V_1A describe arcs to intersect in A_1 A_2 A_3 and A_4. These lie on the required ellipse. Take other points such as A and plot a succession of points on the curve. Note that $BF_1 = CV_1$.

For large ellipses we may anchor at F and F_1 the ends of a cord (dotted lines) of length VV_1 and plot the ellipse by moving P while keeping the cord taut. This is the gardener's method for marking elliptical plots.

(*b*) Given the major and minor axes and using auxiliary circles.

Circles are drawn about the axes, fig. 2. Any radial line is taken to intersect the circles in D and E. Lines drawn through D and E parallel to the axes intersect in P, a point on the ellipse.

(*c*) Given the circumscribing parallelogram.

The method is shown in fig. 3. Semicircles on the sides are divided into an equal number of parts, here 6, and perpendiculars are dropped from these points to the sides. Parallels to the sides from the feet of the perpendiculars intersect in points on the required ellipse.

The parabola (continued)

It is a property of the parabola that V_1M_1, fig. 4, varies as $PM_1 \times P_1M_1$. If we are given the double ordinate PP_1 and the length of the axis VM, we can readily plot the curve by setting off to scale distances representing $PM_1 \times P_1M_1$. In fig. 5 PP_1 has been divided into 10 equal parts, and VM is represented by 25 units.

The construction for inscribing a parabola in a rectangle is shown in fig. 6, and needs no explanation.

Exercises

1. The foci of an ellipse are 70 mm apart and the major axis is 100 mm long. Draw the ellipse and measure the length of the minor axis. Ans. 36 mm.
2. The axes of an ellipse are 130 mm and 100 mm. Draw the ellipse using auxiliary circles. Mark the foci. (Note, fig. 1, $BF_1 = CV_1$.)
3. A parallelogram has sides of 110 mm and 80 mm and an included angle of 60°. Draw the inscribed ellipse.
4. The sides of a rectangle are 100 mm and 80 mm. Inscribe two parabolas with axes perpendicular to each other.
5. Taking the double ordinate as 100 mm and the length of the axis as 130 mm, draw a parabola as in fig. 5.

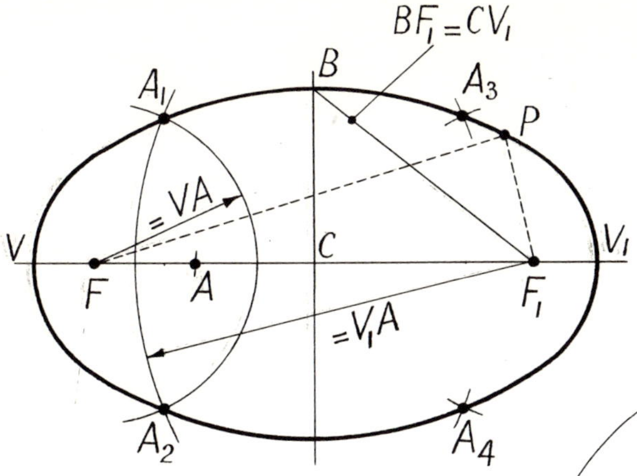

Fig 1

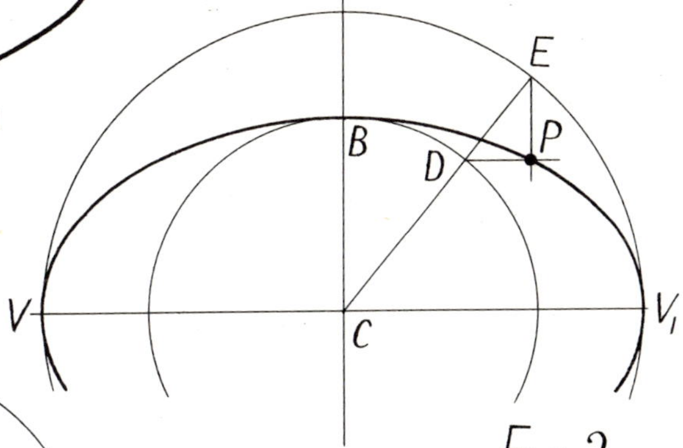

Fig 2

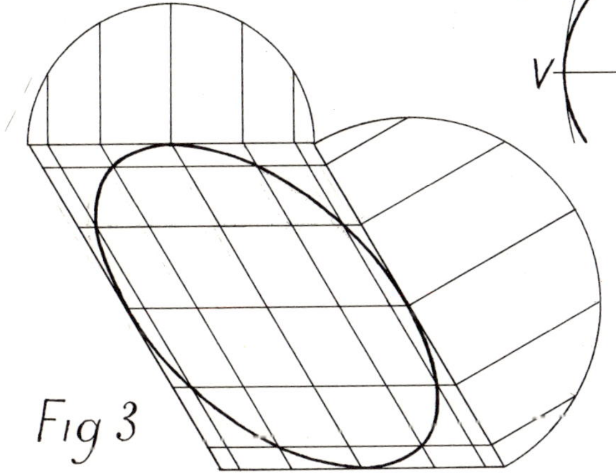

Fig 3

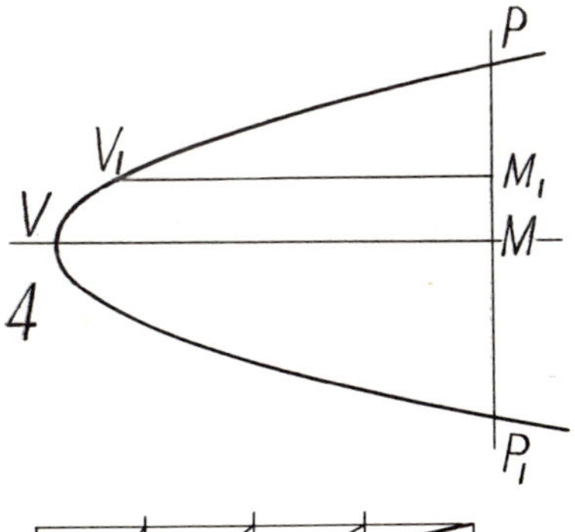

Fig 4

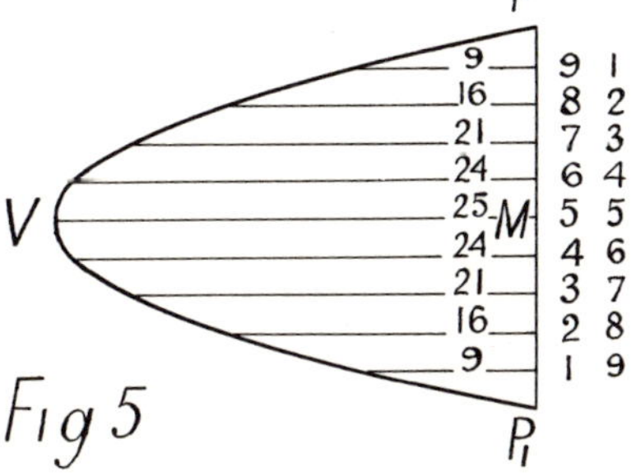

Fig 5

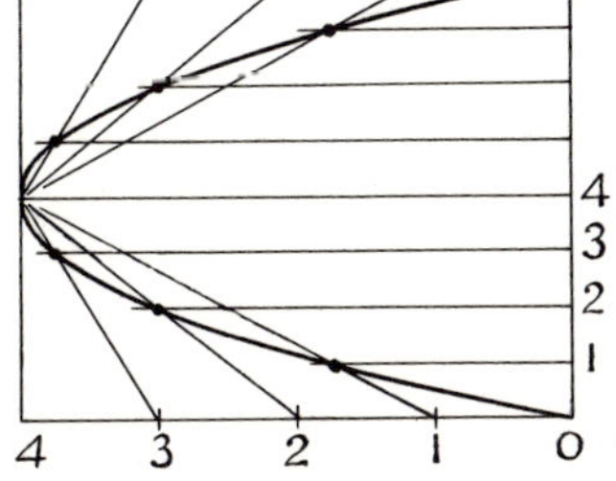

Fig 6

Parallel Curves, Cams

We may draw a parallel to any line, at a distance *d* from it, by first drawing a series of circular arcs of radius *d* from points on the given line and then drawing a line to touch the arcs. If the given line is straight, the required parallel is another straight line; if the given line is a circular arc, fig. 1, the required parallel is a concurrent circular arc, of greater or lesser radius; but if the given curve is not a circle, then the parallel curve and the original differ in type.

Parallels to an ellipse

The middle curve in fig. 2 is part of an ellipse. Two parallel curves are drawn at a distance *d* on each side of the ellipse. *Neither of these is an ellipse*. The shape of the inner curve is obviously not elliptical (note the crossed lines at the ends), nor is the outer curve, although it seems to be.

Cams

A simple radial cam is shown in fig 3. The cam is a disc. It revolves about O and displaces a sliding piece AB which moves vertically and is held against the cam. A roller is provided at the end of AB to enable the parts to work smoothly. As the cam turns from the position shown it lifts AB, holds it for a while, and then allows it to drop gradually to its original position.

We are usually given the extreme positions of the centre of the roller on the sliding piece and the kind of movement it is to have. Let us suppose that, as the cam turns uniformly through 120°, the *centre* of the roller, fig. 4, is to move uniformly from C to D. Let us divide CD into say 6 equal parts and let us set off radial lines, from O, at 20° intervals, giving 6 angular divisions. If we mark off the distances 01 02 . . . along the radii we obtain points C1 C2 . . . which lie on an Archimedean spiral (see page 100). We require a line parallel to this curve and we therefore draw circles, representing the roller, at centres C1 C2 . . ., as in fig. 5. The outline of the required cam is a curve drawn to touch these circles.

Exercises

1. Construct an ellipse, major axis 200 mm, minor axis 90 mm. Draw curves parallel to it and distant 25 mm.
2. Draw a cam to give the following motion to a roller-ended slider. Rise of 50 mm at uniform speed during one-third revolution; rest during the next third; uniform fall during the last third. Diameter of roller 25 mm. Least distance of the centre of the roller C to the cam axis O, 50 mm. The solution is partly shown in fig. 5.
3. Repeat Ex. 2 but take a roller diameter of 50 mm. Draw only one half of the cam.
4. Draw the cam in Ex. 2 if the rise of 50 mm is to take place during the entire revolution with a sudden drop at the end.

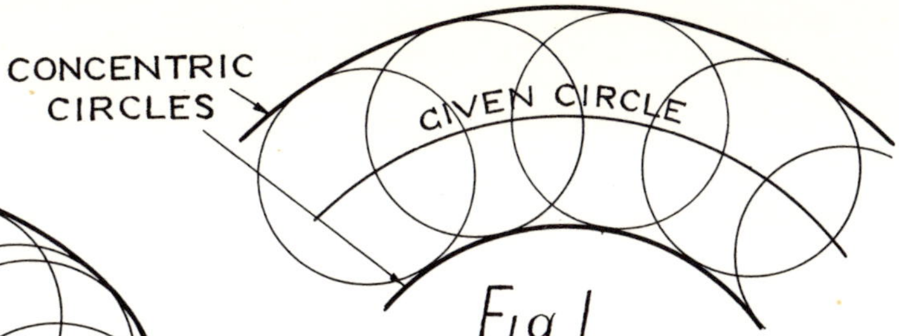

Fig 1

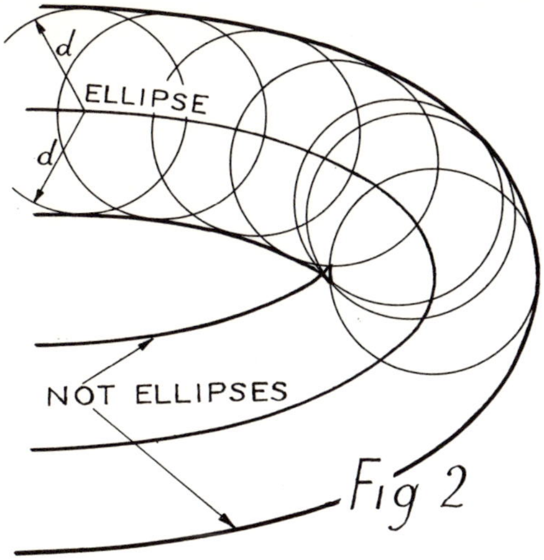

Fig 2

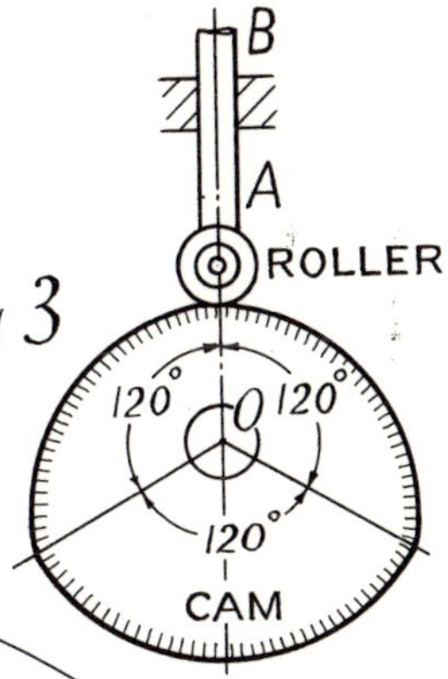

Fig 3

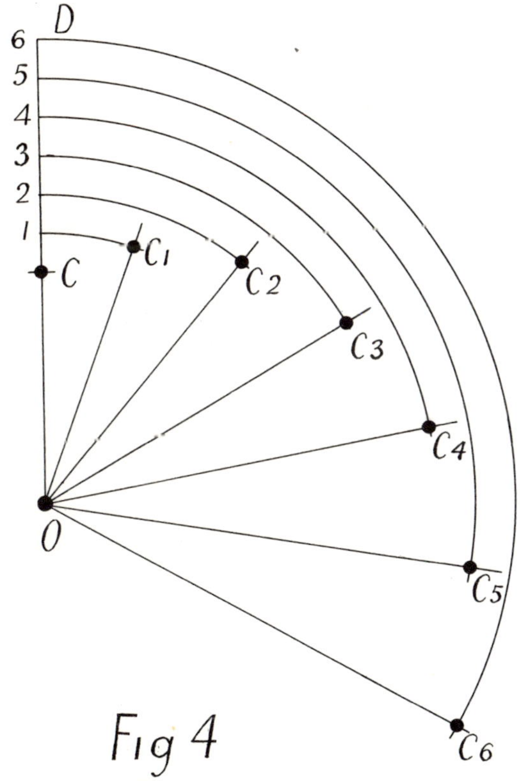

Fig 4

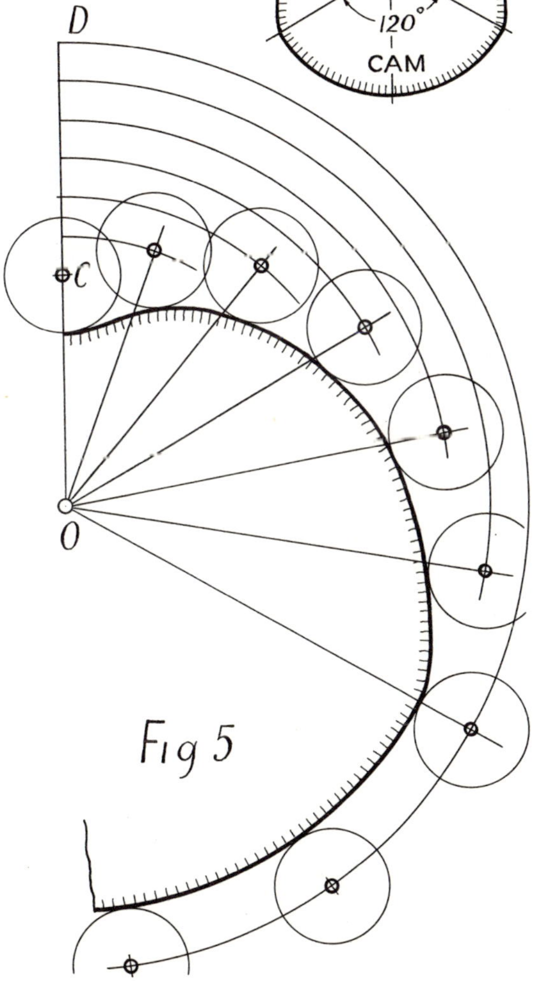

Fig 5

Loci of Points on Moving Mechanisms

The curves traced out by points on moving mechanisms may be plotted point by point, either by drawing an outline of the mechanism in a series of positions, or by using a paper trammel. The trammel method is usually preferable because it is quicker and less confusing.

It will be found that it is often not easy to guess the form of the locus; that points are required more closely together at some sections of the locus than at others—compare *a* and *b*, fig. 2; and that there are usually key positions for the mechanism which set clear limits to the outline of the locus. Two typical cases are dealt with below and solved opposite.

Locus of point of intersection of revolving links

The two revolving links OA and O_1A_1, fig. 1, are hinged at their ends to fixed points O and O_1. One link O_1A_1 revolves twice as quickly as the other. We require to find the locus of P, the point of intersection.

Method.—Join OO_1 and with centres O and O_1 draw equal circles. Mark off a distance *ab*, on circle O, and mark off this distance *twice*, giving a_1b_1 on circle O_1. Join O*b* and O_1b_1 and produce them to intersect in a point on the required locus. Repeat the construction for other points above and below OO_1 and draw a fair curve through them.

Locus of ends of rod moved by a crank

The crank OA, fig. 2, revolves about O. Hinged to A on the crank is a rod PQ which is constrained always to pass through a fixed point C. We require the loci of the ends of P and Q as the crank OA revolves about O.

The peculiarities of the paths of P and Q could hardly have been foreseen. A simple trammel may be used carrying a line $Q_1A_1P_1$ (in which Q_1A_1 = QA and A_1P_1 = AP). This is applied so that A_1 lies always on the circle centre O and the line always passes through C. Note that extreme points on each locus are given when the line Q_1P_1 is a tangent to the circle, as shown in the figure.

Exercises

Choose a suitable scale for each drawing.

1. The points O and O_1, fig. 1, are 1 metre apart and the link OA revolves twice as quickly as O_1A_1. Plot the locus of P, the point in which they touch. Limit your locus to about 1 m on each side of OO_1.
2. Take the following dimensions for the mechanism in fig. 2. OA = 0·5 m, PQ = 1·6 m, QA = 1·2 m. C is 0·7 m to the left of a vertical and 0·23 m below a horizontal, through O. Plot the loci of P and Q.
3. In fig. 3 OA is 0·3 m, AB is 0·7 m, AP = PB. Plot the locus of P.
4. The two rods OA and O_1B, fig. 4, oscillate about centres O and O_1. The connecting link AB is vertical when the other links are horizontal. A point C divides AB so that AC : CB :: BO_1 : AO. Plot the locus of C as the links move as far as possible in each direction. (This is Watt's parallel motion. The central part of the locus of C is very nearly straight and vertical.)
5. The equal cranks in fig. 5 are 0·3 m long and the centres OO_1 are 1·2 m apart. The rod AB is 1·2 m long. Find the locus of a point C on AB 0·4 m from A for a complete revolution of the crank.

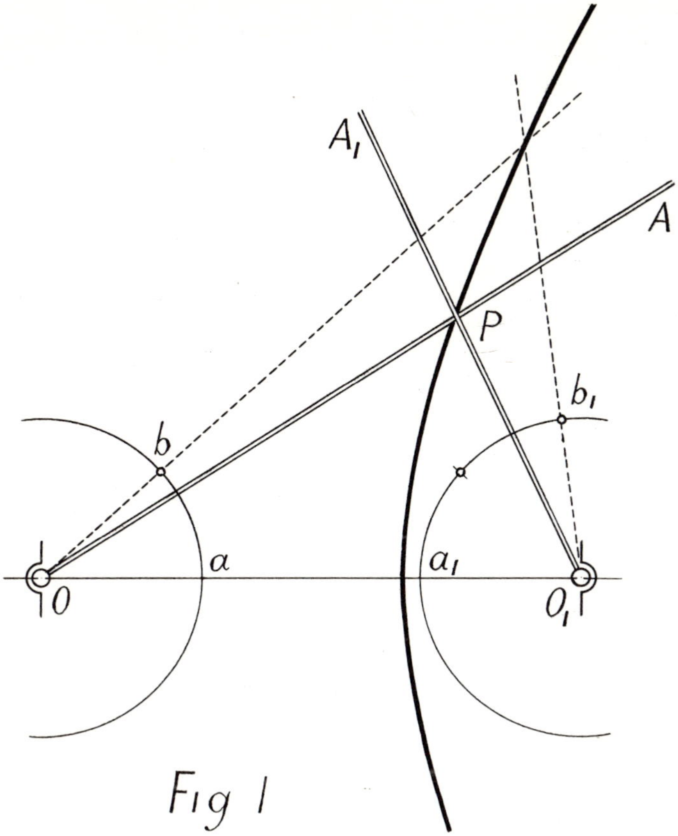

Fig 1

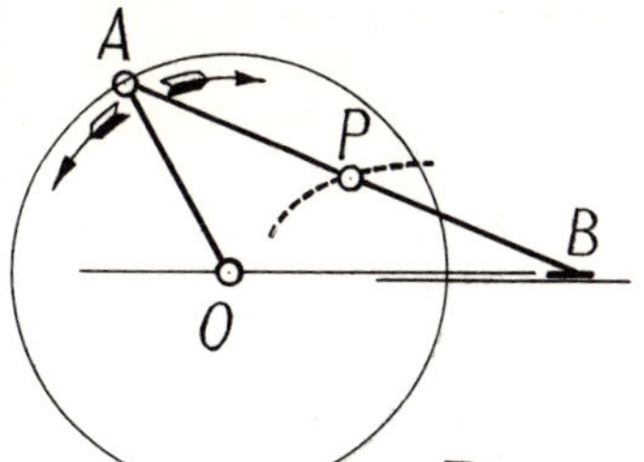

Fig 3

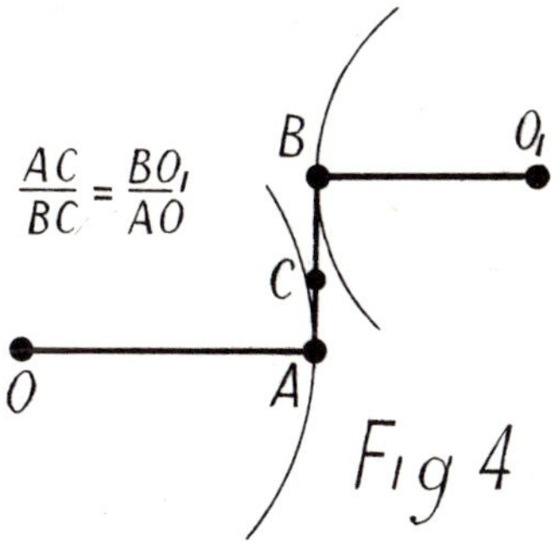

Fig 4

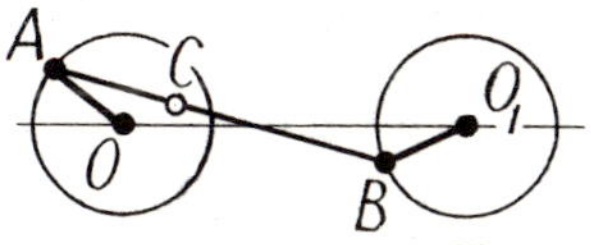

Fig 5

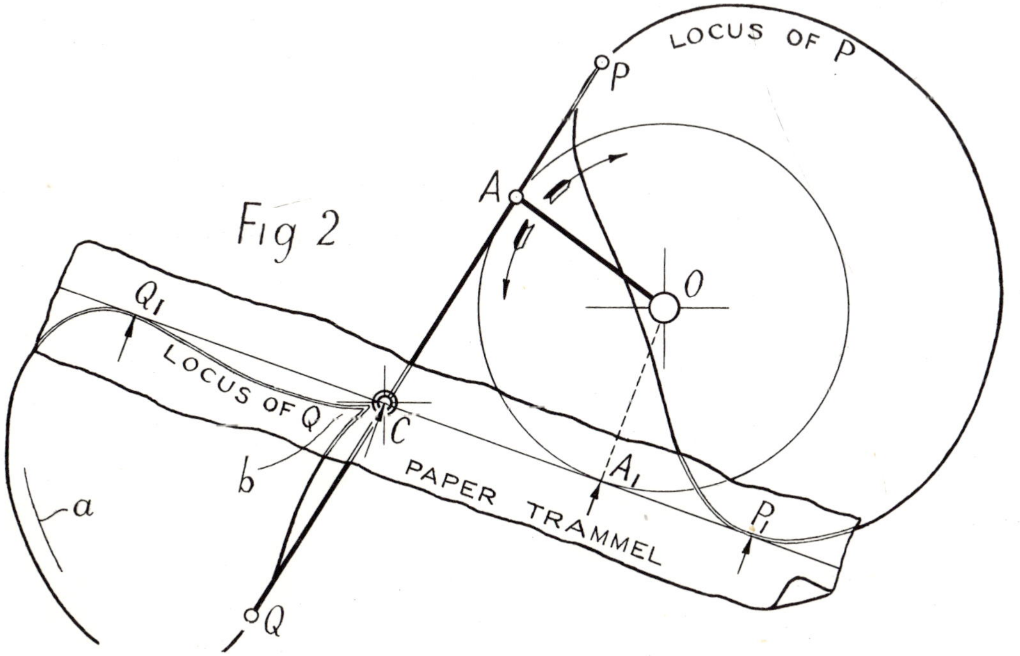

Fig 2

Approximate Constructions

Length of a circular arc

Let AB, fig. 1, be the given circular arc subtending an angle Q at its centre O. Draw a tangent to the arc at B. Join AB and produce it to D making BD = $\frac{1}{2}$AB. With centre D and radius DA describe an arc to intersect the tangent in E. Then the length of the straight line BE is almost equal to the length of the arc AB.

Given length marked along a circular arc

The converse of the above construction is as follows, fig. 2. Let AB be the given arc. Draw a tangent to the arc at B, and mark off BE equal to the given length. Take a point F on BE, such that BF = $\frac{1}{4}$ BE. With F as centre and FE as radius describe an arc to intersect the given arc in A. The length of the arc AB is almost equal to that of the line BE.

The construction in figs. 1 and 2 are due to Rankine. They are discussed more fully in the author's *Practical Geometry and Engineering Graphics;* the errors involved are also examined.

Circumference of a circle

A close approximation to the length of the circumference of a circle is given by the very simple construction in fig. 3. The student may like to calculate the error involved.

Let AB be the diameter of the given circle. Draw a line perpendicular to it at B and mark off BE = 3 × AB. On AB describe a semicircle. From O, its centre, draw a radius OC at 30° to AB. From C draw CD perpendicular to AB intersecting it at D. Join DE. The length of DE gives a close approximation to the length of the circumference.

Diameter of a circle from the given circumference

Let AB, fig. 4, equal the given circumference. Bisect it and draw a semicircle on AB. Using the same radius and with B as centre draw an arc cutting the circumference in D. Draw DE perpendicular to AB. With E as centre and ED as radius describe an arc cutting AB in F. The length of AF is almost the diameter of a circle, of which AB is the circumference. The construction gives diameter = 0·317 circumference; actual diameter = 0·318 circumference.

Note.—In all the above the most careful construction is necessary for an acceptable result.

Exercises

1. An arc subtends an angle of 50° at the centre of a circle of 100 mm radius. Calculate its length and compare with it the length found graphically.
2. Find the angle subtended at the centre of a circular arc 100 mm radius by a circumferential length of 90 mm. Check by calculation.
3. Find the length of the circumference of a circle 80 mm diameter using the methods of both fig. 1 and fig. 3.
4. The circumference of a circle is 6·28 m. Find its diameter graphically and check by calculation.

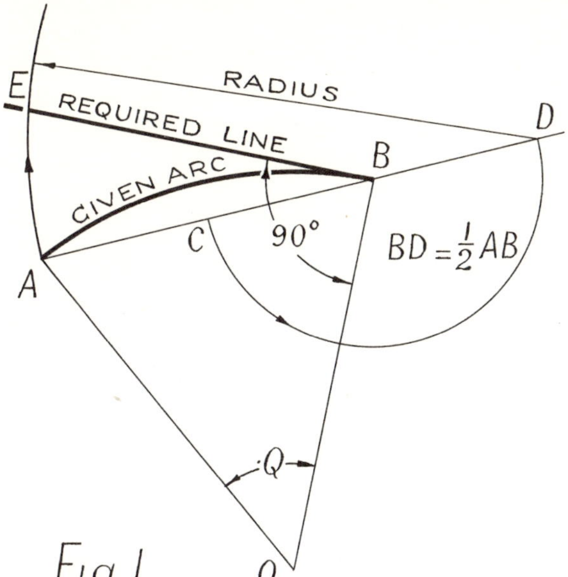

Fig 1

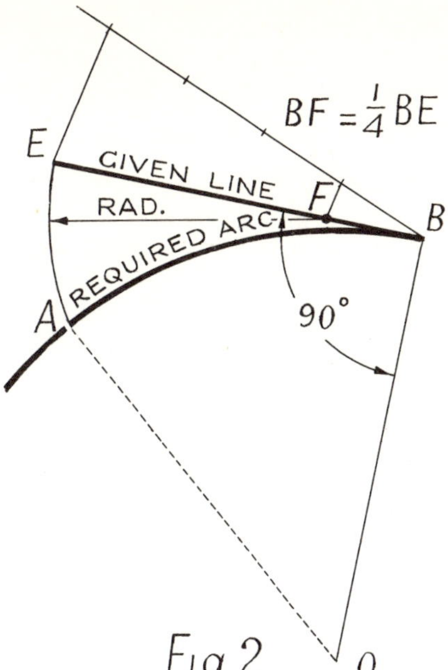

Fig 2

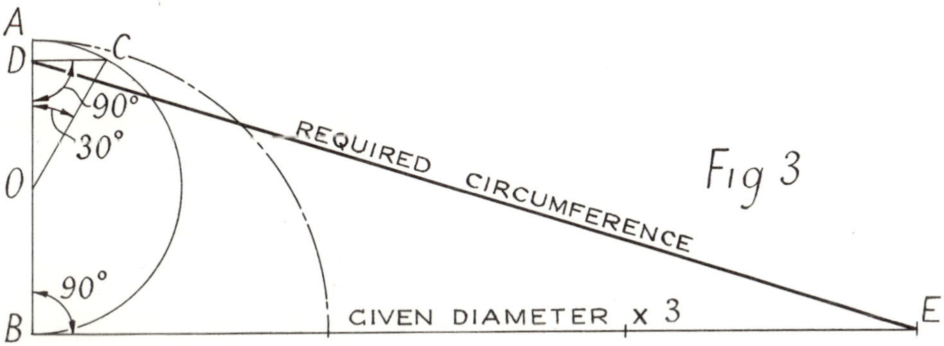

Fig 3

Fig 4

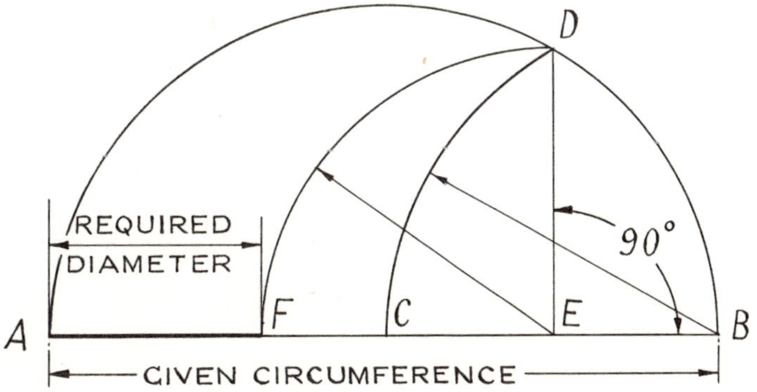

Roulettes, Involutes and Cycloids

If one curve rolls without sliding upon another curve, any point on the rolling curve traces out a *roulette*. The locus of P, fig. 1, as the curve rolls from one position to another, is a roulette.

Involute and evolute

If the rolling or generating line is straight, the locus of a point in the line is called the *involute* of the base curve. The locus of P, fig. 2, as the line AB rolls from one position to the other, is the involute of the base curve shown.

The point of contact C is the centre, for an instant, about which P turns; and the locus of C, i.e. the *original base curve*, is called the *evolute* of the involute path of P.

Cycloids

If the rolling line is a circle and the base line is straight, then the locus of a point P on the circumference, fig. 3, is a *cycloid*. But if the base line is also a circle we may have two curves depending on whether the rolling circle rolls on the outside or the inside of the base circle. If the circle rolls on the *outside*, fig. 4, the locus of P is an *epicycloid;* but if the circle rolls on the *inside* of the base circle, the locus of P_1 is a *hypocycloid*.

These curves were at one time of importance in engineering because they were used for the shapes of the teeth in gearwheels. They have been supplanted by teeth which are involute in form — see page 98.

Tracing-paper solutions

Close approximations to the true roulette forms can be achieved by the use of tracing paper. The method is as follows.

Draw the rolling curve on tracing paper and apply it over the base curve, as in fig. 5. P is a point on the curve, and A the point of contact. Using a fine needle, mark the position of P and then transfer the needle to A. Allow the tracing paper to roll about A until the curve (dotted line) *overlaps* the fixed curve by a small amount, cutting it at B. Transfer the needle to B and revolve the paper until the curve *touches* the fixed curve at B (chain line). The point P has now moved to P_1. Mark it, and plot a succession of points P_1P_2 . . . A fair curve through them gives the required roulette.

Exercises

1. A circle 50 mm diameter rolls along a straight line. A point P on the circumference is at first and last in contact with the straight line. Draw the cycloidal path of P.
2. A circle 50 mm diameter rolls once around (*a*) the outside, (*b*) the inside, of a circular arc 80 mm radius. Plot the complete epicycloid and hypocycloid traced by a point on the circle. Increase the diameter of the rolling circle to 80 mm and plot the hypocycloid. Note that it is now a diameter of the fixed circle.
3. A parabola CD, fig. 6, rolls on the line AB. Plot the roulette of its focus F.
4. A line AB, fig. 7 touches an ellipse tangentially at C, and rolls until it is in contact with V_1. Plot the roulette of P.

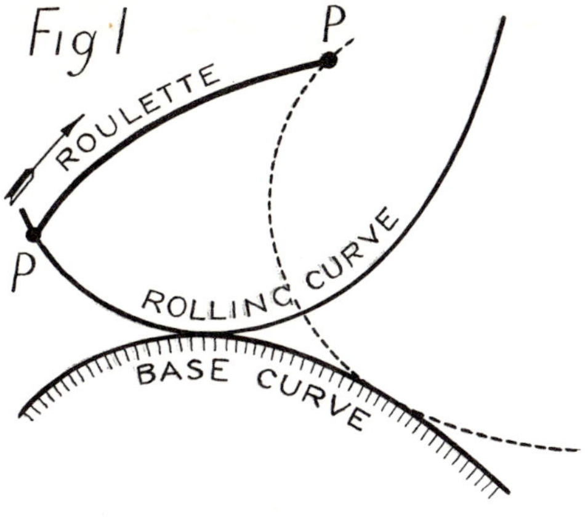

Fig 1

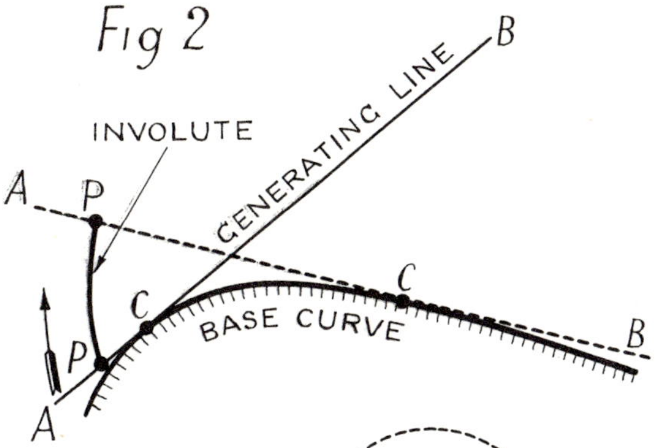

Fig 2

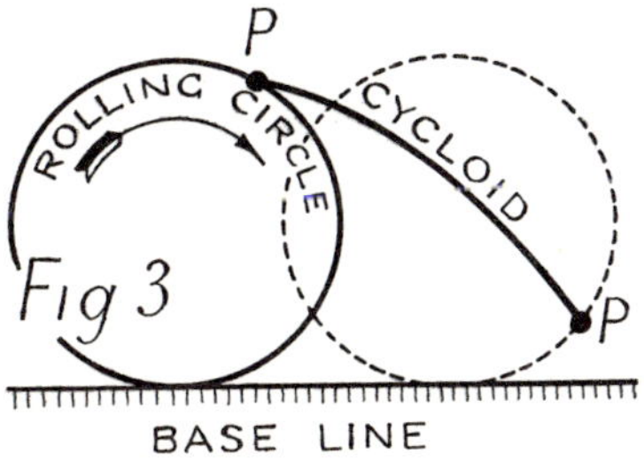

Fig 3

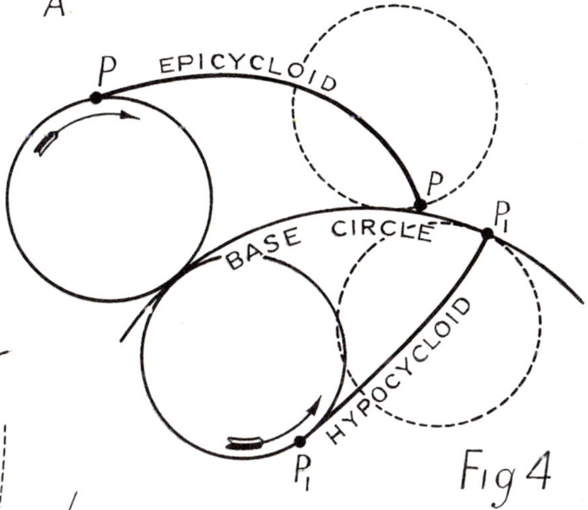

Fig 4

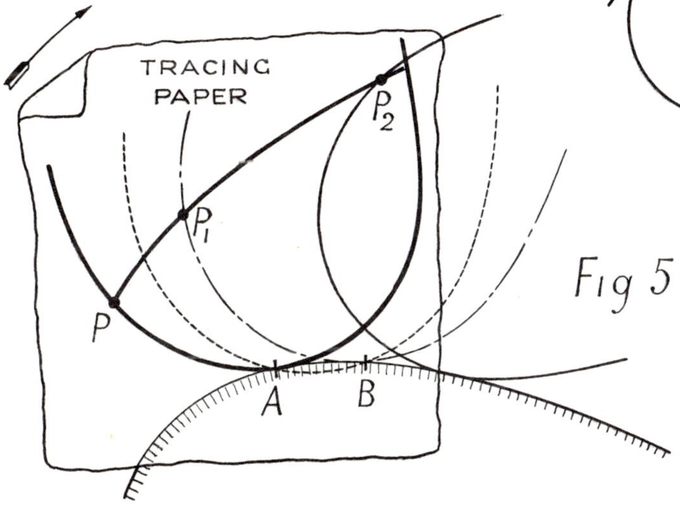

Fig 5

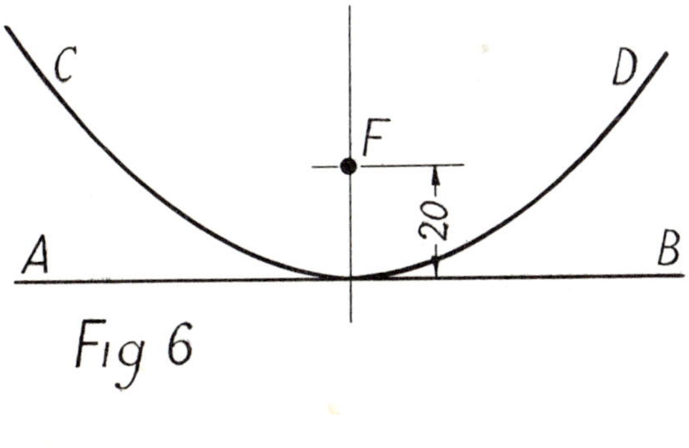

Fig 6

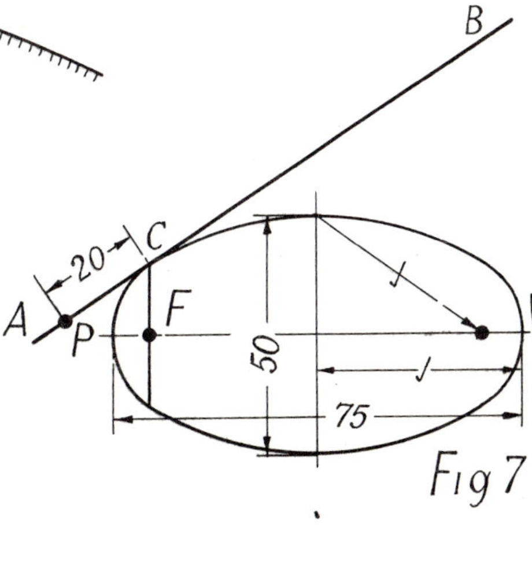

Fig 7

Involute to a circle

The involute may be determined by using tracing paper, as described on page 96, and the method discussed here is an alternative.

We first require the length of the circumference of the base circle, centre O, fig. 1. This may be obtained graphically as on page 94. Divide this line, and the circle, into the same number of equal parts (say 12) 1, 2, 3 . . . At the points on the circle draw tangents and mark off along them, successively, lengths equal to $\frac{1}{12}$ $\frac{2}{12}$ $\frac{3}{12}$ . . . of the circumference, giving points p_1, p_2, p_3 . . . on the required involute. The *normal* to the involute at N is given by the tangent NM to the base circle; a line at N drawn at right angles to NM gives the *tangent* at N.

Toothed gearing

The two circles in fig. 3 represent discs in contact at P. If one disc turns about O, friction between the discs will cause the other to turn about O_1. To prevent slipping at P when power is to be transmitted, grooves may be cut in the rubbing surfaces, and projecting strips added between the grooves, forming the *gear teeth* shown in fig. 4. The imaginary circles in fig. 4 corresponding to those in fig. 3 are called the *pitch circles;* P is the pitch point.

The profiles of the teeth will be correct if the motion between the wheels is the same as that given by the plain discs, fig. 3. For this to be so, the profiles may be either *cycloidal* or *involute curves*, and involute curves are now almost always used.

To draw the involute teeth (may be deferred)

Exercises

1. The student should actually set out the profiles of two mating teeth in cardboard, as indicated by the strips in fig. 4 and fig. 2. Take pitch circle diameters of 250 and 400 mm.*, i.e. $OO_1 =$ 325 mm. Draw the common tangent TT_1, fig. 2, and set off NN_1, through P at 20° to TT_1. Draw circles centres O and O_1 to touch NN_1; these are the *base circles*. Take any point Q on NN_1 and, using tracing paper applied to the two base circles, plot short lengths of involute curves passing through Q. These give the profiles of mating teeth. Complete each tooth, making the width on the pitch circle = 20 mm and limiting the height to clear the opposite base circle.

   Cut out these two radial strips in stiff cardboard and apply them over a diagram as in fig. 2, using pins at centres O and O_1. Move one strip against the other and note the kind of motion between the two profiles—a mixture of rolling and sliding. Note also that the point of contact Q lies always upon NN_1, which is called the *path of contact*. The angle *A* is the *pressure angle* and is usually 20°.
2. Repeat the construction in Ex. 1, making the angle $A = 14\frac{1}{2}°$.

* If compasses for these radii are not available, plot the circular arcs by using a thread anchored by a pin at O and O_1.

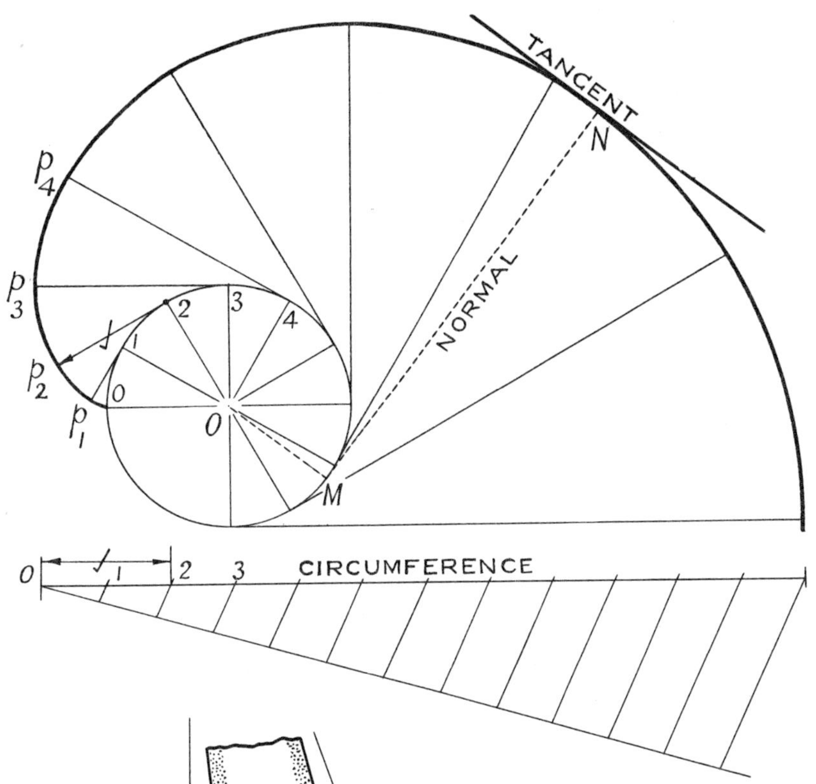

Fig 1

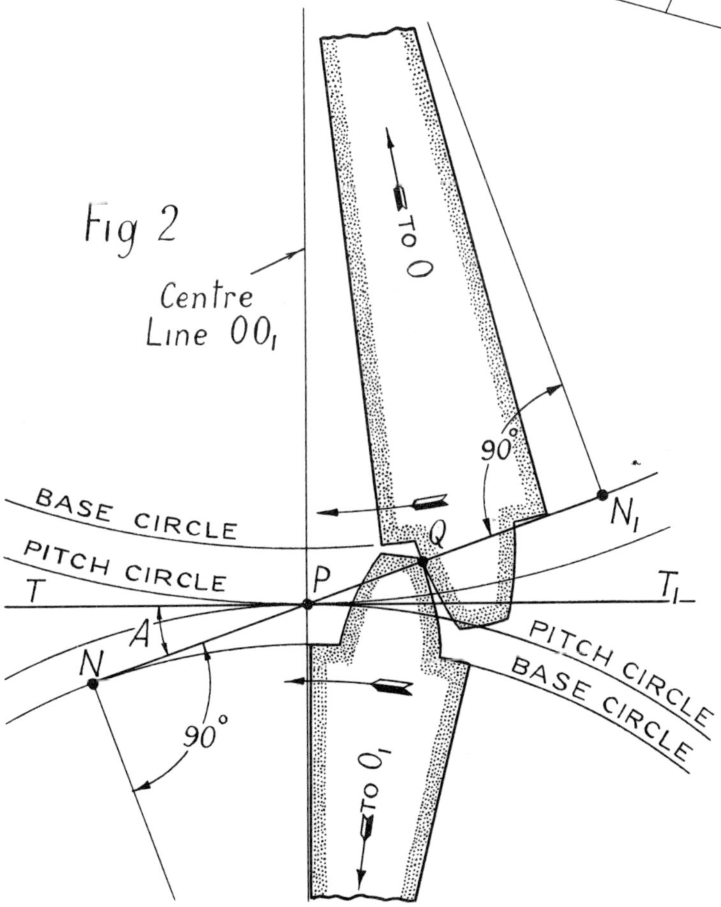

Fig 2

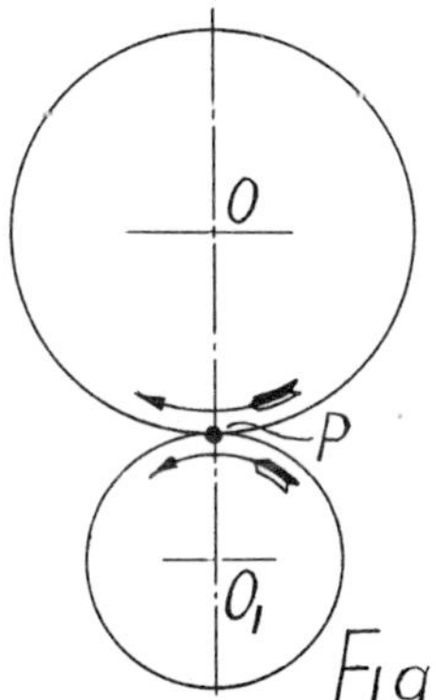

Fig 3

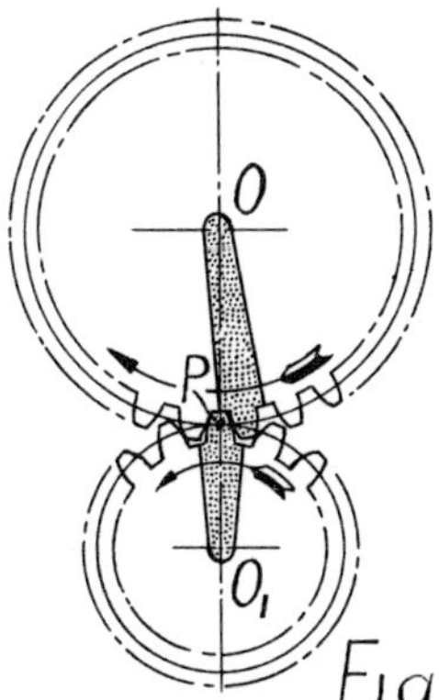

Fig 4

Spiral and Helix

Archimedean spiral

If we imagine a line to rotate in a plane about one of its ends, and, at the same time, a point to move in one direction along the line, then the path traced out by the moving point is a spiral. There are many spirals, depending on the way the point moves, but we shall consider only one, the spiral of Archimedes, fig. 1. In this the point moves equal distances along the line for equal angular movements of the line.

Let OP and OQ be the initial and final positions of the point for one turn of the spiral. Divide PQ into, say, 12 equal parts and set out 12 equidistant radii from O. With centre O and radii O1, O2, O3 . . ., draw arcs to cut successive radii in points 1, 2, 3 . . . A fair curve drawn through the points gives the required spiral. In fig. 1 the curve has been taken onwards into the second convolution.

Helix

In this curve a point moves around a cylindrical surface and at the same time advances in the direction of the axis. The axial and the rotary movements are both uniform.

The right-angled triangle ABC in fig. 2 is cut from paper and wrapped around a cylinder, B and C coinciding. The length of BC is equal to the circumference of the circle. The hypotenuse AB becomes a helix. The axial advance in one complete turn is AC, and this is called the *pitch* of the helix. To obtain the projection A_1B_1 of the helix, divide AB and the circumference of the cylinder into a similar number of equal parts, say 12. Number them as in fig. 2. Draw horizontals and verticals to give points on the helix.

Helical spring of square section (may be deferred)

We may imagine the square *abcd*, fig. 3, to move around and along the axis, thus generating the spring. To project the spring, draw helices from the four corners *a b c* and *d*, as shown in the lower part of the drawing, and then thicken in only those lines which will be seen, as shown in the upper part. The divisions of the pitch distance should be chosen to divide equally the side of the square; the same construction lines then serve for the upper and lower pairs of helices. The construction should be clear from the drawing.

Exercises

1. Draw *two* convolutions of an Archimedean spiral, least radius 10 mm, greatest radius 100 mm.
2. Plot two complete turns of a helix 100 mm diameter, 50 mm pitch. Using the same axis and pitch, plot, within the first, a second helix of 60 mm diameter.
3. Project one complete turn of a spring: section 25 mm square, outside diameter 130 mm, pitch 100 mm.
4. Repeat Ex. 3 for a spring of *circular* section, 25 mm diameter. Hint. Regard the spring as the envelope of a sphere 25 mm diameter the centre of which moves along the helical centre line.

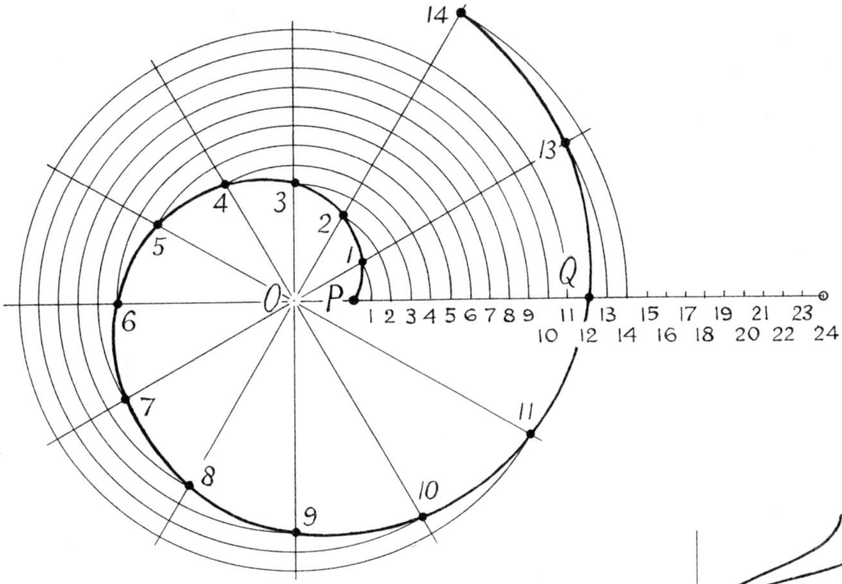

Fig 1

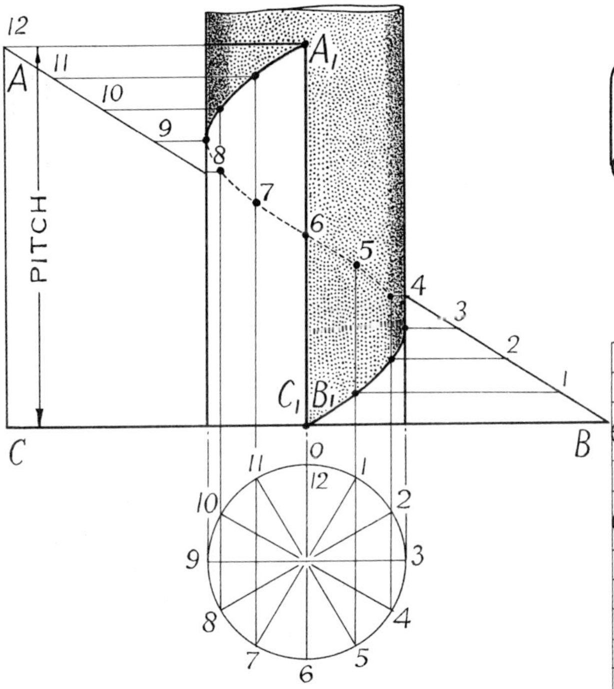

Fig 2

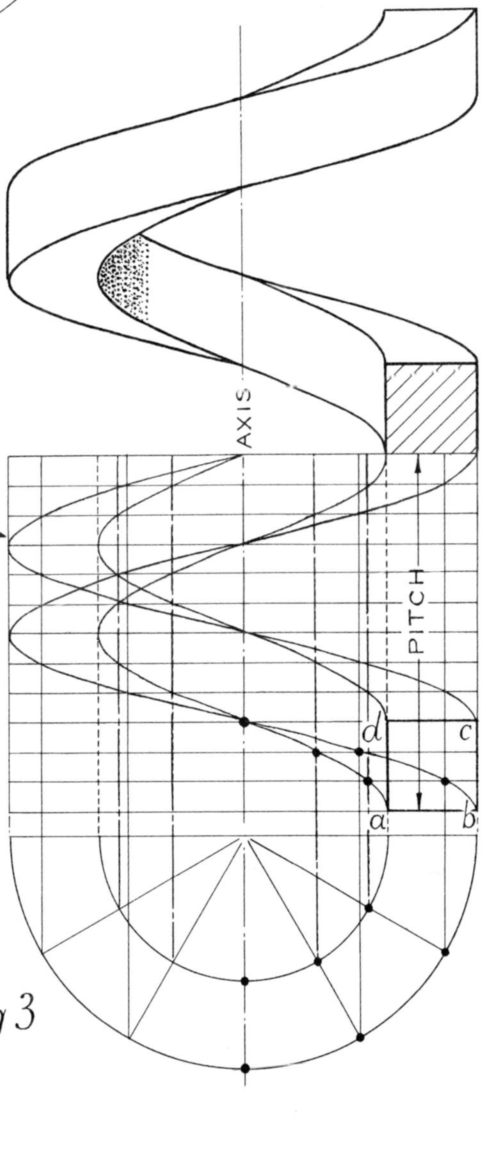

Fig 3

Areas of Irregular Figures

There are many ways of finding the approximate values of areas included within irregular curves. One of the most acceptable is now described. It is based upon a simple construction used for graphical integration.

Method

Axes OX and OY, and an irregular curve, are shown in fig. 1. We wish to find the area between the curve and the axes.

Take any pole p on OX produced. Divide the area into vertical strips (only two are shown). Transfer the mid-ordinate height h to OY giving Oc. Join cp. Repeat for each strip; the next gives dp. From O_1 on O_1X_1 in fig. 2 draw O_1c_1 between the ordinates, parallel to pc. Then draw c_1d_1 parallel to pd. Build up thus the stepped curve in fig. 2. The required area is given by:

Area $= H \times P$ (*i.e. the product of the final height and the pole distance*).

Note.—(1) It will be clear that the length of P influences the size and accuracy of fig 2. Again, ordinates should be spaced to suit the curve, being closer where the curve is changing most. (2) An area bounded wholly by curves, as shown in fig. 3, can be converted into the kind shown in fig. 1, as indicated in figs. 3 and 4.

Proof

First area shaded = mid ordinate × width of strip $= h \times w$

Because the triangles Opc, fig. 1, and $a_1O_1c_1$, fig. 2, are similar, then $Oc/Op = a_1c_1/a_1O_1$; i.e. $h/P = a_1c_1/w$

$\therefore h \times w = P \times a_1c_1$

Hence, first area shaded $= P \times a_1c_1 = P \times h_1$

By similar reasoning, the second area shaded (dotted lines) $= P \times h_2$

Hence the sum of the two shaded areas $= P(h_1 + h_2)$

It follows that the Total Area $= P \times H$, i.e. Pole Length × Final Height H.

Exercises

1. Redraw the curve in fig. 1 twice the size given. Measure the area under the curve. Ans. 7200 mm².

2. The major and minor axes, a and b, of an ellipse are 150 mm and 100 mm in length. Find its area (plot only one quadrant). Check by calculation. Area $= \frac{1}{4}\pi\, a.b.$

3. The aerofoil shown in fig. 5 is defined by distances on ordinates above and below a centre line. The overall length of the aerofoil is 100 units and the value of the ordinates are given below. Find the area enclosed by the curve. (Ans. 1436 units²).

Station % Length	Upper Surface	Lower Surface	Station % Length	Upper Surface	Lower Surface
2·5	6·14	3·14	40	11·49	8·83
5·0	7·93	4·52	50	10·40	8·14
10	10·03	6·32	60	8·90	7·07
15	11·19	7·51	70	7·09	5·72
20	11·80	8·30	80	5·05	4·13
25	12·05	8·76	90	2·76	2·30
30	12·06	8·95	95	1·53	1·30
			100	0·22	0·22

(Hint. It is probably best to find each area separately, one construction lying above and the other below a horizontal. Suitable scales are : OX and OY 1:10, Op = 2 units. Then area $= (H_1 + H_2) \times 10 \times 10 \times 2$ square units.)

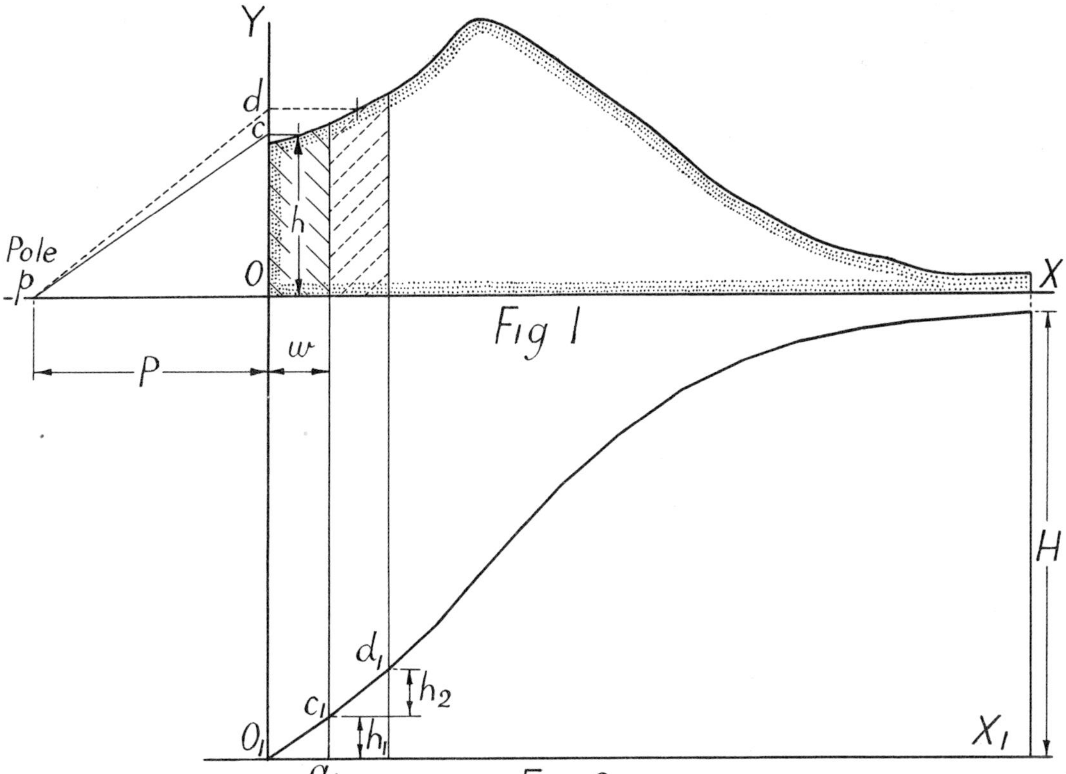

Fig 1

Fig 2

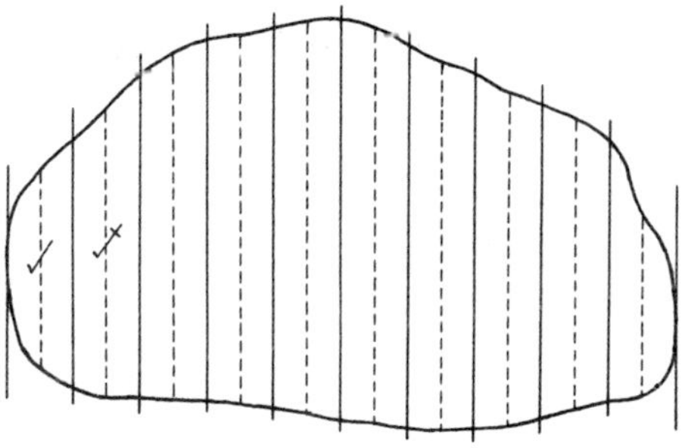

Fig 3

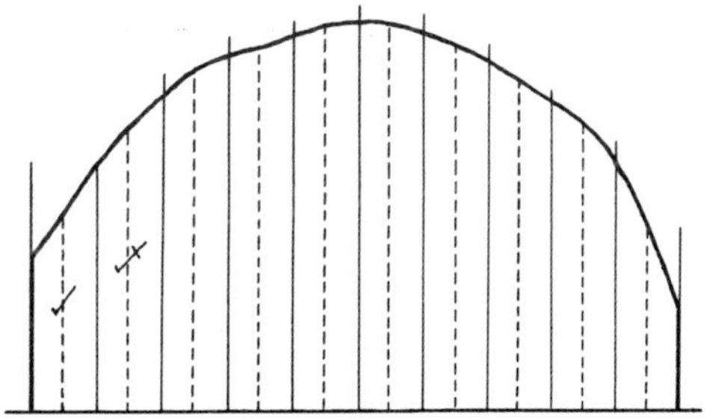

Fig 4

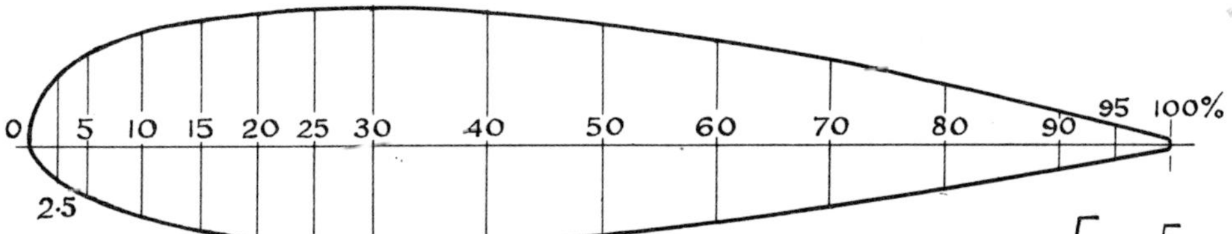

Fig 5

Principles of Projection

In Part I we made drawings of various objects by recording views taken in two or three directions and arranging the views in line. We also saw that it was necessary to label the views in order to avoid misinterpretation.
We shall deal with the matter a little differently now, using systems known as First Angle and Third Angle projection.

Orthographic projection (two views)

Consider the object shown in fig. 1. From selected points on its outline, parallel lines such as AC have been drawn to meet a plane surface. Such lines are called *projectors,* and the resulting drawing is called a *projection* of the object. If the projectors are perpendicular to the plane, then the projection is called *orthographic* or *orthogonal*, i.e. a correct or true view.
The plane surface in fig. 1 is horizontal (HP). If we take another plane at right angles to HP, as in fig. 2, intersecting it in *xy*, and if we make another orthographic projection on this vertical plane VP, we obtain a second view. If the two planes are opened out until they are in the same plane, we get the views one above the other, as in fig. 3.
Another arrangement of the projections is given if we suppose the planes to be transparent and to come between the observer and the object, as in fig. 4. If these are supposed hinged at *xy*, then, when they are opened out, the arrangement of views is that given in fig. 5.
The projection on the HP is commonly called a *plan* and the projection on the VP an *elevation*. It will be seen that in one arrangement the plan comes below the elevation, and in the other above it. A very useful way of looking at it is to say that in the first system, fig. 3, one view shows what would be seen by looking on the *far* side of the other view; while in the second system fig. 5, one view shows what would be seen by looking on the *near* side of the other view.
The lines connecting the views in fig. 3 do not appear in fig. 2. They also are called projectors, and they are at right angles to *xy*. That this is so will be clear from figs. 4 and 5; two projectors are shown in fig. 4, linked to the same lines as in fig. 5.

Definitions

1. ***The orthographic projection*** **of a point on a plane is given by the foot of the perpendicular from the point to the plane.**
2. ***The trace of a line*** **on a plane is the point of intersection of the line, produced if necessary, with the plane. In fig. 1, C is the trace of AB.**
3. ***The trace of a plane*** **on another plane is the line of intersection of the two planes. The line *xy* in figs. 2 and 4 is the trace of one plane on the other.**

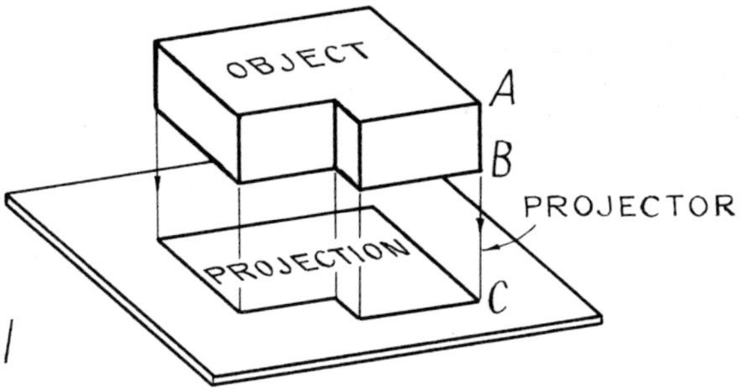

Fig 1

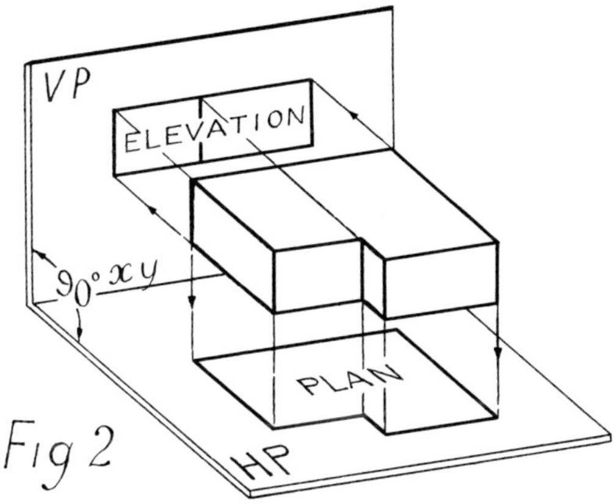

Fig 2

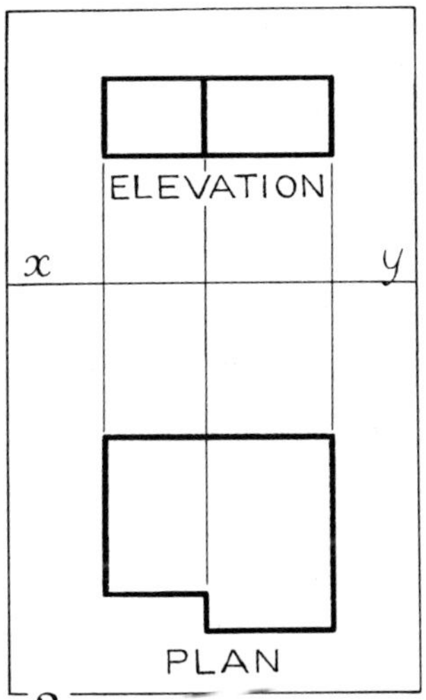

Fig 3

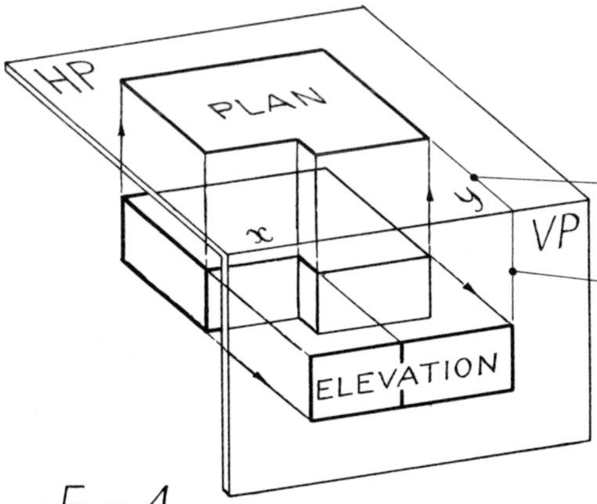

Fig 4

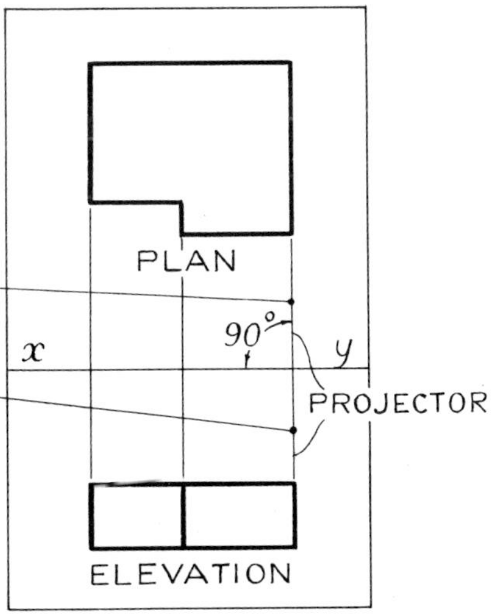

Fig 5

Orthographic projection (three views)

We shall take the work of page 104 a stage further by introducing a third plane at right angles to both the horizontal plane HP and the vertical plane VP. Groups of drawings are shown opposite: figs. 1 and 3 give the arrangement of an object within these three *planes of reference*, as they are called; figs. 2 and 4 give the orthographic projections of the object on these planes, after opening out. The views in fig. 2 and 4 indicate that the new plane is opened out to coincide with the VP, before the HP and VP are opened out.

We can adopt the terms *plan, elevation, end view* for the projections, but an alternative, having much to commend it, is to label the views as was suggested in Part I and as shown opposite. We have used the elevation as a reference view, and the other views are defined from it by arrows and letters. This method makes it perfectly clear that in the First Angle system, figs. 1 and 2, a view shows what would be seen by looking on the *far* side of the adjacent view; and that in the Third Angle system, figs. 3 and 4, a view shows what would be seen by looking on the *near* side of the adjacent view.

First and third angle projection

The names First and Third Angles refer to the quadrants shown in fig. 5. Space is divided by vertical and horizontal planes into four 90° angles. The object to be described can be placed in either the 1st or the 3rd Angle, these being represented by figs. 1 and 3 respectively. Then, when the planes are opened out, we get the arrangement of views in fig. 2 or 4. If we place an object in the 2nd or 4th Angles the views overlap when the planes are opened out.

Comparison between projections in 1st and 3rd angles

The views in fig. 2, or fig. 4, will, when dimensioned, describe the object completely. One system has no natural advantage over the other. Projection in the 1st Angle is traditional, and was at one time universal. It is still current in Europe. The 3rd Angle system is exclusively used in N. America; it is widely used in Britain, where its general acceptance is probably only a matter of time—because of the overriding need for uniformity of practice in matters concerning defence.

The advantages of 1st Angle *pictorial* views in revealing the solutions to geometrical problems will be clear from the examples in this book; but in the actual solution of the problem either system may be used. Both are used in these books but the 3rd Angle system is given a preference.

A misapprehension arises in thinking that an object should be specified by a drawing of two elevations and a plan view as shown in figs. 2 and 4. There is no limit to the number of orthographic views which may be projected from a plan and elevation. It is often helpful to project two side elevations, one to the left and one to the right, of the front elevation, in order to clarify points which otherwise could only be shown with difficulty by dotted lines. Such an arrangement is required in Ex. 3 and 4. When applying dimensions to these and other drawings, a more pleasing result is obtained if they are distributed around the various views without repetition.

Exercises

1. Measure the dimensions of the object opposite and quadruple them. Draw the three views in each system and add dimensions.
2. Fig. 6 shows a socket. Draw, full size, in 1st angle projection:
   (*a*) an elevation seen in direction of arrow A.
   (*b*) a side elevation seen in the direction of the arrow B.
   (*c*) a plan view below view (*a*).
3. A corner bracket is shown in fig. 7. Draw, scale full size, in 3rd angle projection:
   (*a*) an elevation seen in the direction of the arrow A.
   (*b*) a side elevation seen in the direction of the arrow B.
   (*c*) a side elevation seen in the direction of the arrow C.
   (*d*) a plan view above view (*a*).
4. A two-forked bracket is shown in fig. 8. Draw, scale full size, in 1st angle projection:
   (*a*) an elevation seen in the direction of the arrow A.
   (*b*) a side elevation seen in the direction of the arrow B.
   (*c*) a side elevation seen in the direction of the arrow C.
   (*d*) a plan view below view (*a*).

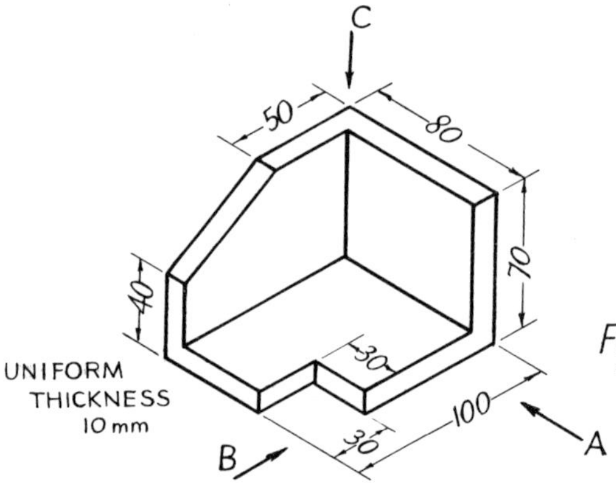

Fig 6

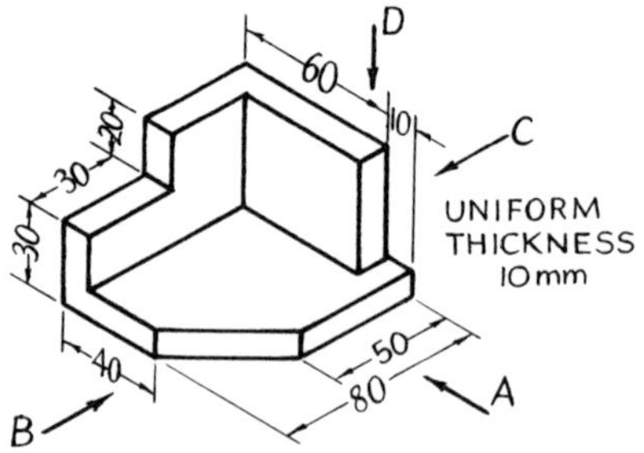

Fig 7

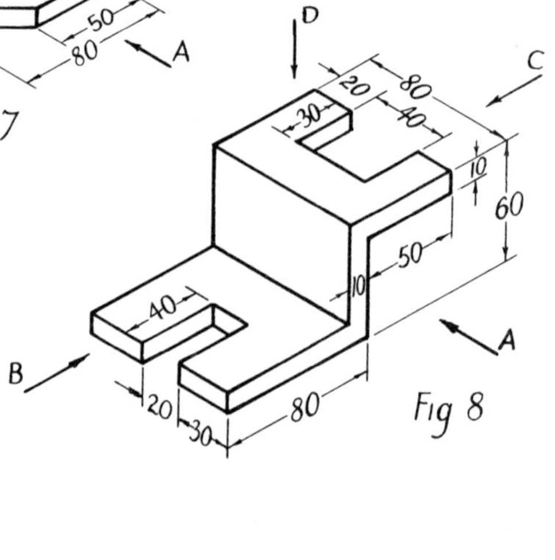

Fig 8

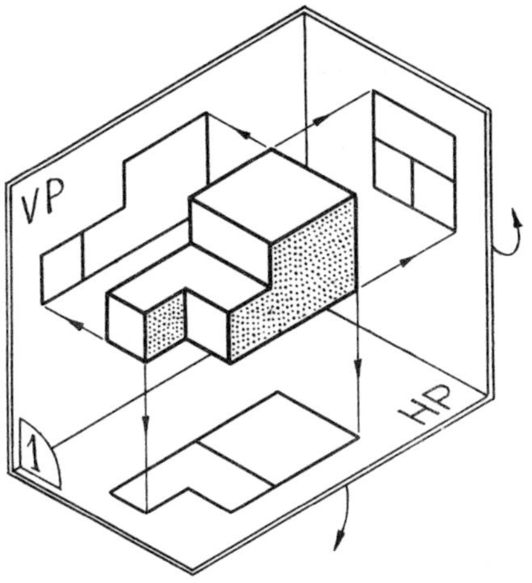

Fig 1

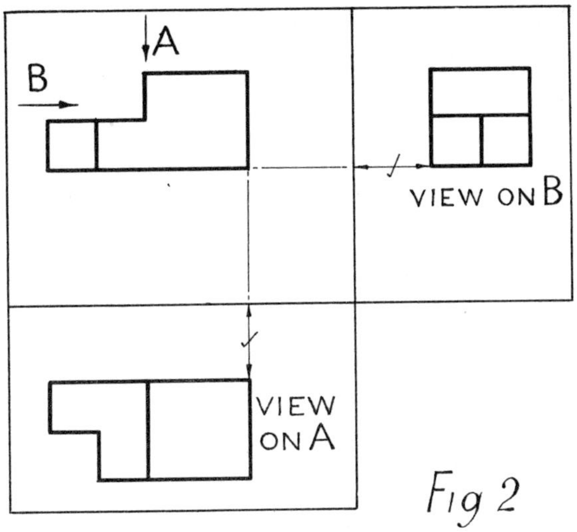

Fig 2

FIRST ANGLE

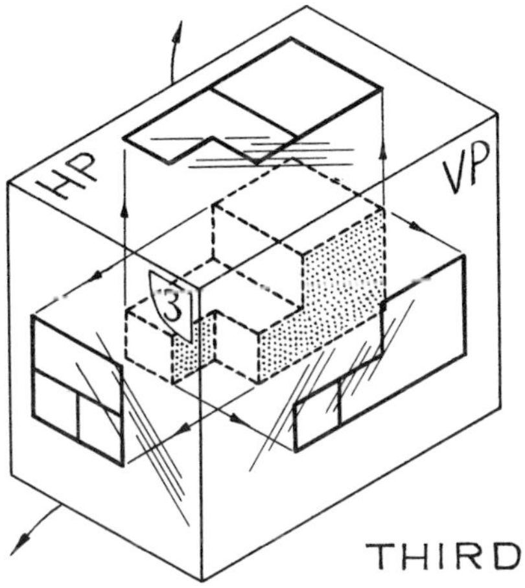

THIRD ANGLE

Fig 3

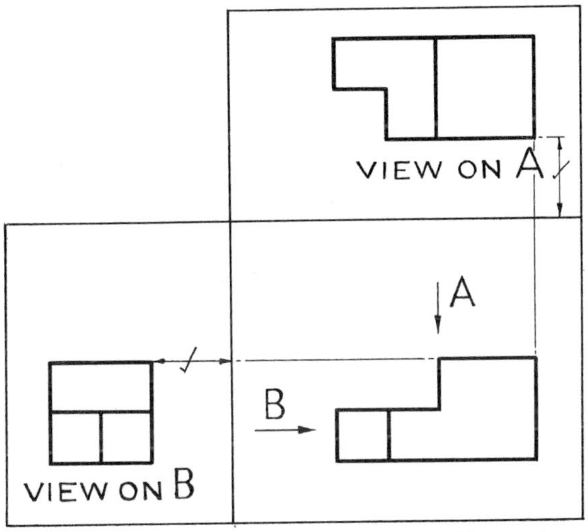

Fig 4

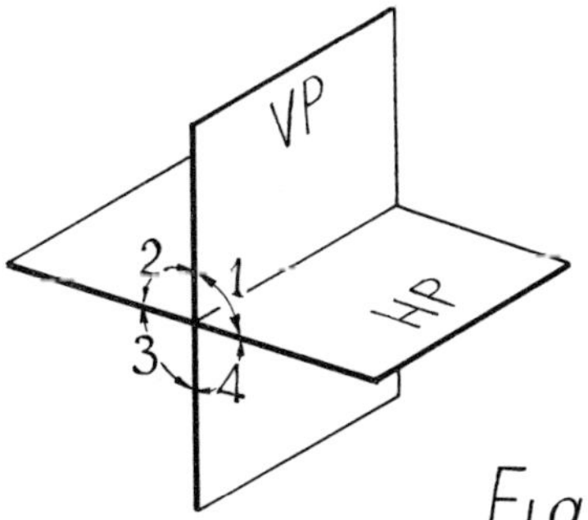

Fig 5

The Straight Line

Fundamental projections

There are some elementary operations concerning the projections of straight lines which are of first importance. Although the student may perform them easily, it is very necessary that he should fully appreciate the reasons for the operations.

We shall now consider some projections which will give us: (*a*) a view in which a line shows its true length; and (*b*) a view in which a line appears as a point.

Consider the pictorial view in fig. 1. This shows a line AB and its projections *ab* on the horizontal plane (I), and a_1b_1 on the vertical plane (II). These projections are shown in the usual way in fig. 2. Notice that neither *ab* nor a_1b_1 gives the true length of AB. If we were at liberty to show the projections of the line AB as we liked, we should not have arranged them as in fig. 1. We should probably have made the line vertical, as in fig. 7, or horizontal. Let us convert the view in fig. 1 to that in fig. 7 by stages.

First, let us interpose another vertical plane (III), behind the line AB and parallel to it, as in fig. 3. Note that *ab* is now parallel to the new line x_1y_1. Note also that we have not changed the position of AB and that the heights of A and B above Plane I remain the same. We may therefore obtain the projection a_3b_3 from *ab* as in fig. 4. We draw x_1y_1 parallel to *ab* and then mark off, from x_1y_1 along the projectors from *a* and *b*, distances which are equal to the distances of a_1 and b_1 from *xy*. *The length* a_2b_2 *is the true length of* AB. The thickened lines in fig. 4 give projections which replace those in fig. 2 in what follows.

Let us now take a view *along* the line AB. We proceed by using another plane (IV), fig. 5, *taken at right angles to* AB. This gives x_2y_2, fig. 6, and the projection a_3b_3 results. This is a point, as far from x_2y_2 as *a* or *b* is from x_1y_1. The student may get a clearer idea of these last two projections if he turns his book around until AB, fig. 5, is vertical. He will then see that we now have two simple projections of the line AB, as shown in fig. 7; one view gives the true length of the line and the other gives its position.

These transformations are of great importance because they enable us to reduce difficult problems to simpler ones.

The Rule for these operations may now be stated:

If we have two projections P and Q of an object (on *xy*), we may obtain a new projection R (on x_1y_1) by projecting either from P or Q along lines perpendicular to x_1y_1. If we project from P, distances in R from x_1y_1 are the same as distances of the same points in Q from *xy*. If we project from Q, distances in R from x_1y_1 are the same as distances in P from *xy*.

The true length of a line is given by one projection if the other projection is parallel to *xy*.

Exercises

See next page.

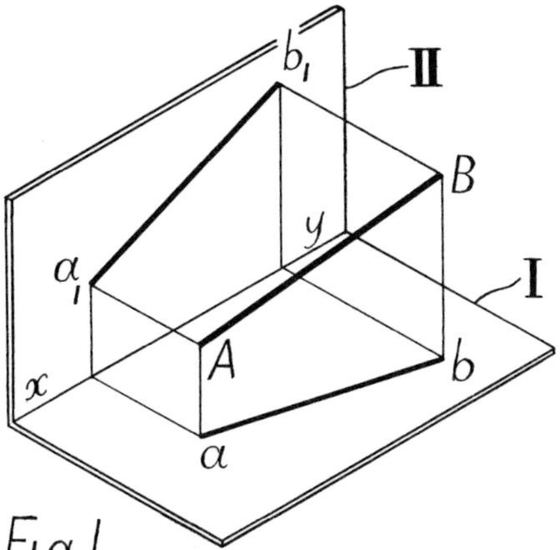

Fig 1

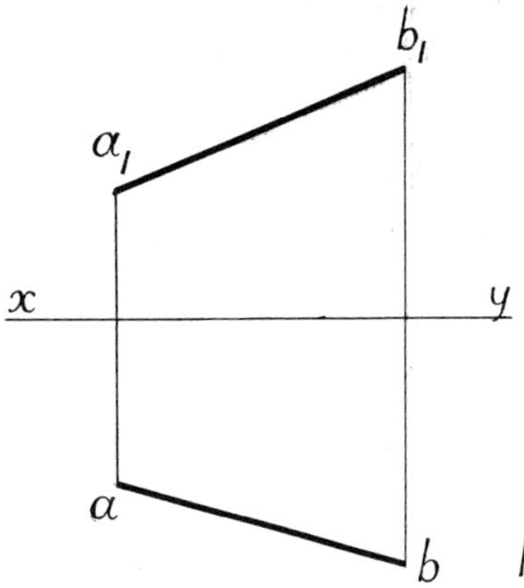

Fig 2

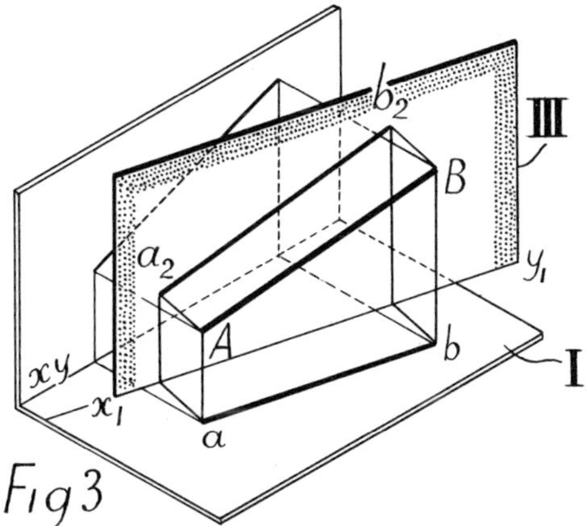

Fig 3

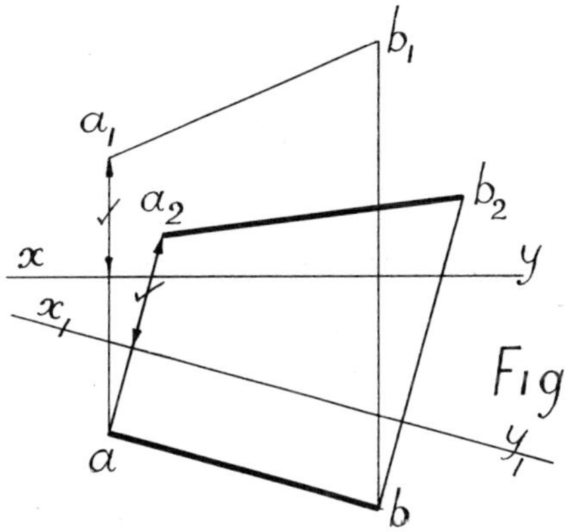

Fig 4

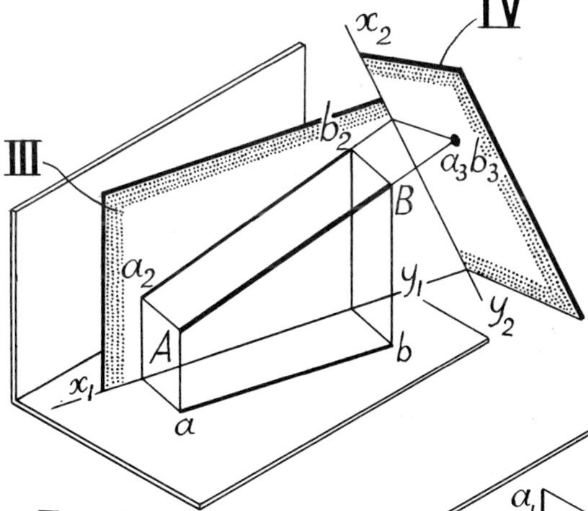

Fig 5

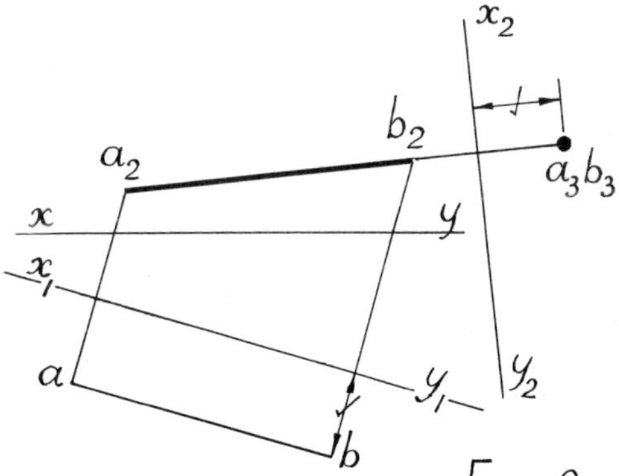

Fig 6

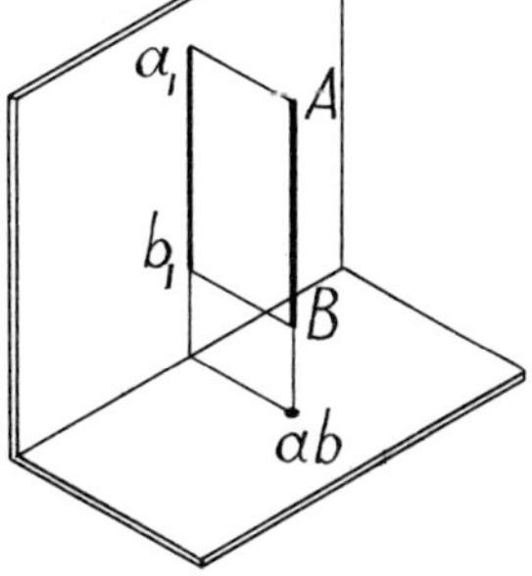

Fig 7

Inclinations to the planes of reference (may be deferred)

It is frequently necessary to obtain the inclination of a line to the HP or the VP. If a line is parallel to one plane, then its projection on that plane gives its inclination to the other.

This gives us a clue for the solution of the general case in which the line is not parallel to the HP or to the VP. We choose another plane, parallel to the line and at right angles to either the HP or the VP. A projection on this plane gives the required inclination.

Figs. 1 and 2 are very similar to figs. 1 and 3 on the previous page. The line AB, fig. 1, is not parallel to the HP or the VP. Let us now insert a vertical plane parallel to the line, as in fig. 2; x_1y_1 is parallel to *ab*. The new projection is a_2b_2 and the angle between a_2b_2 and x_1y_1 is the required inclination Q of the line AB to the HP. The solution is given in fig. 3.

To find the inclination of AB to the VP, we use another plane, perpendicular to the VP and parallel to the line, as in fig. 4; x_2y_2 is parallel to a_1b_1. The new projection is a_3b_3 and the angle between a_3b_3 and x_2y_2 is the required inclination Q_1 of the line AB to the VP. The solution is given in fig. 5.

The actual constructions in figs. 3 and 5 are very easy to make. They should be applied to both 1st Angle and 3rd Angle systems

Exercises (dimensions are in millimetres)

1. A line AB is shown by its projections *ab* and a_1b_1, fig. 7, in the 1st Angle System. Find the length of AB and its inclinations to HP and VP. Ans. Length 62.6 mm; Q, 18.5°; Q_1, 31°.
2. The projections *ab* and a_1b_1 of a line are given in fig. 8, using the 3rd Angle System. Find the length of the line and its inclinations to HP and VP. (Note: *ab* is the plan, a_1b_1 the elevation.)
   Ans. Length 57.4 mm; Q and Q_1, 20°.
3. The plan of a line 60 mm long is shown in fig. 9. The elevation of one end is at b_1. Complete the elevation and measure the inclinations of the line to HP and VP.
   Ans. Q, 41.5°; Q_1, 19.5°.

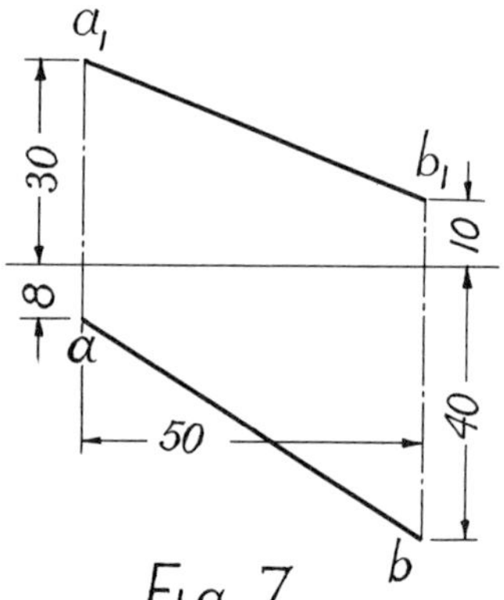

Fig 7

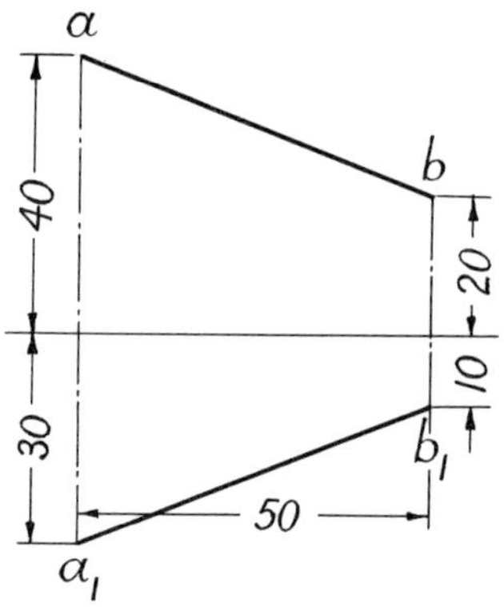

Fig 8

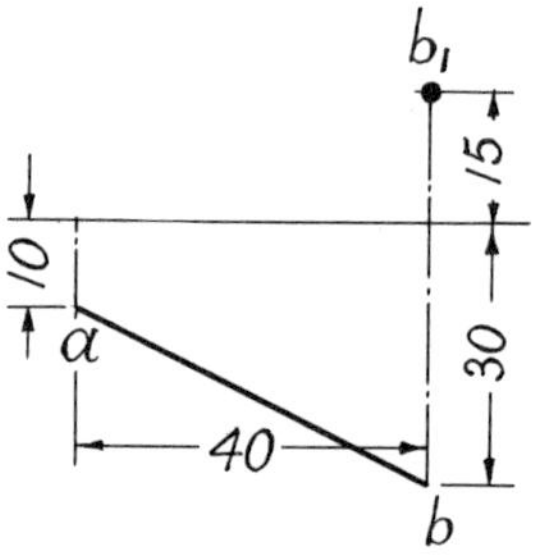

Fig 9

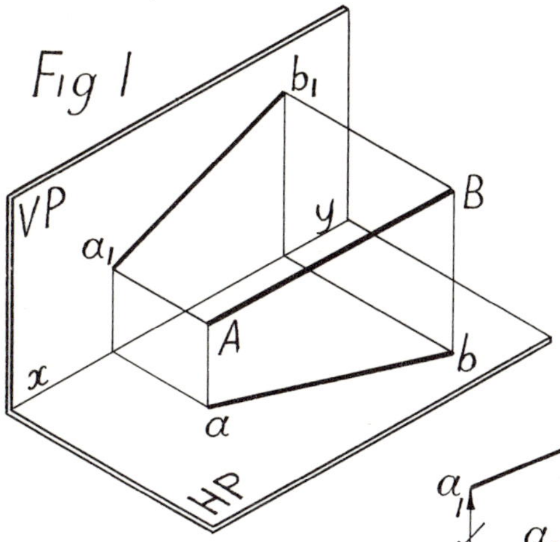

Fig 1

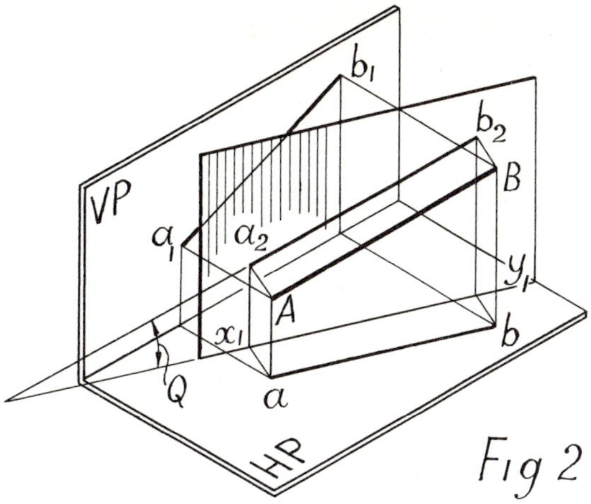

Fig 2

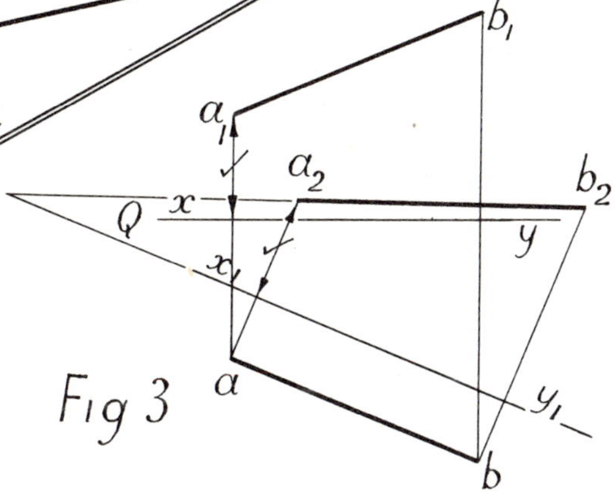

Fig 3

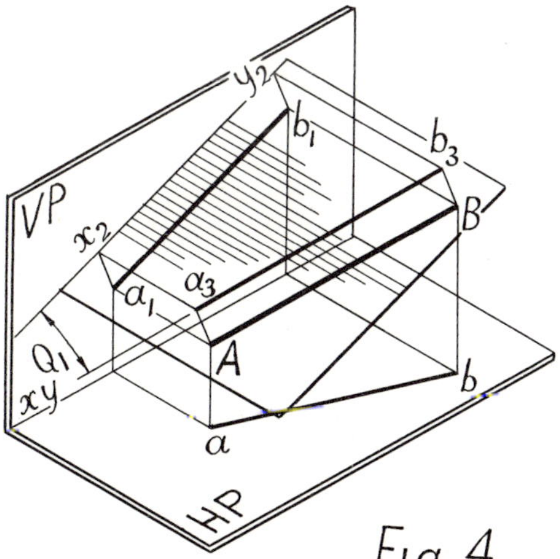

Fig 4

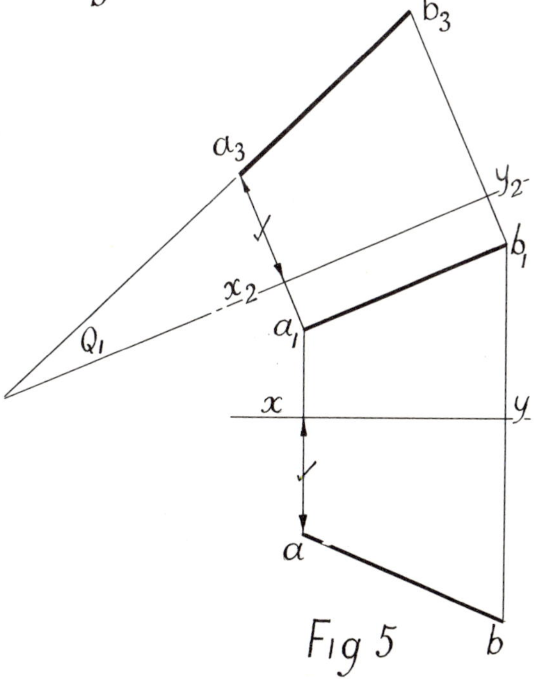

Fig 5

Auxiliary Projection of Solids: Cube

There is often little reason for showing solids in unusual positions but if these projections are required we are now able to make them, beginning with the simplest projections of the solid. We shall consider some illustrations of the method.

Cube: View along a diagonal of the solid

Such a diagonal is shown by AB in fig. 1. We require a view of the cube in which this line AB appears as a point.
Let us first draw the simplest views I and II of the cube, given by the two squares. The projections of the diagonal AB are *ab* and a_1b_1. We now proceed exactly as described on page 108.
Take x_1y_1 parallel to *ab*, and project the new view III of the cube; the diagonal is now a_2b_2. Note that all points in this view are as far from x_1y_1 as the corresponding points in I are from *xy*.
Now take x_2y_2 perpendicular to a_2b_2. Ignore view I. Project a new view IV, transferring distances from x_1y_1 in view II. We now have the diagonal AB as a point, a_3b_3. Hence this view, IV, is the view required.
The outline in fig. 2 IV is a regular hexagon, as may be tested. All the lines represent edges of the cube *and all have the same length*—but less than the true length of the edge of the cube. The view is a true isometric projection of the cube (*iso* meaning same, and *metric* meaning length). The length of the shortened edges in IV, compared with the true length of these edges in I, enables us to prepare a scale for the construction of isometric drawings. As mentioned in Part I, isometric drawings in which *true distances* are marked off along the three isometric axes show the objects a little larger than real size; if they are made *to scale*, the isometric views match the ordinary projections of the object. This is fully discussed on page 128.

Exercises (dimensions are in millimetres)

1. Draw a view of the rectangular prism in fig. 3 taken along the diagonal AB.
Each of the hexagonal prisms in figs. 4 and 5 have the same dimensions.
2. Draw a view looking along the line AB, fig. 4.
3. Draw a view looking along the line CD, fig. 5.
4. A cube has an edge of 40 mm. Draw a view of the cube taking x_2y_2 in fig. 2 *parallel* to a_2b_2. Note that although the projection is similar to IV the outline is not now a regular hexagon.

5. Four views of a cube are shown in fig. 6, drawn in 1st angle projection. Draw the given view full size and add a section on the line AB.
6. An irregular hexagonal prism is shown on page 42, fig. 8. Draw, scale full size, in 3rd angle projection, the following views:
   (*a*) the front elevation.
   (*b*) the plan view.
   (*c*) the side elevation.
   (*d*) the auxiliary projection showing the true shape of the surface created by the cutting plane.

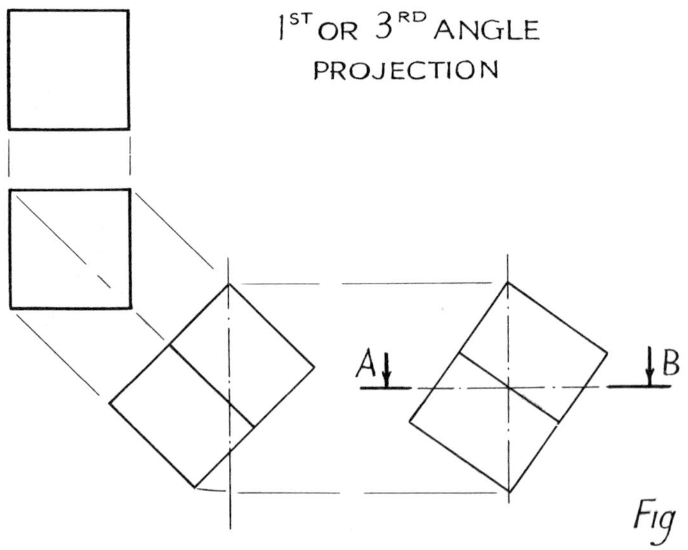

Fig 6

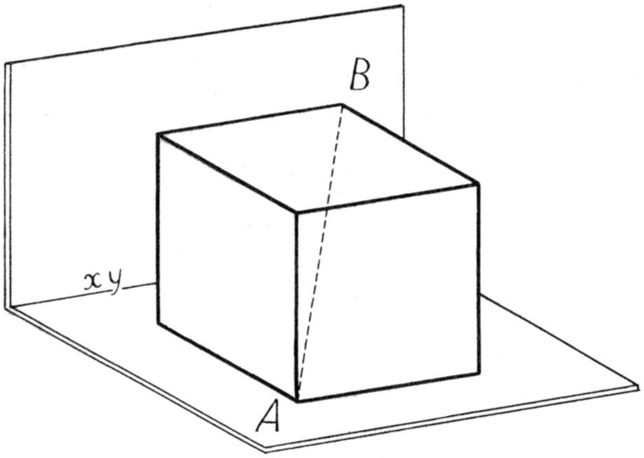

Fig 1

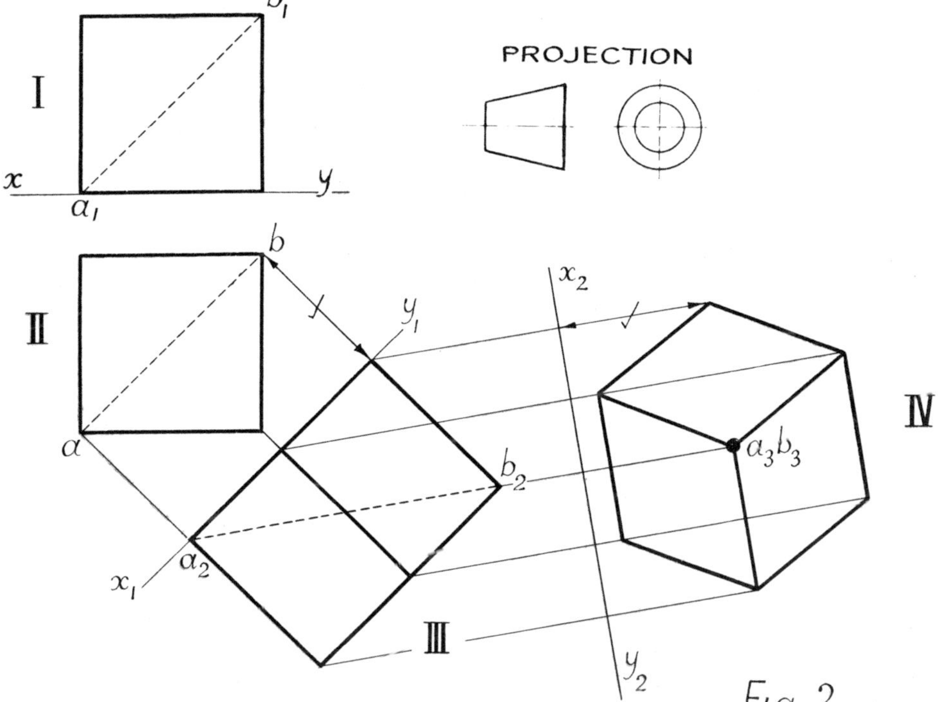

PROJECTION

Fig 2

Fig 3

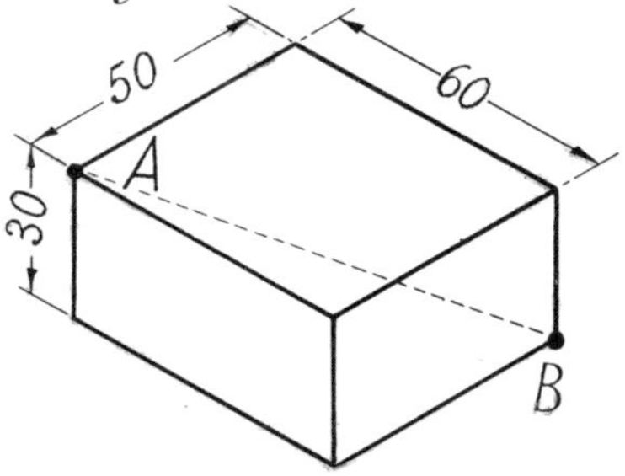

Fig 4

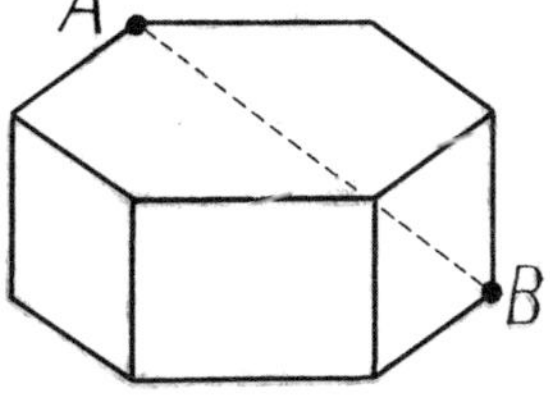

Fig 5

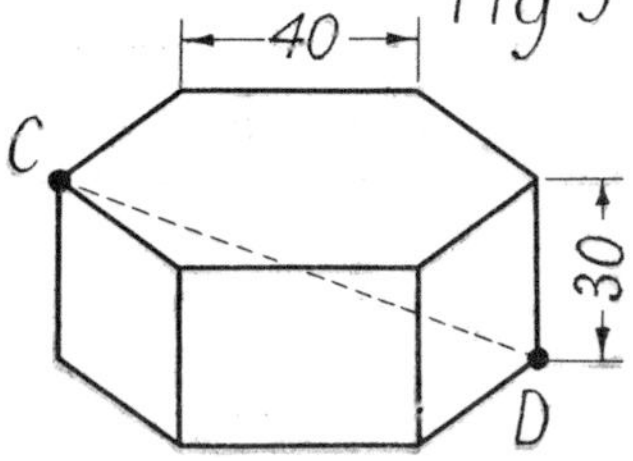

Projection on a plane parallel to a given surface

It is sometimes desirable to obtain the projection of an object when one of its surfaces is parallel to the plane of projection. For example, we may require views of a pyramid lying on one of its side faces. The procedure to be adopted is invariably the same and is as follows:

(*a*) Chose the simplest possible projections of the solid, and then:

(*b*) Arrange the solid so that the surface in question is represented *in one view by a line.*

(*c*) Project a new view on a plane parallel to this line. The view gives the true shape of the surface and the required view of the object.

Hexagonal pyramid

The pyramid in fig. 1 is shown by two simple views I and II, and a side face *abc* (shaded) in view I appears as a line $a_1b_1c_1$ in view II. This means that in view II we are looking along the surface, seeing no width. If we take a line x_1y_1 parallel to $a_1b_1c_1$ and project a new view of the pyramid, applying the rules of projection set out on page 108, we shall obtain the view required. Points such as a_2, b_2 or c_2 in the new view III are as far from x_1y_1 as are the corresponding points *a*, *b* or *c* from *xy*. If these distances are transferred methodically the process becomes almost automatic.

If we turn the book until x_1y_1 is horizontal, and if we ignore view I, the two views assume then their normal positions; they have been transferred, but rearranged, in fig. 2.

It is important to notice that the triangle $a_2b_2c_2$ gives the true shape of the side face. *The true shape of any surface is revealed by a projection on a plane parallel to that surface.*

If a projection of the pyramid in a more general position is required this may be obtained from projections II and III, shown separately in fig. 2 for convenience. Suppose we require a view on a vertical plane making 45° with the edge a_2c_2. We draw x_2y_2 at 45° to a_2c_2 and project the new view IV from III, taking distances from II.

Exercises

1. Assume the pyramids in figs. 1 and 2 to have a base hexagon of 30 mm side and a vertical height of 70 mm. Obtain views III and IV but with the angle marked 45° changed to 30°.
2. The hexagonal pyramid shown in fig. 3 has its top cut off. Views are required when the lower part is resting, upside down, on the cut surface; take the views in the directions of the arrows, one of which is parallel and the other perpendicular, to the cut surface. (Hint: draw a view from the top; then take x_1y_1 parallel to the top and project; finally take x_2y_2 perpendicular to x_1y_1 and project again.)
3. Two elevations of a rectangular pyramid with a cutting plane are shown in fig. 4. Draw, full size, in 1st angle projection:
   (*a*) the given view A.
   (*b*), (*c*) and (*d*) views in the directions of arrows B, C and D, with the portion above the cutting plane removed.
4. Fig. 5 is the elevation of a hexagonal pyramid with one face resting in the horizontal plane. Part of the pyramid is removed by the cutting plane CP. Draw full size in 3rd angle projection, the given view, a view to the right, and a view giving the true shape of the surface created by the CP.

Note.—Two preliminary views are required in order to construct the given view.

5. A regular hexagonal pyramid is shown on page 44, fig. 6, drawn in 1st angle projection. Draw, scale full size, in 1st angle projection, the following views of the truncated pyramid:
   (*a*) the elevation.
   (*b*) the plan view.
   (*c*) a side elevation seen from the left hand side of view (*a*).
   (*d*) the auxiliary view showing the true shape of the surface created by the cutting plane.

PROJECTION

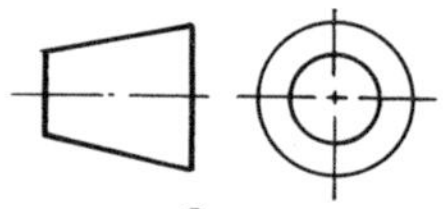

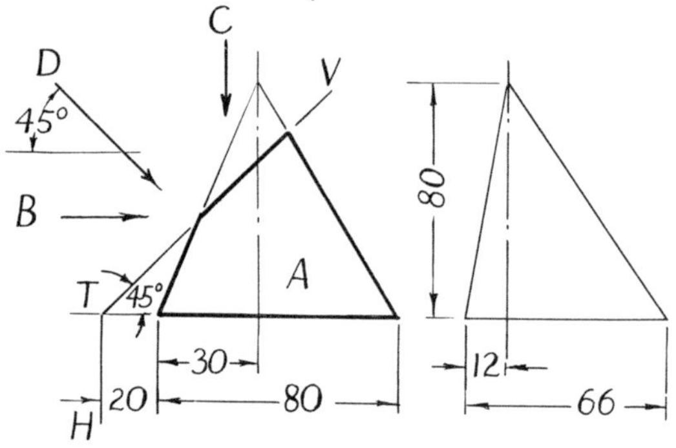

Fig 4

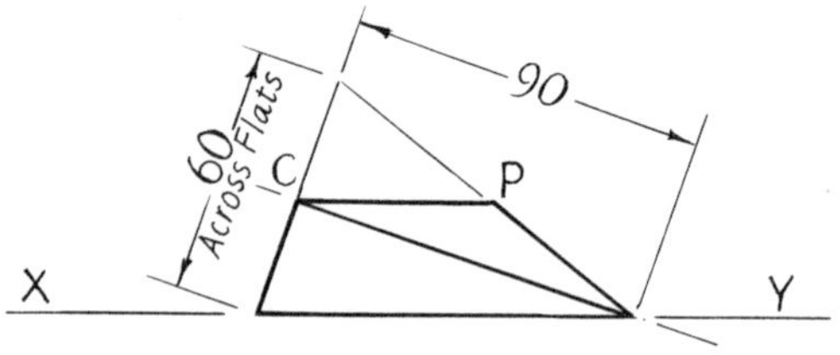

Fig 5

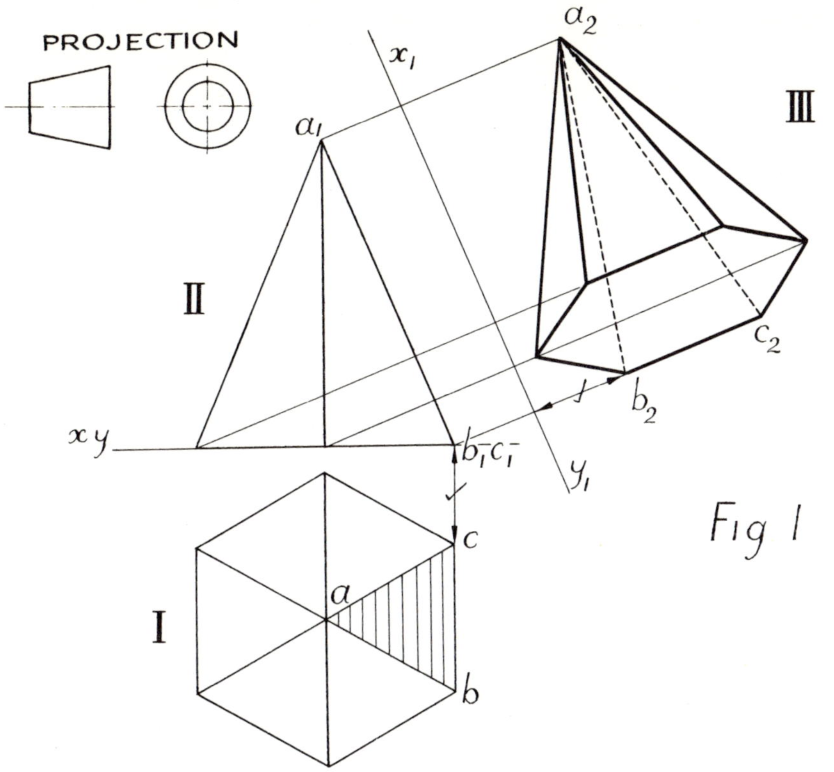

Fig 1

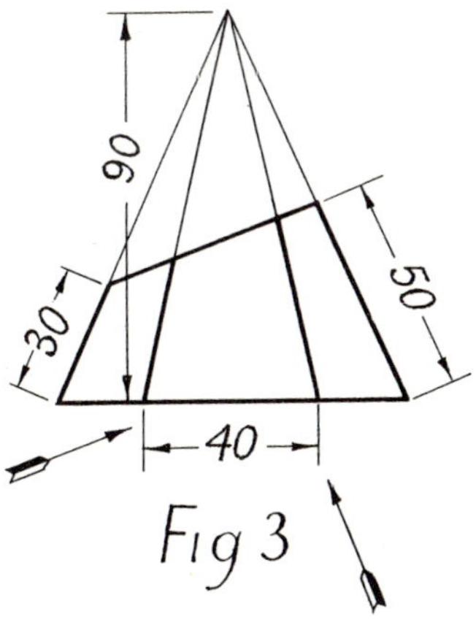

Fig 3

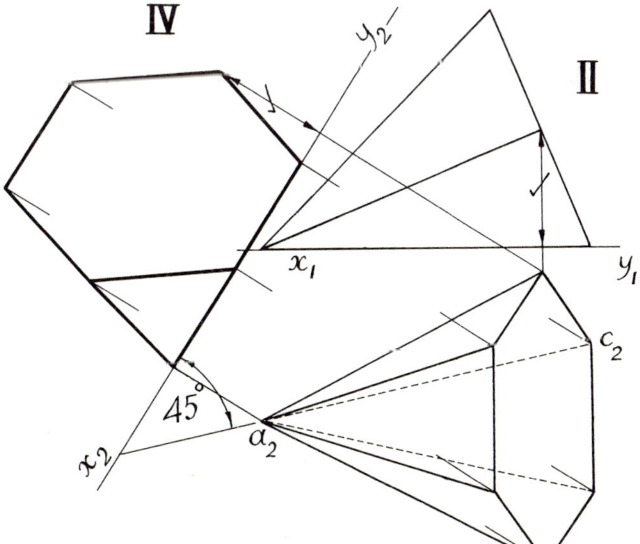

Fig 2

Auxiliary Projections of Solids: Cylinder and Cone

Cylinder and cone

These surfaces can be regarded as those traced out by a rectangle in the one case, and a triangle in the other, revolving about a side. The 'fixed' side is the *axis*, and the side tracing out the surface is the *generator*.

Projections are sometimes required when the position of the solid is defined by giving the position of the axis or generator in relation to the planes of reference. It is usually best to begin with the most easily drawn projections of the solid, and to obtain the required views by auxiliary projections.

Cylinder

Let us suppose that projections of the cylinder are required when its axis is inclined at a given angle Q to one of the planes of reference. First draw the simple views I and II in fig. 1. Then take x_1y_1 making the given angle Q with the axis, and draw the auxiliary view III. In this, the circular ends are ellipses. We can project these point by point: four points a_2, b_2, c_2 and d_2 are readily marked, and intermediate points such as p_2 and q_2 can be projected in the usual way. If desired the ellipse may be drawn by the aid of a trammel, because a_2c_2 and b_2d_2 give us the minor and major axes respectively.

Cone

The auxiliary projection of the cone in III, fig. 2, is readily obtained from I and II. A generator of the conical surface is given by the line ab, a_1b_1. We require a view in which this generator lies in the horizontal plane. We have taken x_1y_1 along a_1b_1, and projected the required view III in the way already described. (See Ex. 2 for view IV.)

Exercises

1. A cylinder has a diameter of 50 mm and a length of 70 mm. It lies with its curved surface on the HP and with its axis making an angle of 35° with the VP. Draw the projections of the cylinder on the HP and VP.

2. A cone has a diameter of 75 mm and a vertical height of 100 mm. It lies with its curved surface on the HP, the generator in contact making an angle of 30° with the HP. Draw the projections of the cone on the HP and VP. (Hint: proceed as in fig. 2. Then take x_2y_2 at 30° with a_2b_2 and project view IV; take distances from x_2y_2 in IV equal to corresponding distances from x_1y_1 in I.)
3. Consider the two views I and II of fig. 2 as constructed views of a right circular cone, 70 mm diameter and 100 mm height. These two views should be drawn with very thin lines. Rotate the cone clockwise about b until the generator ab is in the horizontal plane with its axis parallel to the vertical plane. This is the new elevation. In 1st angle project a plan view and a side elevation to the right of this new elevation.
4. Repeat Ex. 3 but with the axis of the cone inclined to the vertical plane by 30° when in its final position.
5. A shaft guide is shown in fig. 3. It is used to support a shaft inclined at 30° to the horizontal plane. In 3rd angle projection draw the given view full size and also views seen in the directions of the arrows L, M and N.

Note.—An attempt should be made to build up the views L, M and N simultaneously, since many measurements are common to all views. This practice will help to speed up drawing.

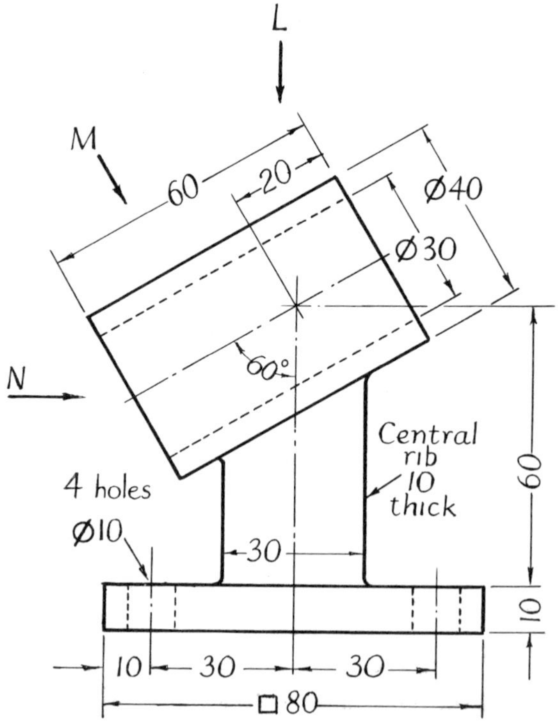

Fig 3

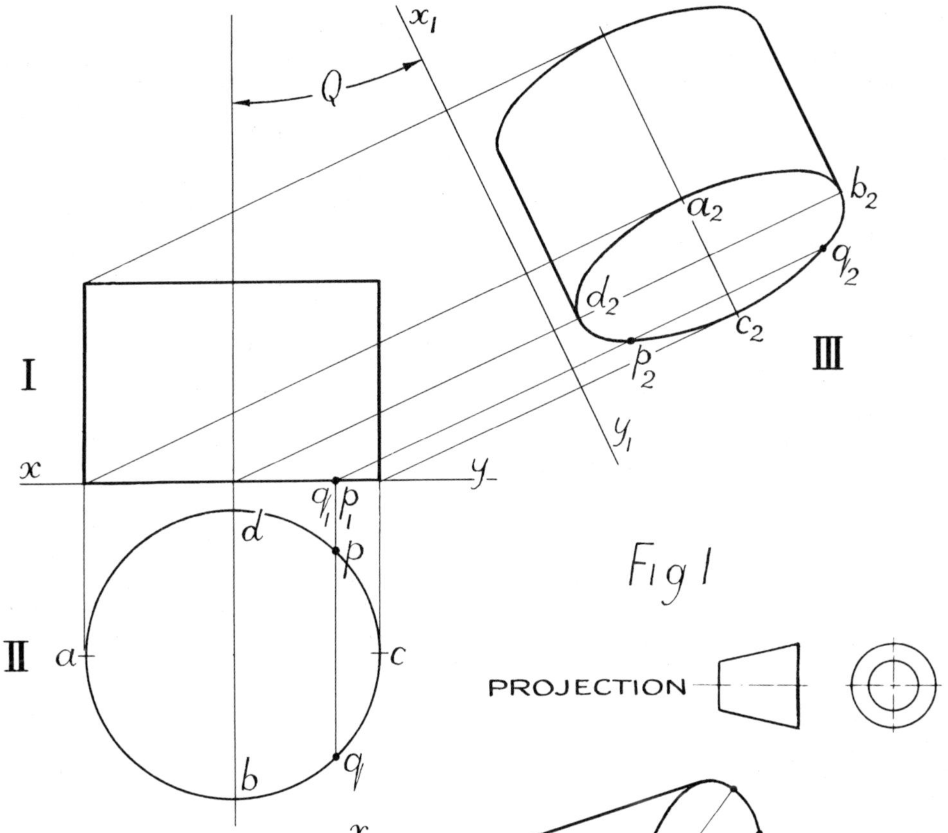

x1
Q
b2
a2
q2
d2
c2
III
p2
y1
I
x
y
q1 p1
d
p
II
a
c
q
b
PROJECTION

Fig 1

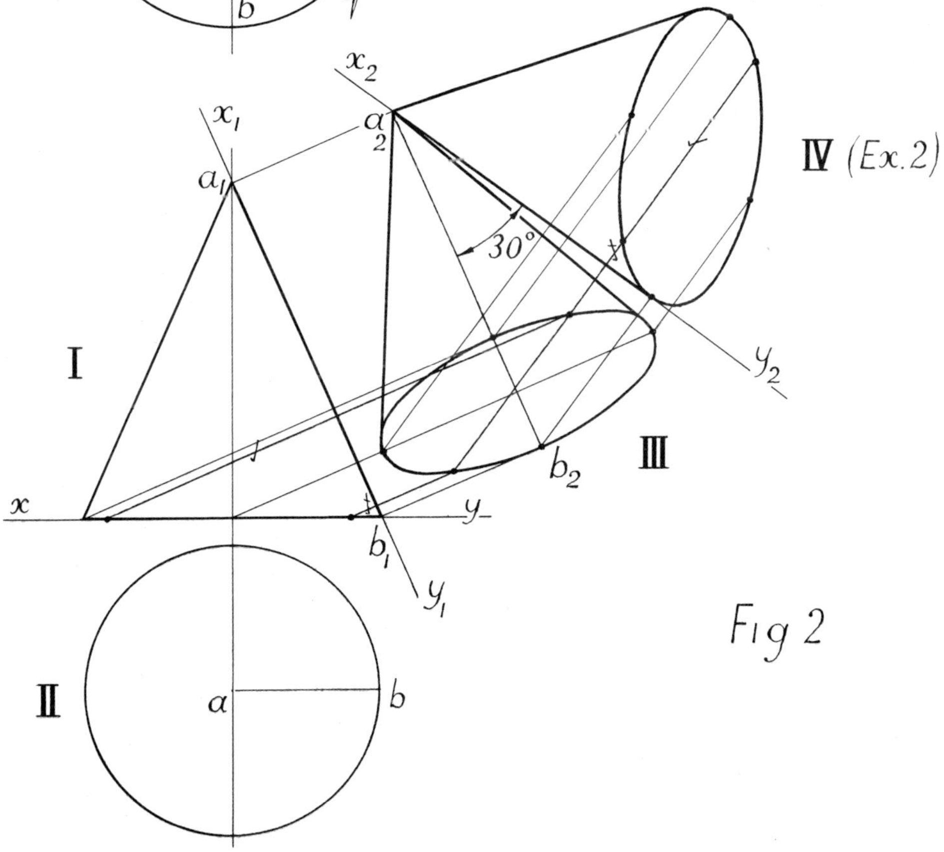

x2
a2
x1
a1
IV (Ex. 2)
30°
y2
I
III
b2
x
y
b1
y1
II
a
b

Fig 2

Sections of Solids: Cube and Pyramid

Sections of solids have been dealt with in Part I, the cutting planes being either parallel or perpendicular to the planes of reference. We shall now consider sections produced by cutting planes which are inclined to one plane of reference and perpendicular to the other. The plane in fig. 1 is inclined to the horizontal plane HP, and that in fig. 3 is inclined to the vertical plane VP. The traces, i.e. the lines in which the plane intersects the VP and the HP, are usually marked VT and HT.

Cube: plane inclined to the HP

A cube is shown by two views I and II in fig. 2. It is cut by a plane represented by VTH. The arrangement is shown pictorially in fig. 1. The line VT gives an edge view of the section and the sectioned area in view II is obtained by simple projection. The *true shape* of the section is given by taking x_1y_1 parallel to VT and projecting view III (of the section only). Distances of points in III, from x_1y_1, are the same as distances of corresponding points in II from xy.

Hexagonal pyramid: plane inclined to the VP

The problem is shown pictorially in fig. 3; careful application of the principles of projection is required. The pyramid lies with a sloping side in the HP and with its axis parallel to the VP. We begin by drawing the simple projections in I and II (I being unsectioned) and then obtaining those required by taking x_1y_1 coincident with a sloping face. View III is easily obtained, and views I and III give the required projections (the page may be turned to bring x_1y_1 horizontal). The traces VT and HT of the section plane may now be drawn. HT now gives an edge view of the section in view III and the sectioned area in I is obtained by projection from III, taking points where HT cuts the edges of the pyramid. The *true shape* of the section is given by taking x_2y_2 parallel to HT and projecting view IV from III, using distances in I.

Both of these examples should be worked out by the student.

Exercises

1. Obtain the true shape of the section of the cube in fig. 2 III using the following particulars. Edge of cube 50 mm. One edge of cube is inclined at 30° to xy. The plane is inclined at 30° to HP (i.e. VT makes 30° with xy) and HT is 20 mm from the near edge of the cube.
2. Obtain a view of the sectioned cube given by the arrow A.
3. Obtain the true shape of the sectioned pyramid in fig. 4 IV using the following particulars. Edge of base 25 mm, vertical height 90 mm, inclination of plane to VP 30°. The plane bisects the axis of the pyramid. (The view will differ slightly from that shown—the difference is important.)
4. Complete view II, fig. 4, with the front sectioned part of the pyramid removed (see dotted outline).
5. An oblique pentagonal pyramid is shown in fig. 5. It is cut by the plane VTH. Draw, scale full size, in 1st angle projection, the following views of the pyramid with the part above the cutting plane shown by thin lines:
   (*a*) a front elevation corresponding to the one given.
   (*b*) a plan view.
   (*c*) a view seen from the right of view (*a*).
   (*d*) a true view of the surface created by the cutting plane.

Note.—(i) The pyramid is said to be oblique when, with its base in the horizontal plane, the axis is not vertical.
(ii) If the oblique pyramid is cut by any plane parallel to the base, it will produce a regular pentagon.

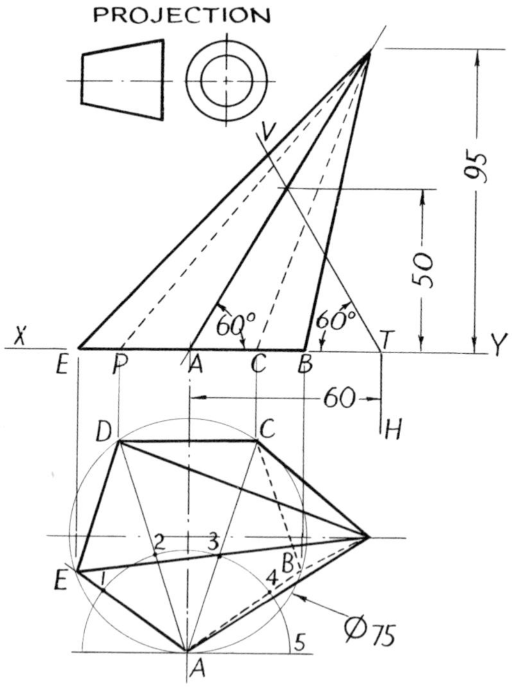

Fig 5

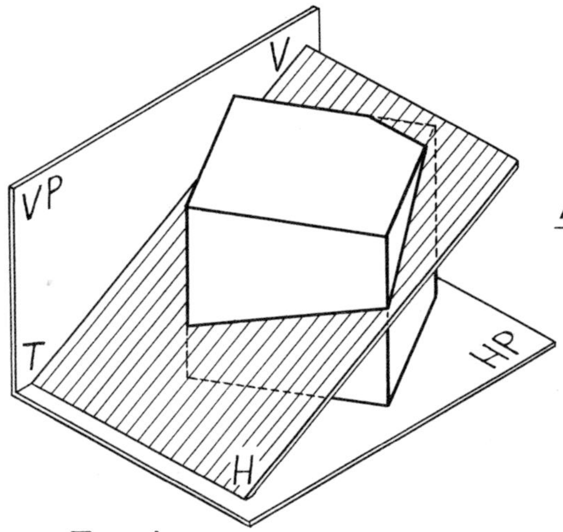

Fig 1

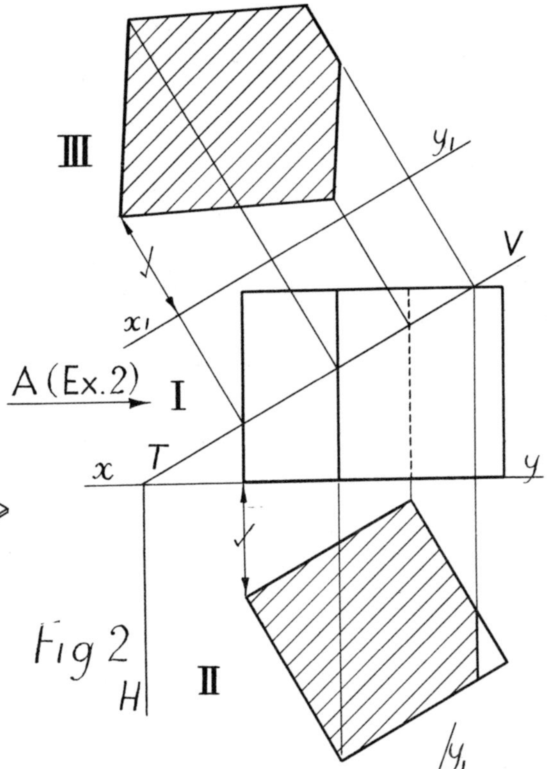

Fig 2

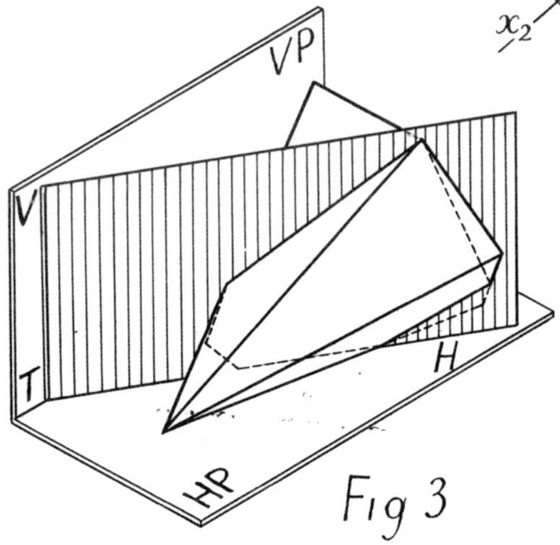

Fig 3

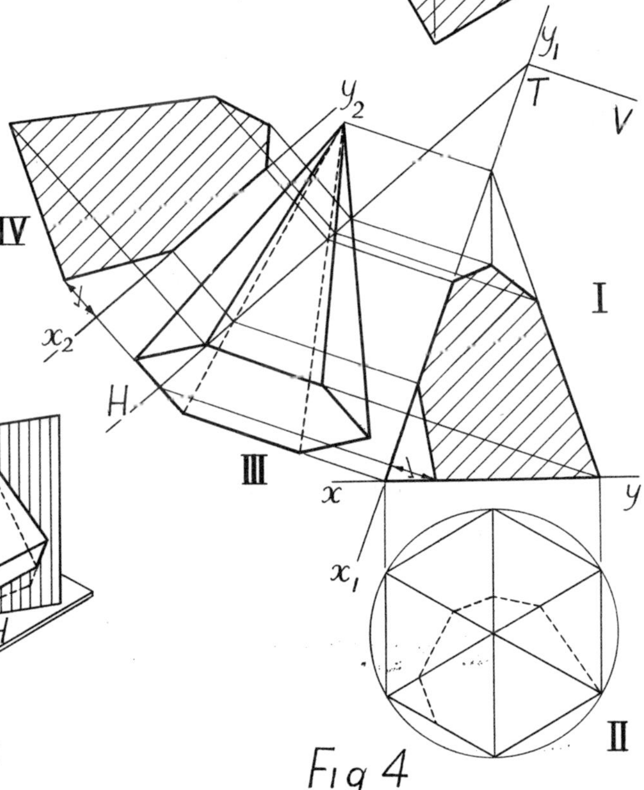

Fig 4

Sections of Solids: Cone (Conic Sections)

Sections of the cone

On pages 86 and 88 the Ellipse, Parabola and Hyperbola were considered as the locus of a point satisfying certain conditions. We shall now obtain the same curves from plane sections of the cone.

Each of the figures opposite shows two simple views of a cone. The cone is cut by an inclined plane VT. In fig. 1, VT is parallel to a generator of the cone; the outline of the section is a *parabola*. In fig. 2, VT cuts two opposite generators of the cone; the outline of the section is an *ellipse*. In fig. 3, VT cuts, on the same side of the axis, both surfaces of the double cone; the outline of the section is a *hyperbola*.

True shape of the conic sections

The following construction applies to each figure.

Draw a centre line *cl* parallel to *xy*. Draw another line c_1l_1 (which will be the centre line of the section) parallel to VT. Take any point p_1 on VT and draw a horizontal through it. Regard this horizontal as the edge view of a plane, cutting the surface of the cone in a circle, as shown. Project from p_1 to cut this circle in *p* and *p*. These are points not only on the cutting plane VT but also on the surface of the cone; hence they are on the outline of the required section. Transfer these points to the auxiliary projection; that is, mark off distances p_2q_2 equal to *pq* on each side of c_1l_1, on the projector from p_1. The points p_2 lie on the required outline. Repeat the construction for other points, and draw a fair curve through them. The position of selected points such as p_1 is a matter for judgment; they should be closer together where the conic section is most curved and less close on the straighter portions.

The construction is very similar to that employed for the sections of other solids but here, because of the lack of edges and plane surfaces, we have used the device of taking cutting planes at right angles to the axis of the cone. We could equally well have used generators—thin lines, fig. 1.

A line joining points such as *p* will give a view of the sectioned cone taken along the axis. This is left as an exercise for the student.

Exercises

1. Draw the three conic sections shown opposite, using a cone of base diameter 100 mm and generator 100 mm.
   In fig. 1 take VT 10 mm from the outline. In fig. 2 take VT at 30° to *xy* and cutting the axis 50 mm from *xy*. In fig. 3 take VT at 75° to *xy* and passing through a point on the base 12 mm from the outline.
2. Draw a sectional plan and elevation of the part of the cone remaining in fig. 1 when the part to the left of VT is removed.
3. The arrangement shown in fig. 4 is a cone joined to a cylindrical pipe, the junction being an ellipse. Draw the given view and the developments of the surface area of the truncated cone and the cylinder.

Note.—Other constructional views will be necessary.

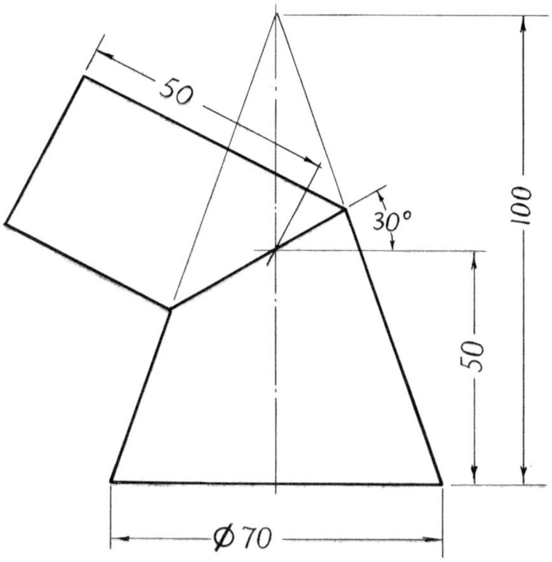

Fig 4

PROJECTION

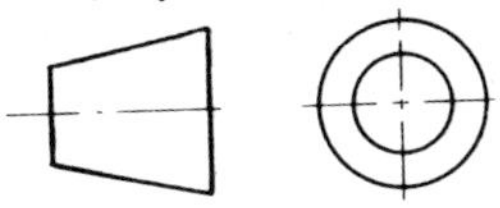

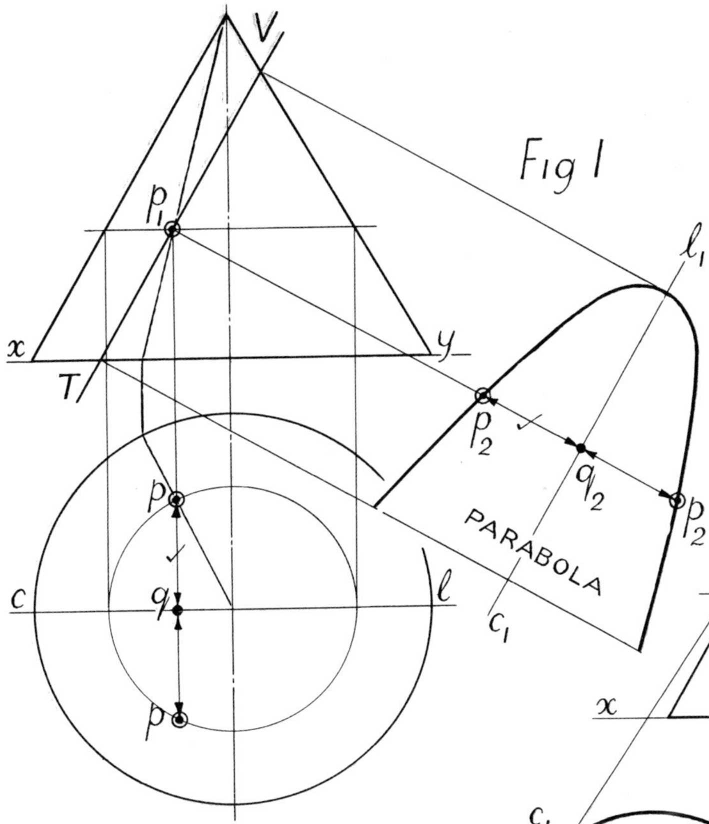

Fig 1

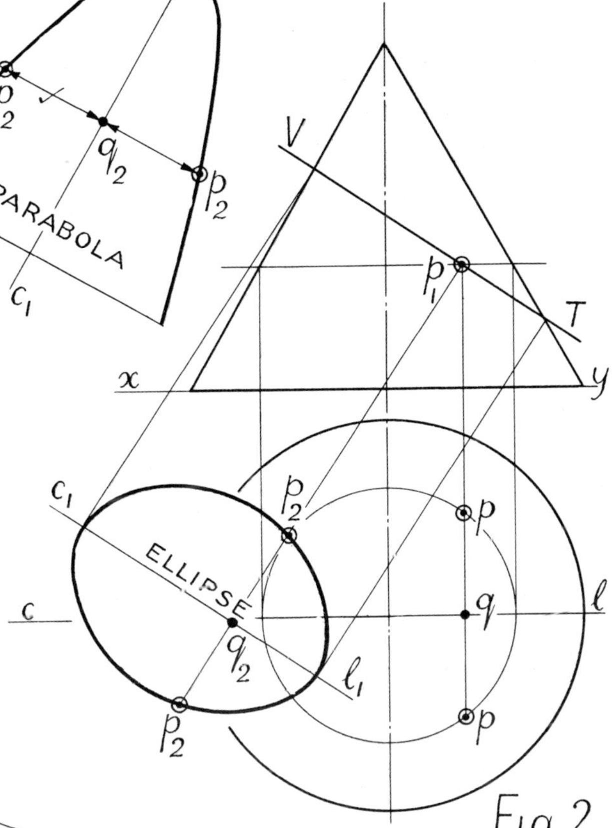

Fig 2

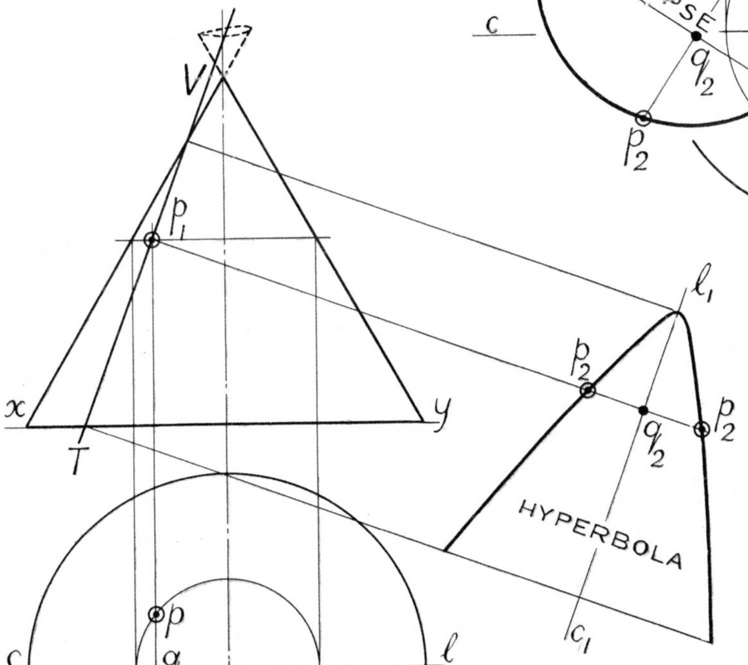

Fig 3

Intersections of Surfaces: Cylinders

General method

If two solids with curved surfaces penetrate each other, the line of intersection is generally a curve which is not contained by one plane. It is plotted by finding points which lie in both surfaces. We shall do this by taking section planes which cut the surfaces in lines which are easily drawn—straight lines or circles. The intersections of these lines give points on the required curve. An example will make the method clear.

Intersecting cylinders

The small drawing, fig. 1, shows cylinders intersecting, the axis of one being vertical and the axis of the other horizontal. A section plane VT1 has been taken which cuts one cylindrical surface in a circle and the other cylindrical surface in two straight and parallel lines. These lines and the circle intersect in P and Q, points which lie on the required curve of intersection.
The construction indicated in fig. 1 is shown in fig. 2. This shows the projections of two cylinders the axes of which do not intersect but are at right angles to each other. The horizontal line VT1 represents the edge view of a section plane cutting the surface of the horizontal cylinder in two lines (or generators) 2*D* apart. If we transfer these two lines to the plan, distance *D* from the centre line, we get the points *p* and *q* where these lines intersect the circular outline. Simple projection gives p_1 and q_1, the elevation of two points on the required curve. It will be evident that points r_1 and s_1 may be marked by considerations of symmetry. The location of other points is a repetition process involving other planes (e.g. VT2); to prevent confusion the planes and corresponding points should be numbered. 'Key points' are *a*, a_1 and *b*, b_1. Usually, not many points are required to define the curve. It is better to use a few well-selected points than to have too many with a confusion of lines leading to error.

Exercises

1, 2, 3. The drawings in figures 3, 4 and 5 represent two intersecting cylinders. The axes are at right angles, intersecting in figs. 3 and 4, but not in fig. 5. Draw projections in the direction A showing the lines of intersection.

4. Turn fig. 5 clockwise through 30°. Project a view showing the line of intersection.
5. If the distance OB in fig. 5 is 60 mm, find the lengths of 12 equidistant generators on the surface of the smaller cylinder, beginning with the shortest. Set these out from a straight line representing the circumference and thus obtain the shape of the branch.
6. Three intersecting cylinders are shown in fig. 6. The smallest, 40 mm diameter, meets the largest, 60 mm diameter, at an angle of 60°, their axes being in the same vertical plane. The 50 mm diameter cylinder intersects the largest at 45°, but their axes, though parallel, are displaced by 5 mm. Project, in 1st angle projection, the given views and add a surface development of the largest cylinder, making the cylinder length 100 mm.

PROJECTION

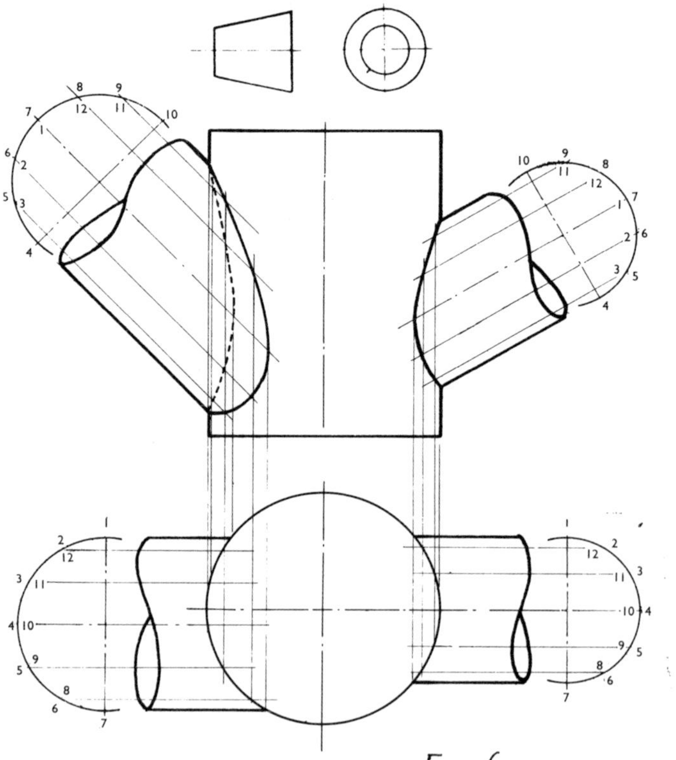

Fig 6

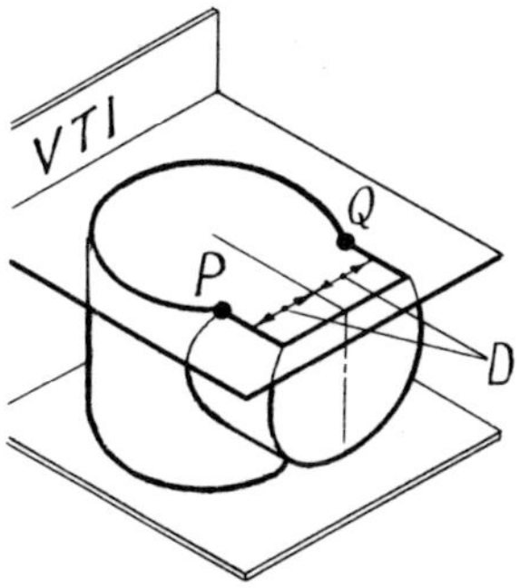

Fig 1

PROJECTION

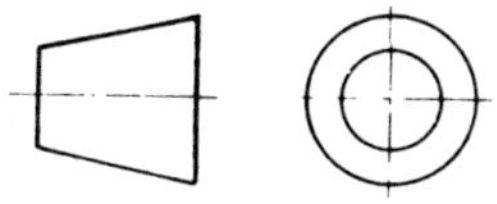

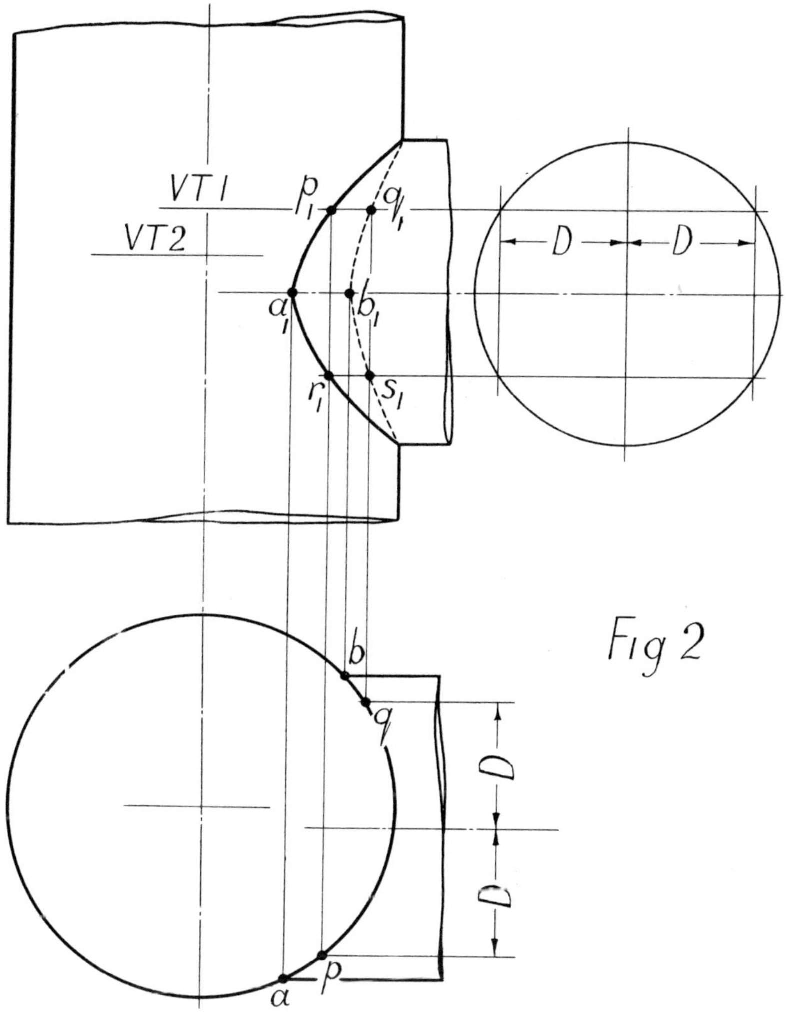

Fig 2

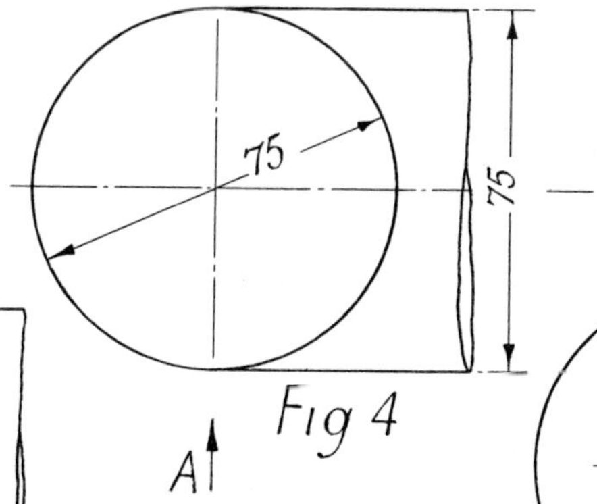

Fig 4

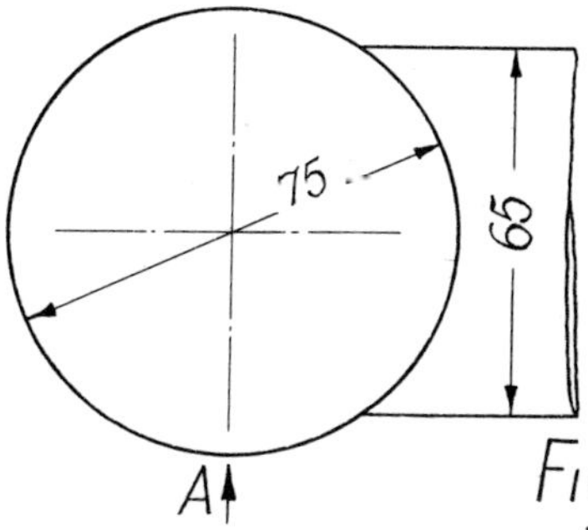

Fig 3

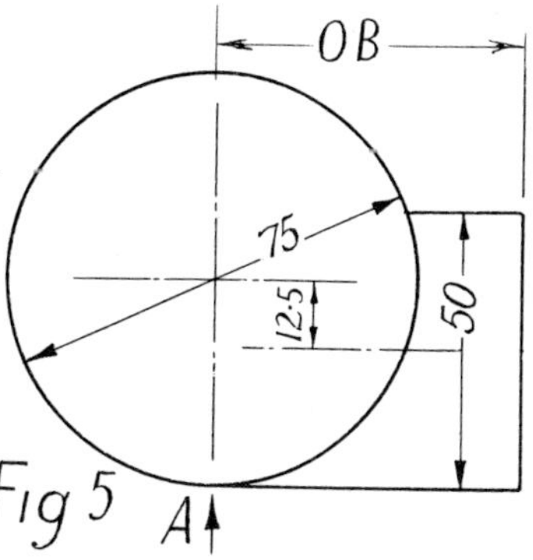

Fig 5

Cone and cylinder

The method of dealing with the simpler examples considered here is similar to that already described.* Section planes will be used which cut the cone in a circle and the cylinder in a rectangle. In other words, the section planes will be parallel to the axis of the cylinder and at right angles to the axis of the cone.

The form of the line of intersection depends upon the relative sizes and positions of the two solids. The cone may envelop the cylinder, as in fig. 1, the cylinder may envelop the cone, as in fig. 2, or neither may envelop the other, when both cone and cylinder envelop a common sphere, as shown in fig. 3. It is a great help in plotting the curves to know which type we have to deal with; hence the first step should be to draw a projection along the axis of the cylinder. This view shows clearly into which class the curve will fall.

The solutions shown opposite should be almost self-explanatory. The cutting plane VT1 gives a circle for the section of the cone and a rectangle for the section of the cylinder. These intersect in points *p, q, r* and *s*, which are on the curves of intersection. The elevations p_1 and q_1 of these points are obtained by projection (r_1 and s_1 are hidden).

Key points are given (*a*) by projecting the points of intersection between the outline of the cone and the outline of the cylinder; and (*b*) by taking a section plane through m_2 and n_2, the points of intersection (or the point of nearest approach) of the circle and triangle in the view along the axis of the cylinder, this giving the points m, m_1 and n, n_1 on the curve.

A part of each curve will of course be hidden from view.

To prevent confusion each section plane and each resulting point (or pair of points) should be given the same number.

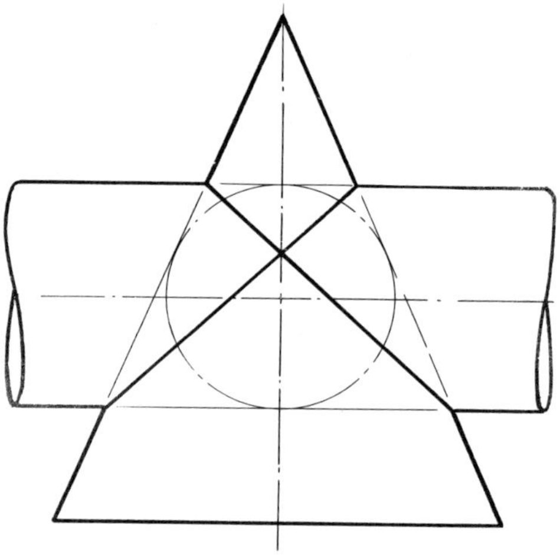

Fig 3

Exercises

1. A cylinder 60 mm diameter rests with its curved surface on the horizontal plane. A cone, height 100 mm, base 100 mm diameter, rests with its base on the horizontal plane. The two axes intersect at right angles. Draw the line of intersection of the surfaces of the two solids.
2. Increase the diameter of the cylinder in (1) to 70 mm and solve the problem.
3. Slightly increase the diameter of the cylinder in (1) until its circular outline touches tangentially that of the cone. Show that the lines of intersection are straight lines in the projection.
4. Displace the cylinder in (1) until its axis is 10 mm in front of the axis of the cone. Draw the lines of intersection of the surfaces of the two solids.
5. A small oil measure is shown in fig. 4. It is formed from three cones. Draw the given view and a development of each truncated cone.

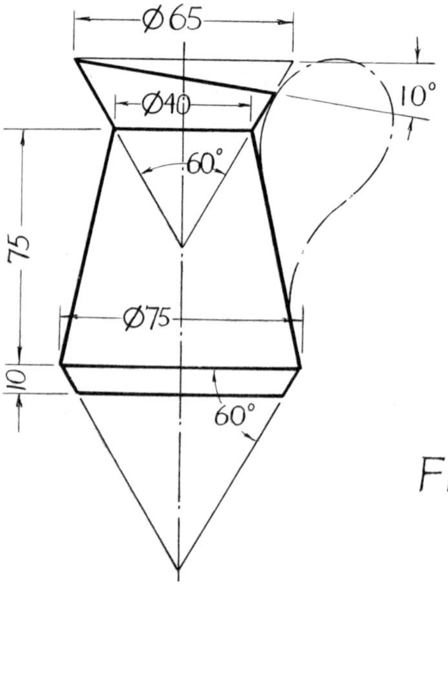

Fig 4

* The *general* method uses section planes which contain the apex of the cone and are parallel to the axis of the cylinder. It is dealt with in the author's *Practical Geometry and Engineering Graphics*.

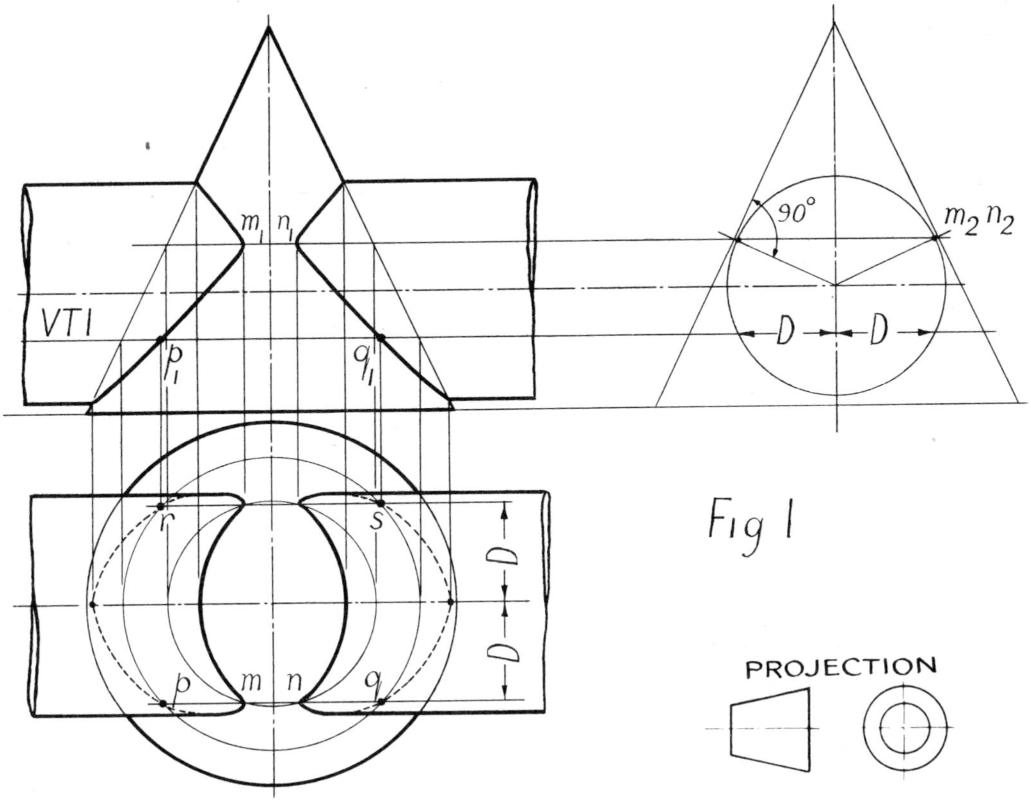
m₁ n₁
VT1
p₁
q₁
90°
m₂ n₂
D
D
r
s
D
D
p
m
n
q
PROJECTION

Fig 1

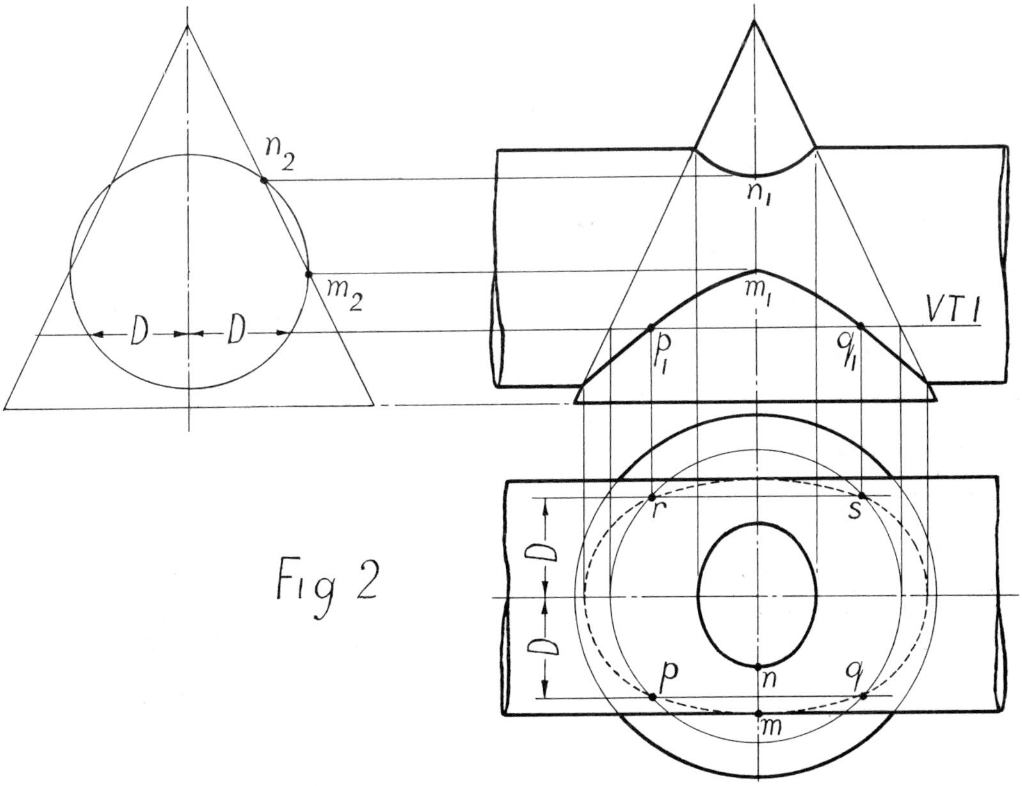
n₂
m₂
D
D
n₁
m₁
VT1
p₁
q₁
r
s
D
D
p
n
q
m

Fig 2

Developments

A development of a surface is the unfolding of the surface on to a plane. Some geometrical solids have surfaces which cannot be developed—the sphere is one—and approximate methods have to be resorted to. We have already considered the development of the surfaces of the cube, prism and pyramid; we shall now consider the cylinder and the cone.

Cylinder

The developed surface of a cylinder is a rectangle, one side of which is equal to the circumference of a circular section and the other equal to the length of the cylinder. The circumference may be calculated, or set off by the approximate method given on page 94. The cylinder in fig. 1, I has been cut by an inclined plane, and the surface of the lower part has been developed. This is a simple matter of transferring the lengths of generators from fig. 1, I to II; we may imagine the surface to be unwrapped. It is necessary to adopt a system of numbering as shown. The length $\pi \times D$ in II has been divided graphically into 12 equal parts corresponding to the 12 circular arcs.

Cylindrical branch

A common problem is that of finding the shape of a sheet-metal pipe which joins another pipe, as in fig. 2, I and II, here shown offset. We first find the projection of the curve of intersection, as on page 122. The rest is a matter of marking and numbering corresponding generators in II and in the development III, and of transferring their lengths.

Cone

The cone shown in fig. 3 has its axis horizontal. The diameter of the base is D and the slant height is H. It is cut by a plane, and we are to find the development of the surface below the plane.
The development of the full conical surface is a sector of a circle of radius H; the length of the arc is πD and the included angle is $180\ D \div H$. We may set off the arc by the method on page 94. Points on the surface may be marked on the development by using generators. Consider the point p_1, on generators 5 and 9. We draw p_1p_2 parallel to the base and draw an arc about a through p_2 to intersect generators 5 and 9 on the development in P_1 and P_2. A fair curve through such points gives the required development.

Exercises

The student is advised to cut out the developed shapes and to satisfy himself of their correctness by bending them into the form of the surface.

1. **Take values of 50 mm for D, 20 mm for l, and 50° for Q for the cylinder in fig. 1. Find the developed shape.**
2. **Take values of 60 mm for D_1 and 50 mm for D in fig. 2, d being 6 mm. Find the developed shape if L is 55 mm.**
3. **A cone has a base diameter of 80 mm and altitude of 90 mm. Find the developed surface.**
4. **The cone in Ex. 3 is cut by a plane making 45° with the base and bisecting the axis. Find the developed shape of the surface below the plane.**
5. **A sheet metal scoop is shown in fig. 4. It is formed from two cylinders. Draw the given view and the development of each cylinder. The handle is at 45°.**
6. **6, 7. Three cylinders intersect in figs. 5 and 6. Draw the given view and the surface development of each pipe.**
8. **Fig. 7 shows the end of a pipe shaped to intersect an elliptical surface. Draw the given view and the surface development of the pipe. It is advisable to draw the semi-ellipse first.**

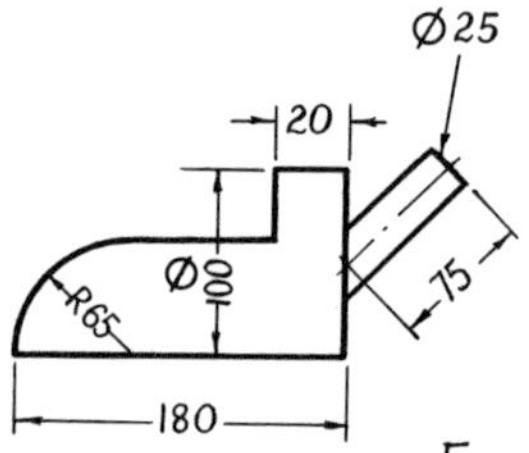

Fig 4

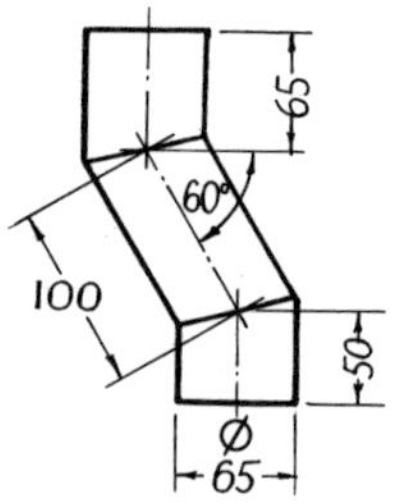

Fig 5

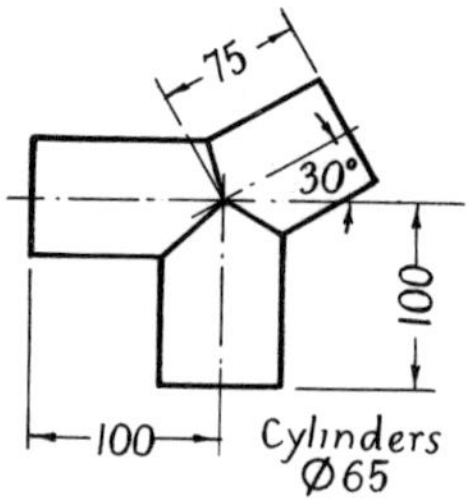

Fig 6

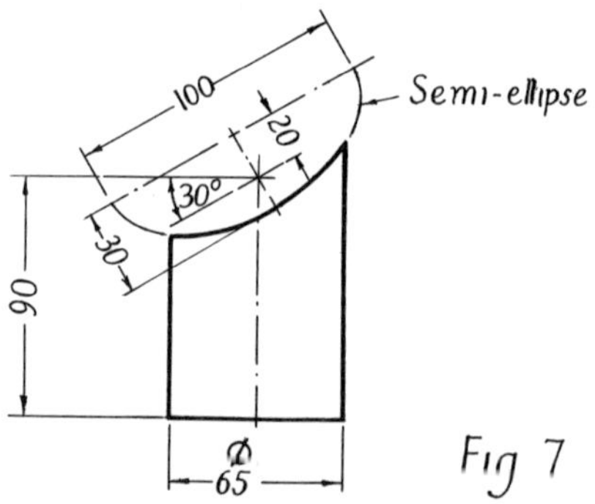

Fig 7

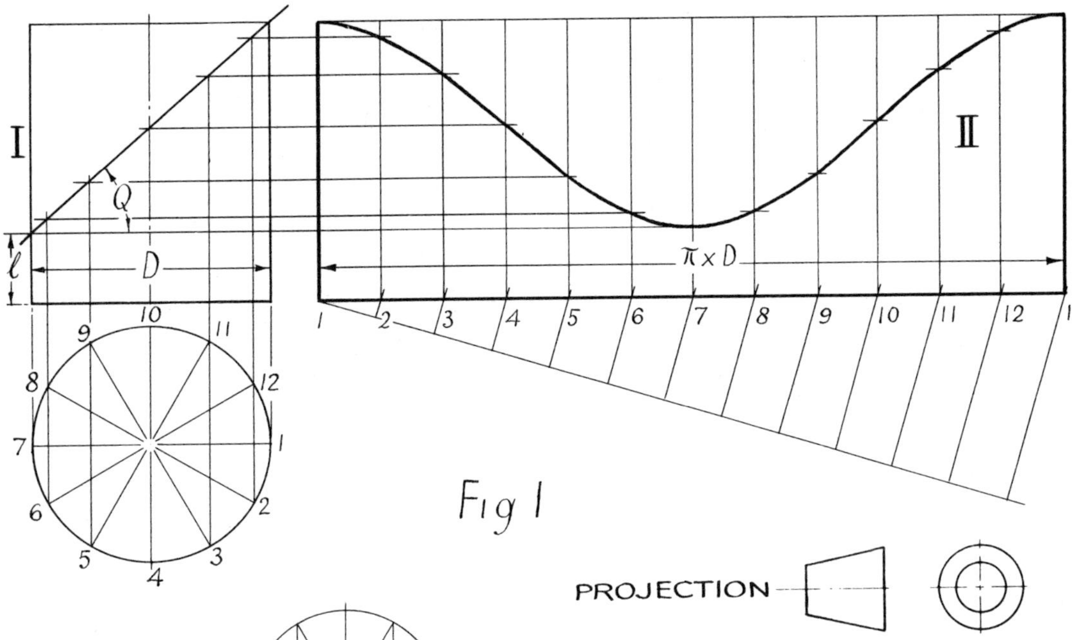

Fig 1

PROJECTION

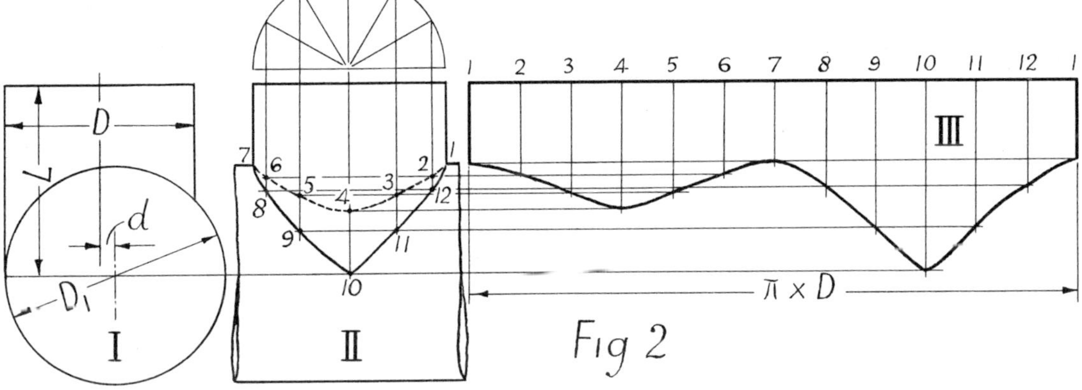

Fig 2

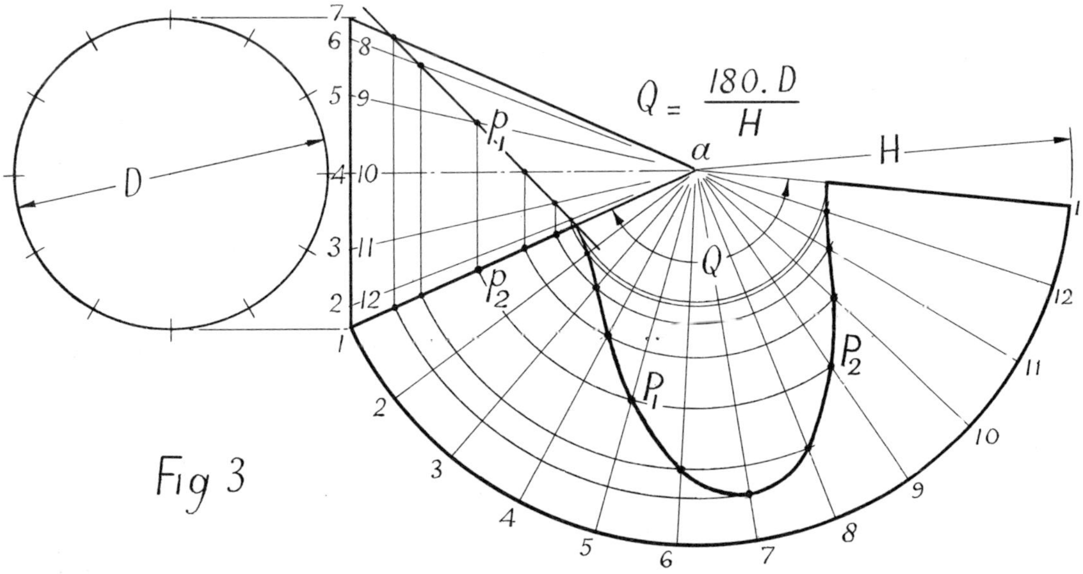

$$Q = \frac{180 . D}{H}$$

Fig 3

Isometric Scale and Isometric Projection

The isometric scale

Although it is convenient to mark off actual distances along the isometric axes (see Part I) so that a natural scale may be used, the practice is not always permissible. For example, in a full-size drawing an object would appear larger than it actually is, in the ratio $\sqrt{3}$ to $\sqrt{2}$. But where an isometric drawing is used to illustrate the solution of an orthographic problem, as on page 119, the use of a scale is necessary to give the correct proportion between the drawings. Again, all projections of the sphere are the same, whether orthographic or isometric; and in a composite solid of which a sphere is part, see opposite, some correction is necessary to give the various parts their relative sizes. This correction is given by using the isometric scale.

Construction of the scale

We have seen that the projection of the cube in fig. 1, obtained as shown on page 112, gives the projections of three edges at right angles, OA, OB, OC, which are all shortened to the same extent. The ratio of their lengths in the drawing, to their true lengths, gives the ratio of the isometric scale to the natural scale. If we draw AE and CE at 45° to the diagonal AC, then the triangle AEC represents the true shape of the triangle ADC, and either AE or EC gives the actual length of the edges AD and DC. This is the basis of the construction in fig. 2.

Set off AE at 45° and AD at 30° to a base line AC. Graduate AE in millimetres to cover the dimensions of the drawings which are to be made, and from each point along AE draw lines perpendicular to AC, thus dividing AD in a similar way. The divisions along AD give dimensions to an isometric scale, and *distances along the isometric axes should be set off to this scale.*

Group of spheres in isometric projection

Three spheres touch and support a fourth, in fig. 3. A moment's thought will show that the distance between any two of the four centres is equal to the sum of two radii, i.e. to a diameter. The centres lie, in threes, at the corners of equilateral triangles. A figure having four sides, each an equilateral triangle, is shown by its projections in fig. 4; it is a regular tetrahedron. If we draw this figure to an isometric scale, we shall have the positions of the centres of the four spheres; these should of course be drawn full size. The completion is left as an exercise for the student.

Exercises

1. **The four equal spheres in the pyramidal stack in fig. 3 are each 50 mm diameter. The base is 100 mm square and 15 mm thick. Make an isometric view of the group.**
2. **Fig. 5 shows the spherical end of an ornamental column. Make an isometric view of this feature to a convenient scale.**
3. **The classic example of the necessity for the use of an isometric scale is illustrated in exercises involving fig. 6. This shows a stone pillar, at a gateway, with a stone sphere on top. First draw, at a scale 1 to 5, an isometric view of both pillar and sphere to a natural scale. Notice that the sphere is always 300 mm diameter from whatever direction it is viewed, whereas the pillar appears to be much larger.**
4. **The application of the isometric scale will be clear when the sphere is drawn to the natural scale and the pillar to the isometric scale.**

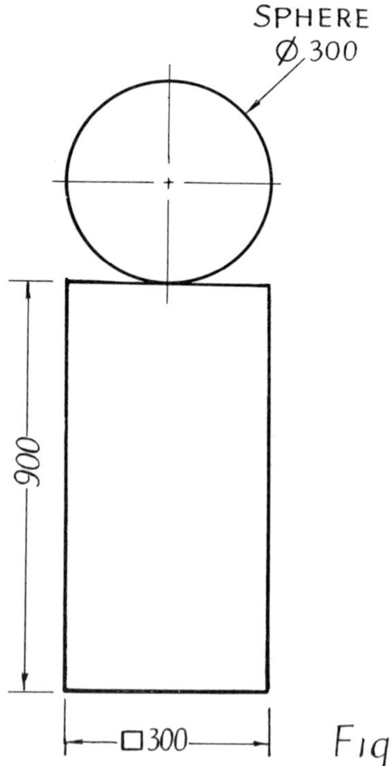

Fig 6

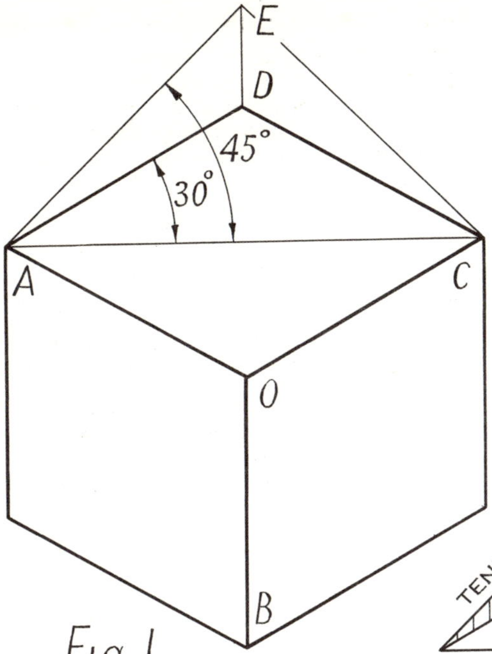

Fig 1

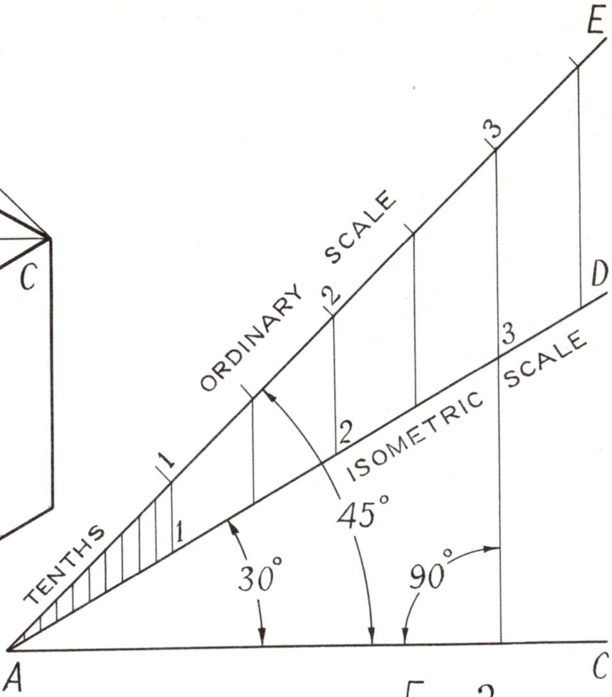

Fig 2

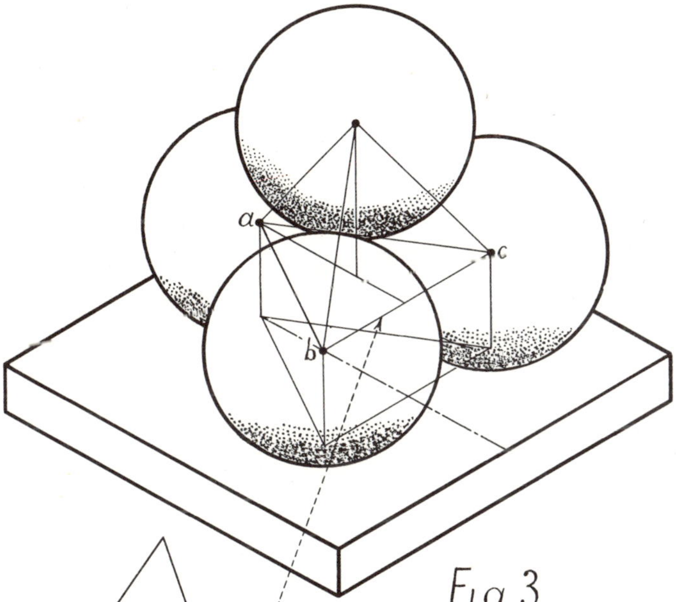

Fig 3

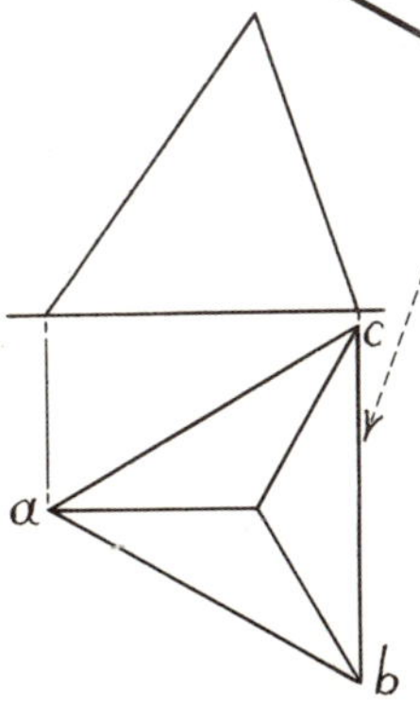

Fig 4

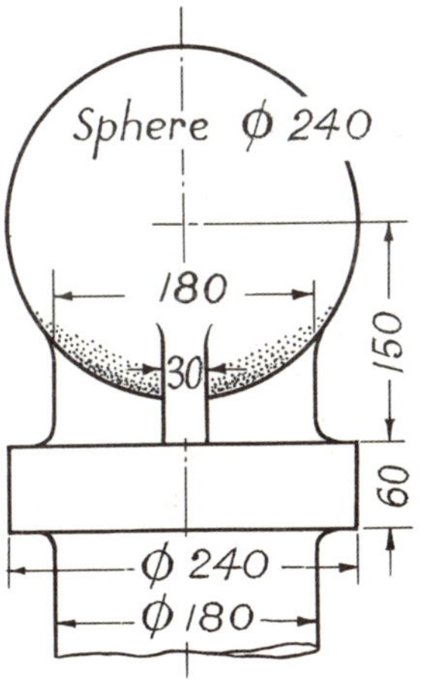

Fig 5

Oblique Projection (Cavalier, Cabinet)

We shall now consider a system of projection in which the projectors, while being parallel, are *inclined* to the planes of reference, as in fig. 1. It will be clear that the oblique projection of a surface, which is parallel to the plane of projection, is an equal figure; the lengths of AB and A_1B_1 are equal.

Drawings in oblique projection

The actual drawing of an oblique projection is often very simple. Consider the object in fig. 2. The principal face has been taken parallel to the plane of projection so that the view of the face is a similar figure. The side edges are set off *to the same scale as that used for the front face*, inclined at a common angle, usually 30°. This is sometimes known as *Cavalier Projection*.

The advantage of oblique projection is that circular parts project as circles, fig. 3—not as the ellipses in isometric projection. A disadvantage is the exaggerated length of the sides of the object. This gives a distorted effect, enhanced by the parallelism of lines which, in a photograph, would converge. We should therefore arrange the longest dimension to be parallel to the projection plane. But as we should also arrange the face with a circular or irregular outline to be parallel to the projection plane, to ease the labour of drawing, we may have conflicting requirements. For example, if we arrange the object as in fig. 4, we have the task of plotting the circular curve. We can reduce the distortion effect by using a scale of one-half for the side dimensions, as in fig. 6. This system is called *Cabinet Drawing*. Figs. 5 and 6 give a comparison of isometric and cabinet projection.

Exercises

1. **Taking the dimensions from fig. 3, draw two projections of the object: (*a*) isometric, using the scale on page 129; (*b*) as in fig. 6.**
2. **Draw an oblique projection of the bearing shown in fig. 7.**
3. **Draw an oblique projection of the bracket shown in fig. 8. Plot the side arc point by point.**
4. Draw an oblique projection of the packing piece shown in fig. 9.
5. Draw an oblique projection of the bracket shown in fig. 10.

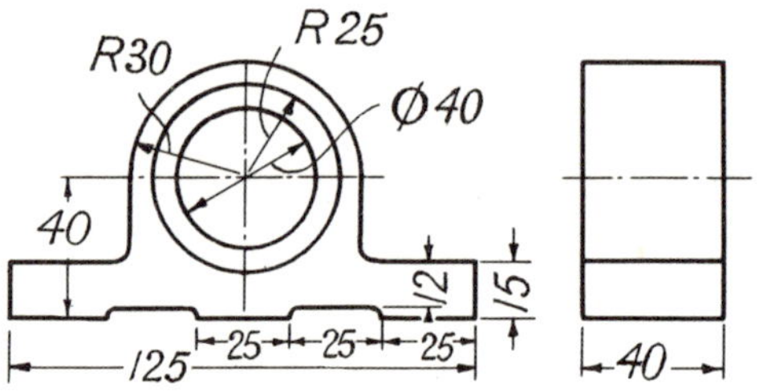

Fig 7

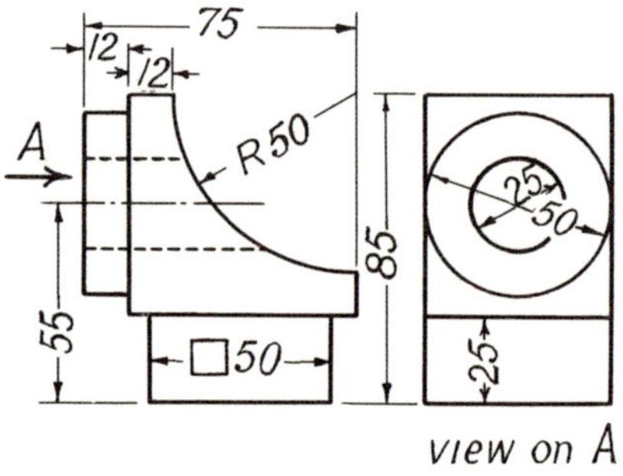

Fig 8

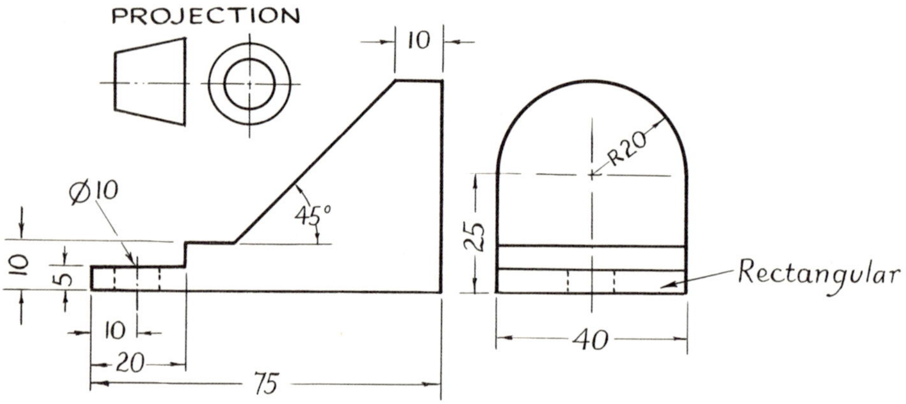

Fig 9

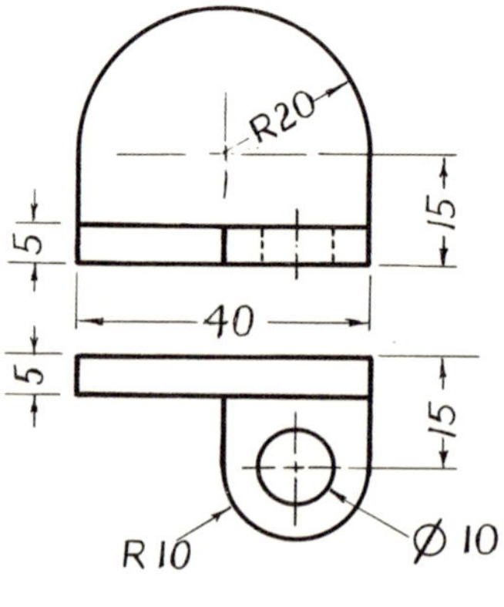

Fig 10

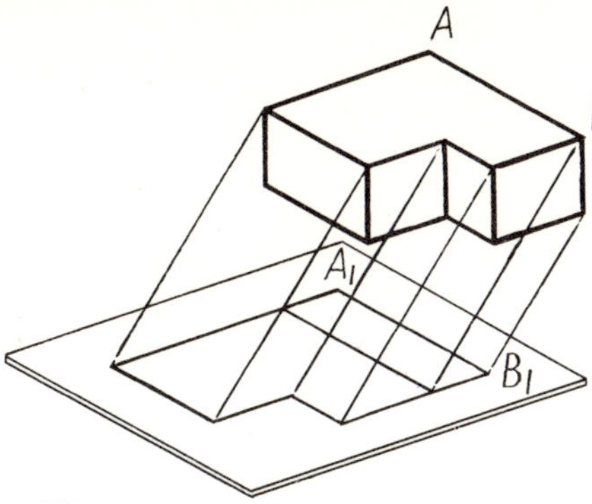

Fig 1

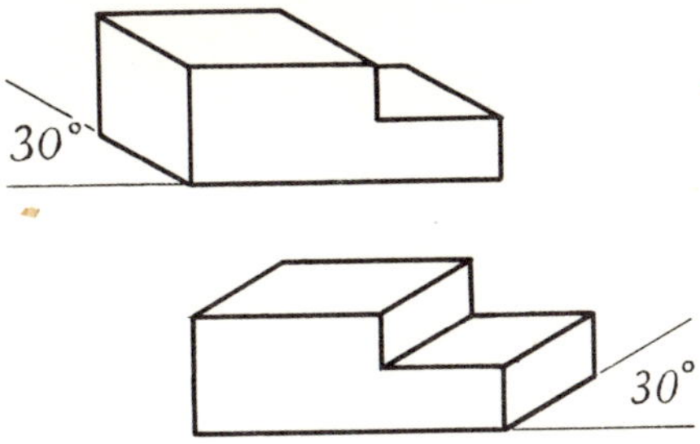

Fig 2

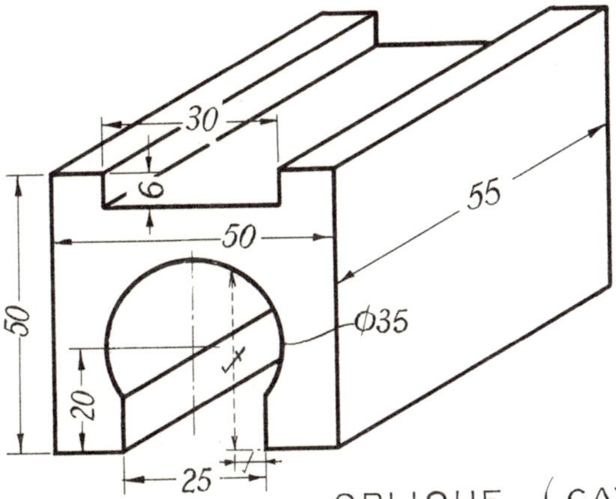

Fig 3

OBLIQUE (CAVALIER)

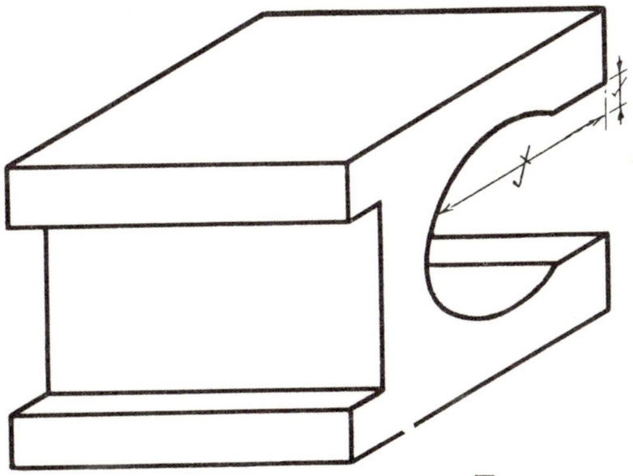

Fig 4

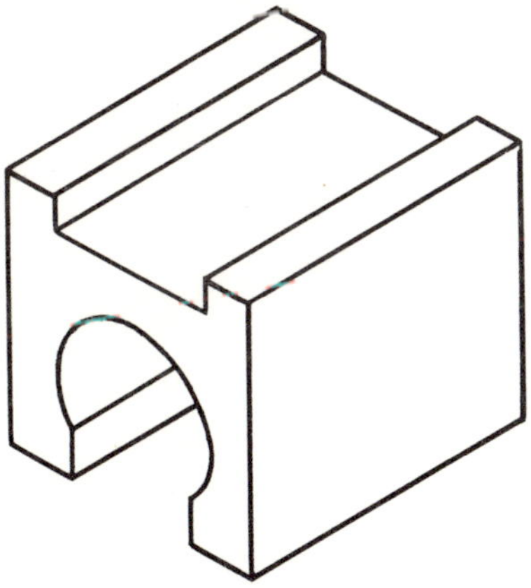

Fig 5

ISOMETRIC

Fig 6

OBLIQUE
(CABINET)

Horizontal Projection or Indexed Plans

In this system of projection only one view is used, a plan. *Points and Lines* are shown by their *indexed or figured plans.* Fig. 1 shows a triangle ABC and fig. 2 shows its indexed plan $a_{20}b_3c_{10}$, this giving the distances from a horizontal plane of the three corners. To describe a curved line the indexed plan of a succession of points along the curve must be given. The indexed plan will only give the true shape of an area if the index figures for points upon it are all the same. Any one contour line on a map represents a true area because the line lies in a horizontal plane. *Planes* are shown by *scales of slope.* A plane VTH is shown in fig. 3. A line AB is drawn in the plane and is perpendicular to TH. This is the *line of greatest slope* and the indexed plan of this line completely fixes the plane. The line is always shown, as in fig. 4, with another thicker line drawn parallel to it and on the left of the line ascending the slope. This thick line has no other significance than to show that the adjacent parallel line SS represents a plane. The unit for the *vertical* scale must of course be stated. Note that a horizontal line in the plane must always be at right angles to the scale of slope.

Solution of problems

As far as this book is concerned, the treatment of problems in Horizontal Projection involves a conversion of the one view to two views, and then a solution in the ordinary way. In other words, we shall regard the Indexed Plan as a convenient way of stating a problem.

True length and inclination of a line

The indexed plan of a line is given, a_9b_3, in fig. 5. Erect perpendiculars to the line at *a* and *b* and set off distances of 9 and 3 units respectively. The line joining these gives the true length, and the angle *Q* between the two lines gives the inclination of the line to the horizontal.

Plane containing three points

The indexed plans of three points are given by $a_{16}b_3c_{10}$ in fig. 6. We require the Scale of Slope.
First take the line joining a_{16} b_3 as an *xy* line and find the elevations of these two points, marked *a* and *b*. Draw *fg* parallel $a_{16}b_3$ and distant 10 units (the index of *c*). Project the point g_{10}. This point must be on the same level as c_{10}, and a line joining c_{10} and d_{10}, being horizontal, must be at right angles to the scale of slope. This is therefore given by SS, and the scale is found by projecting from a_{16} and c_{10}, as shown.

Exercises

1. **Three points are shown in fig. 7. Find the lengths of the sides of a triangle formed by joining them and determine the scale of the plane containing them.**
2. **Fig. 8 shows a plane *ss* and a line *ab*. Find the index of the point of intersection of the line and plane. (Hint: take *ss* as an *xy* line and project the edge view of the plane and the new view of the line. These intersect in the required point.)**
3. **Prepare an isometric view of the solution to Ex. 2.**

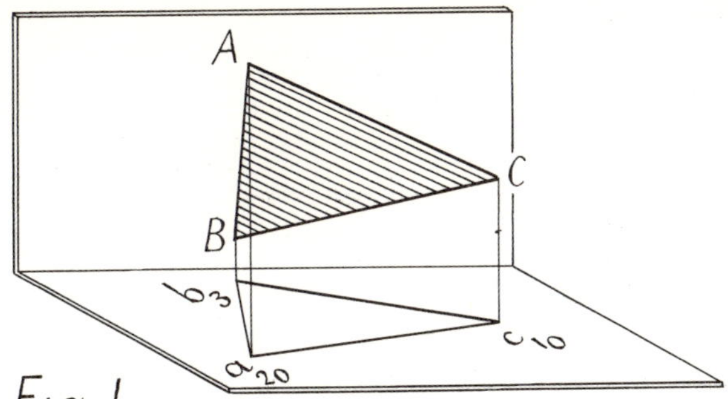

Fig 1

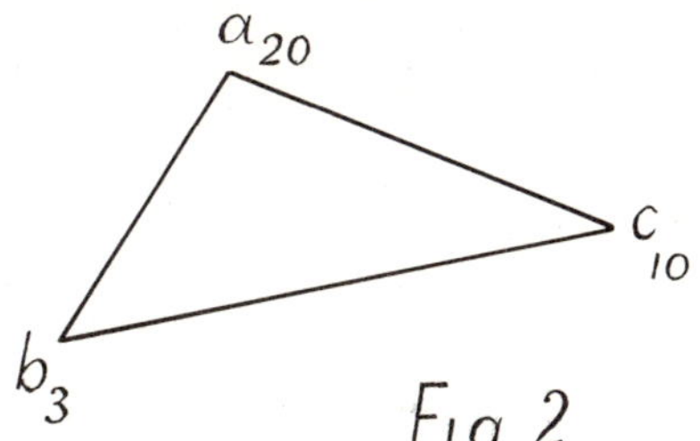

Fig 2

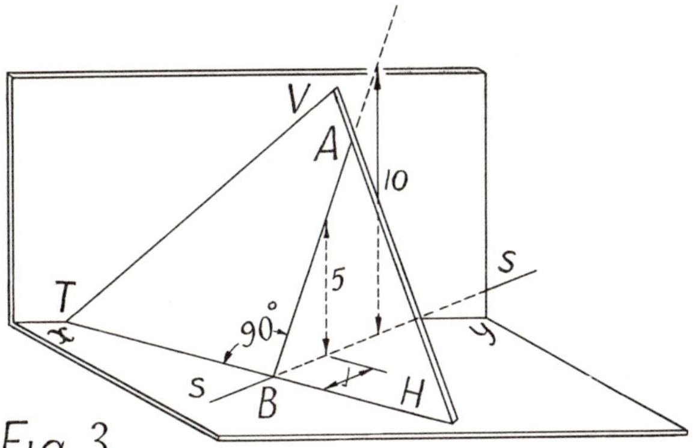

Fig 3

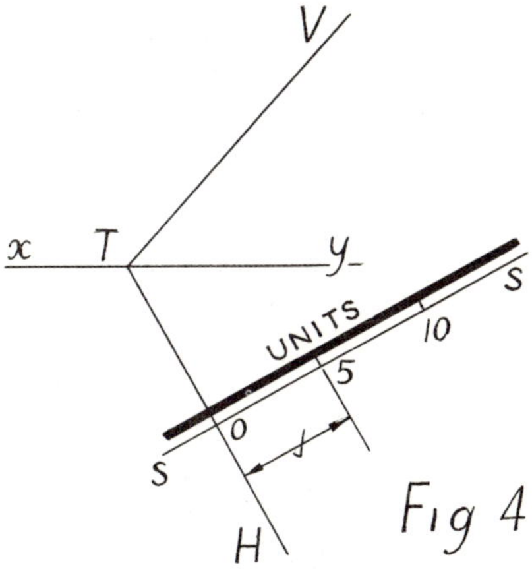

Fig 4

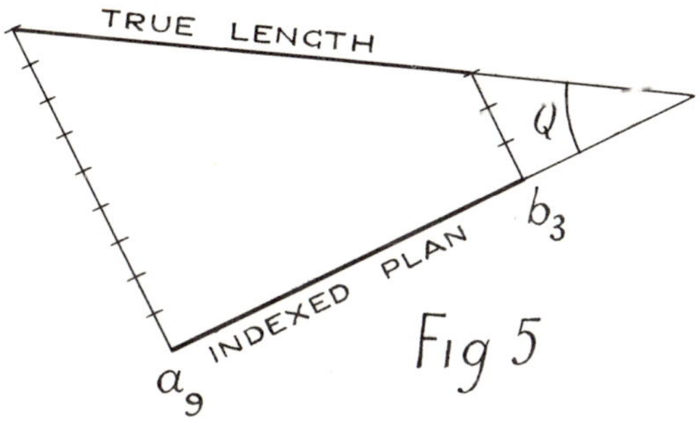

Fig 5

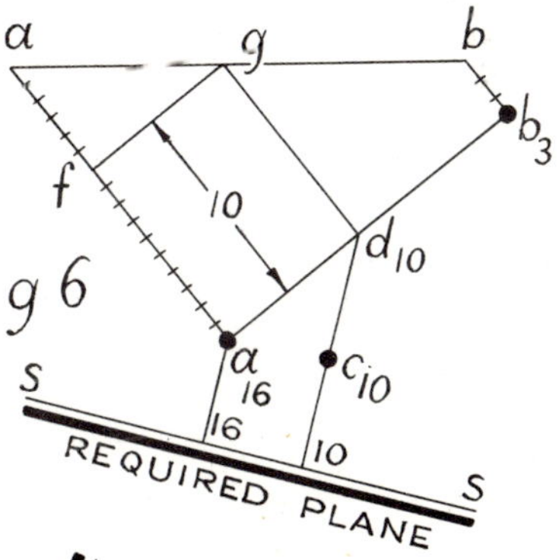

Fig 6

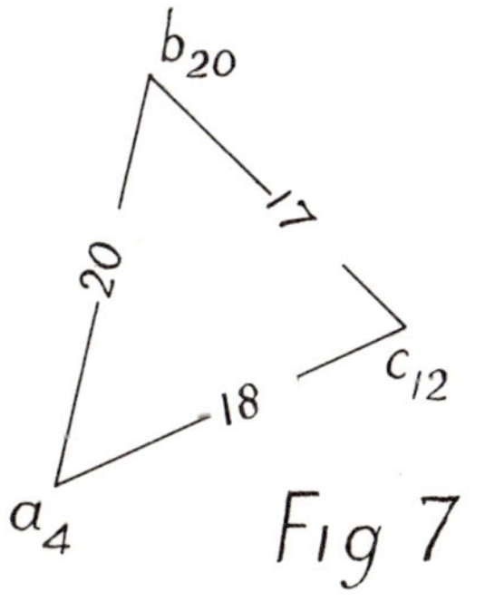

Fig 7

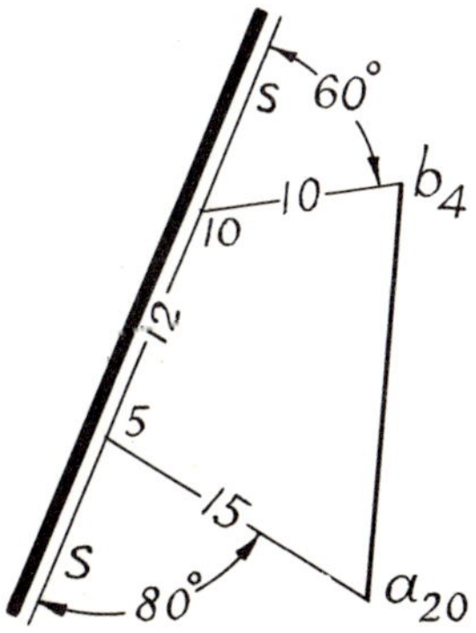

Fig 8

1. Describe a circle touching the three lines in fig. 1. Measure its diameter. (Ans. 53·8 mm.)
2. A triangle has sides 85, 70 and 60 mm long. Construct a rectangle, length 60 mm, of equivalent area. Find, without calculation, the area of the rectangle. (Refer to page 83. Ans. 2060.)
3. Draw an ellipse having foci 130 mm apart and a minor axis 100 mm long. Mark a point P on the ellipse which is 50 mm from one focus. Draw a normal to the ellipse at P and then draw two circles, 40 mm diameter, which touch the ellipse externally and internally at P. (Refer to page 87.)
4. Draw a semicircle 130 mm diameter. Then draw a concentric semicircle having half the area of the first. (Apply the *principle* of fig. 3, page 81.)
5. A triangle ABC has AB = 125 mm, AC = 100 mm, and the angle CAB = 45°. Draw the triangle. Draw (*a*) the circumscribed and (*b*) the inscribed circles and measure their diameters. [(*a*) Bisect two sides and draw perpendiculars; (*b*) bisect two angles. Ans. 126; 56.]
6. Fig. 2 shows a circular disc with a flat part. A cord is fixed at A and is taken to B. Plot the path of B as the cord is unwound taut.
7. A reflector has a mid-section which is a parabola. The open end is 180 mm diameter and the depth of the reflector is 130 mm. Draw the main parabolic section. (Refer to fig. 6, page 89.)
8. Two views of a pyramid are shown in fig. 3. The pyramid has a square base and is held with an edge on a horizontal surface. Obtain a view in the direction B. (First sketch a freehand pictorial view.)
9. The plan of a metal trough is shown in fig. 4. The ends slope at 60° to the plane of the top. Take the dimensions as representing millimetres and obtain the true shape of the front, back and one side.
10. A solid of revolution has an elliptical mid-section, major axis 130 mm, minor axis 80 mm. Each cross-section is circular. The solid is cut through the centre by a plane making 45° with the major axis. Draw the section. (Refer to fig. 1, page 121. An ellipse replaces the triangle. The same method is required.)
11. Draw an isometric view of the object shown in fig. 5.
12. A cone has a base diameter of 80 mm and an altitude of 90 mm. On an elevation of the cone draw the projection of the shortest line which passes around the cone starting from a point on the circumference of the base and returning to the same point. (Refer to fig. 3, page 127. The shortest line is given by a straight line joining points 1 and 1 on the development. Draw this line and project it on to the surface of the cone.)
13. The end of a rod is shaped as shown in fig. 6. The rod is circular and it meets the rectangular end in a rounded fillet. Draw the curve of intersection. (Draw a plan. Take sections parallel to the end face.)
14. The plan of a square flue is given in fig. 7. A lightning conductor is to be taken from A to B around the outside. Find the shortest length and draw the plan of its path to a suitable scale. (Develop the surface and take a straight line from A to B.)
15. Draw the object in fig. 5 in 'Cabinet Projection'.

The following questions relate to illustrations on page 136.

16. A triangular spacer is shown in fig. 8. Draw the two views shown in 3rd angle projection, and views seen in the direction of the arrows A and B.
17. Fig. 9 shows a railway coach sole plate with the front elevation on the left. Draw the two given views full size in 3rd angle projection, and project a plan view which will appear above the front elevation. The centre of the 100 mm radius must be constructed.
18. A flanged elbow is shown in fig. 10, drawn in 3rd angle projection. Draw, scale full size, in 3rd angle projection:
    (*a*) the given front elevation.
    (*b*) the given plan view.
    (*c*) a side elevation as seen from the left of view (*a*).

Note.—It is necessary to construct the plan view first, and then to project into the front elevation.

19. Draw in 1st angle projection, scale full size, the following views of fig. 10.
    (*a*) the given front elevation.
    (*b*) a plan view.
    (*c*) a side elevation seen from the right of view (*a*).
20. Fig. 9, page 130, shows a packing piece in 1st angle projection, the view to the left being the front elevation. Draw, scale full size, in 1st angle projection, the following views:
    (*a*) the given front elevation.
    (*b*) the given side elevation.
    (*c*) a plan view below view (*a*).
    (*d*) an auxiliary view showing the true shape of the sloping surface of view (*a*).

Note.—(i) Views (*a*) and (*b*) should be drawn together, and then the two views (*c*) and (*d*) together.
(ii) Notice the shape of the curve in view (*c*) when projected from views (*a*) and (*b*).

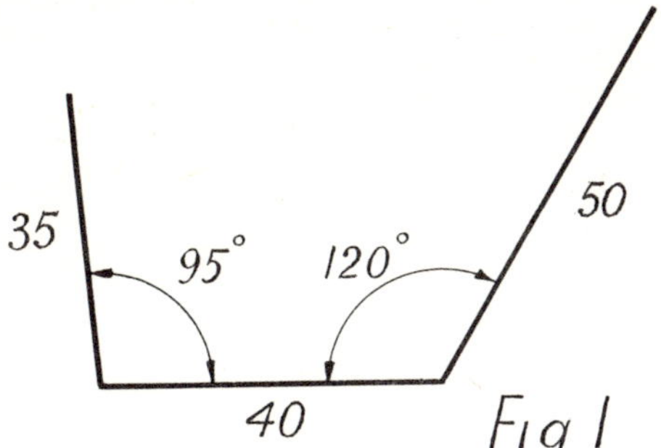

Fig 1

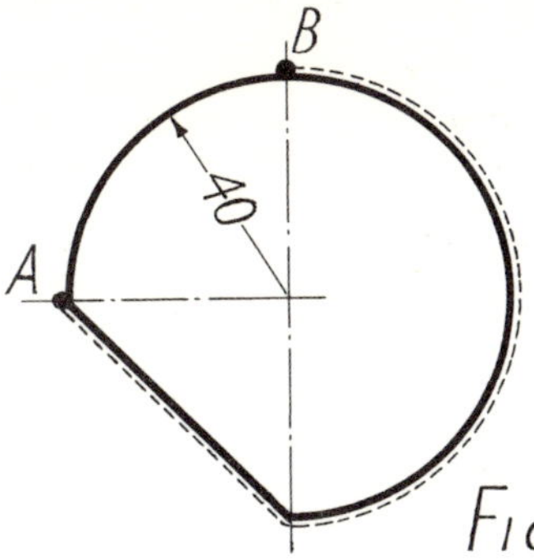

Fig 2

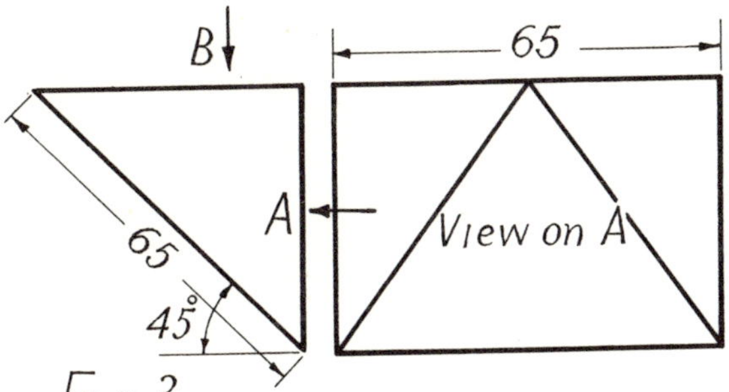

Fig 3

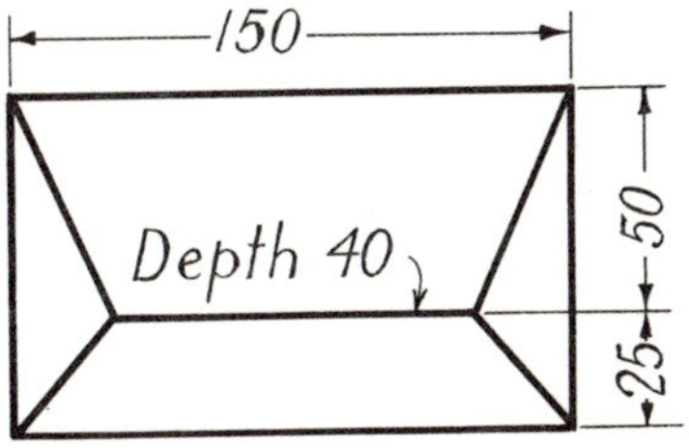

Fig 4

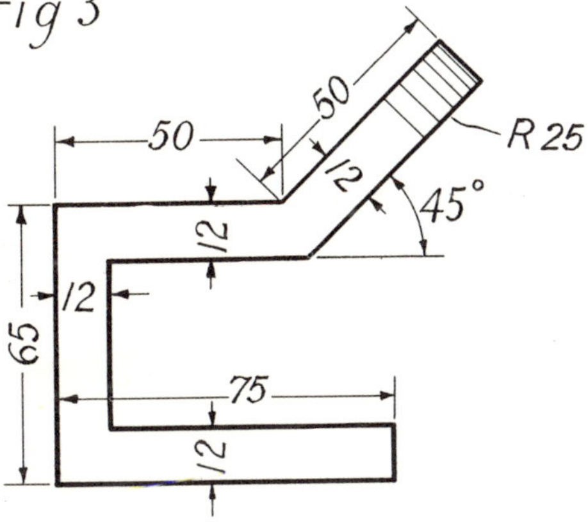

Fig 5

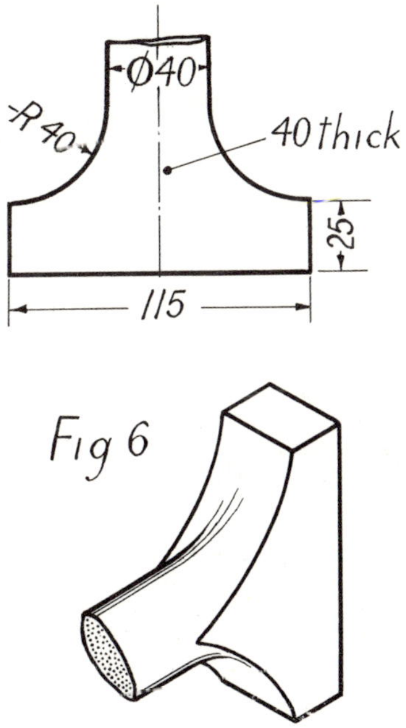

Fig 6

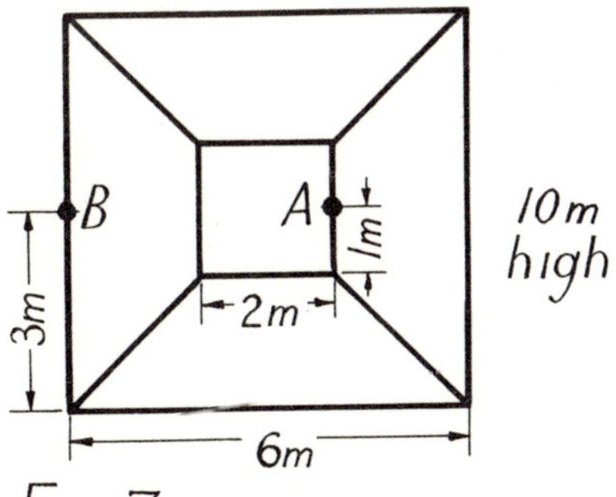

Fig 7

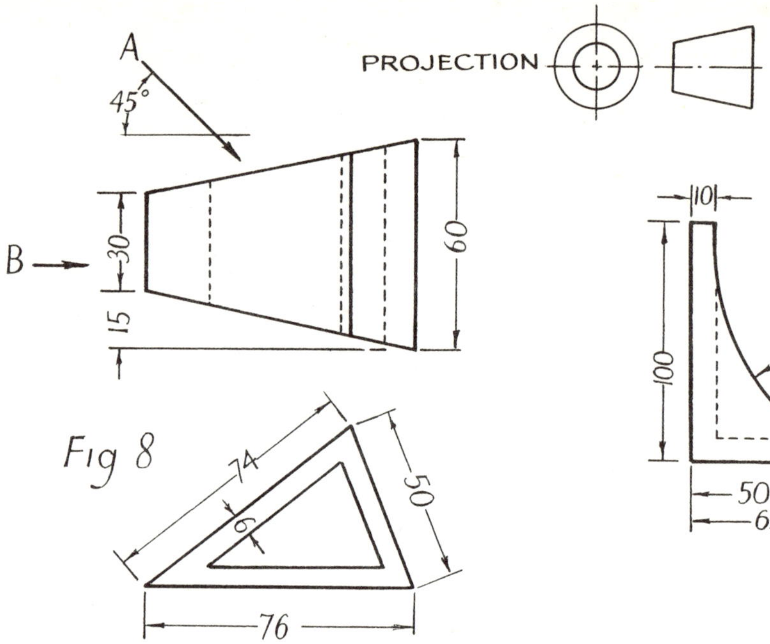

Fig 8

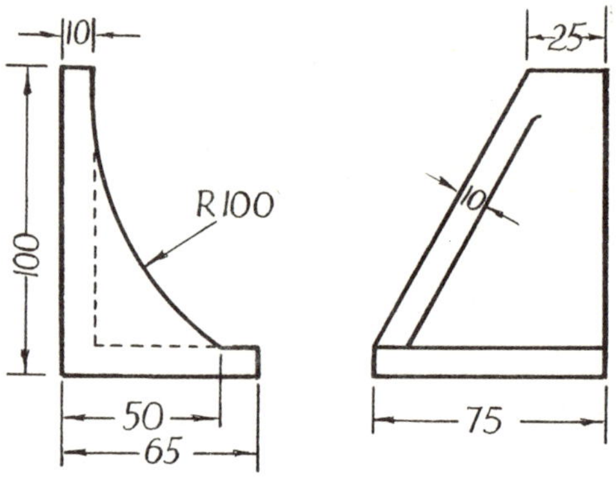

Fig 9

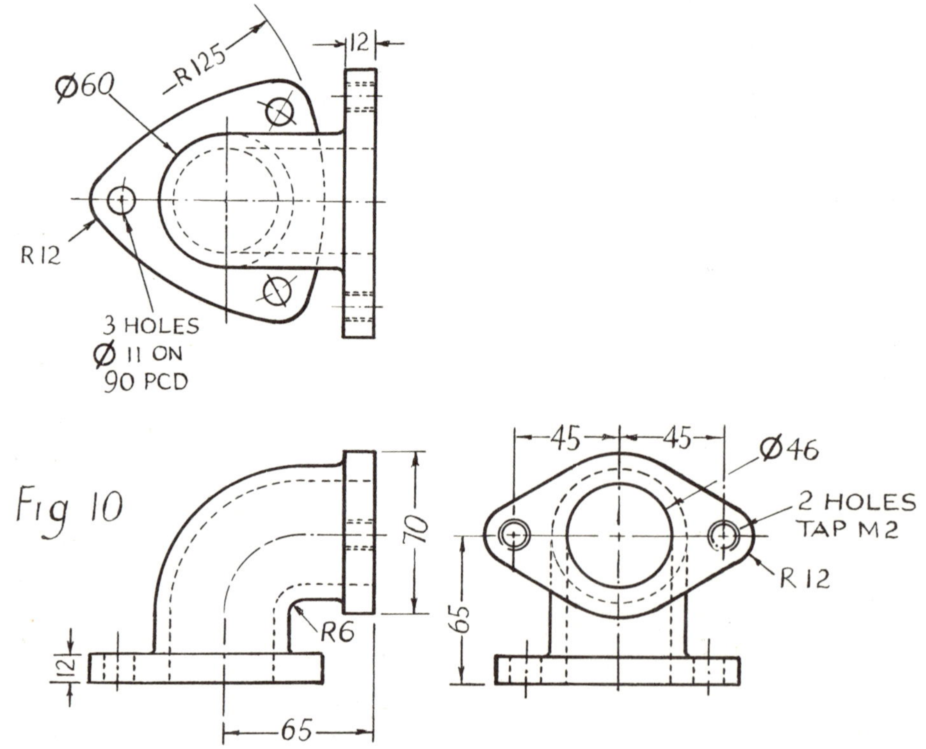

Fig 10

Part Three

Engineering Practice

Standard Practice

Engineering drawing practice

It is of the first importance that the graphical language of signs, symbols, and abbreviations used for the drawings of one firm should be understood by another. It is equally desirable that the conventions should, as far as possible, be common internationally —as, for example, the Plimsoll Line for ships, which is universally known and used.

Each of the principal industrial countries has standardized its Engineering Drawing Practice. Although there are differences between the systems there is much agreement. The practice in Britain is closely in line with that of North America and is set out fully in publication B.S. 308, issued by the British Standards Institution. The Standard was prepared by a committee representing the manufacturers and the users of engineering products and gives full consideration to the practices of other countries. As a result, B.S. 308 has authority and its recommendations should be adopted. They are used and explained in this book.

Systems of projection—Survey of previous work

In Part I drawings were made which represented views of objects taken from several directions. The views were arranged in line and were labelled. In Part II, the object to be described was imagined placed within a group of planes mutually at right angles, and projections on to these planes were taken from points on the object. We saw in Part II that, given two projections, any third projection could be readily drawn. This is further discussed below.

In all earlier work the importance of labelling the views was stressed. This was seen to be necessary because two systems of projection, 1st Angle and 3rd Angle, are in use, permitting two interpretations of the same set of drawings— which could result in the production of different articles. The two systems are illustrated again here, in fig. 1, by drawings of a right-angled bend which explain themselves. Third angle projection has been more generally used in British industry since 1945 in an attempt to co-operate more closely with America and Canada. However, now that closer links have been established with countries of the continent of Europe, a return to 1st angle projection has become necessary in some British industries. Thus it is essential for students and draughtsmen to be able to convert from one system of projection to the other. Examples and exercises in this book use both methods of projection.

Projections of a bracket

The use of the 3rd Angle system applied to a bracket is shown in fig. 2. View III shows the circular outline as a semicircle, and it may at once be drawn from View I. Any other view, e.g. View II, may be obtained from these two views, using any such reference lines as Line 1 and Line 2. By the principles of projection, distances of points in II are as far from Line 2 as are distances of corresponding points in III from Line 1—see distances ticked. Obviously it would be better here to use centre lines (chain dotted) rather than Lines 1 and 2.

Exercise

1. Taking dimensions from the pictorial view, draw three views such as those shown and add two additional views as would be seen in the directions IV and V.

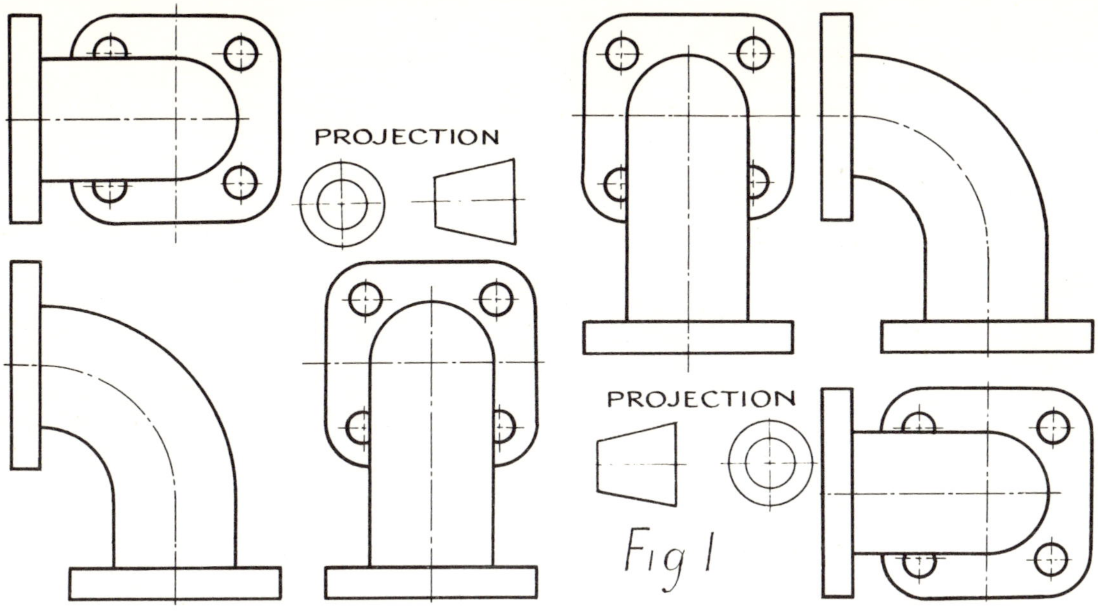

Fig 1

THIRD ANGLE FIRST ANGLE

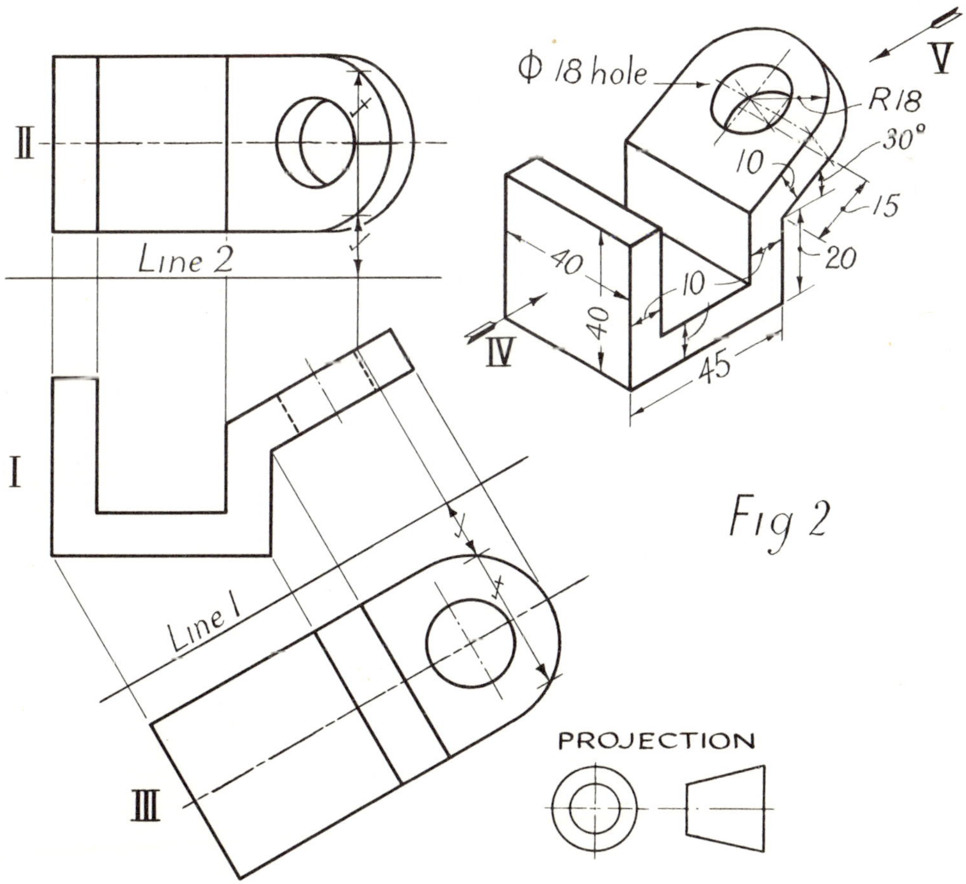

Fig 2

The object shown in fig. 1 is described by the five views in fig. 2. The student should cover fig. 1 and form an idea of the shape of the object from fig. 2 only. It may help to imagine fig. 2 folded to form a box, I giving the top, II the front, V the back and III and IV the sides.

Choice of views

We have now to decide which of these views are necessary fully to describe the object. Views I and II are necessary but insufficient. They do not, for example, reveal the shape of the slot at the back; the important short line *a*, in I, tells us that the slot cannot be a square recess extending from top to bottom. A back view, V, adds little to our knowledge, but a side view, as at III or IV, shows, by the sloping dotted line, that the back of the recess slopes outwards from top to bottom. An underneath view would be similar to I, but without the square at *a*.

The object is completely defined by I and II, with either III and IV, but only if dotted lines are used.

Normally, we should use views corresponding to I, II, III or IV.

What lines represent

A line may represent the intersection of two surfaces, or the profile of a curved surface (such as that of a cone or cylinder). Almost all the lines in the drawings in fig. 2 are formed by surfaces intersecting. They should be identified and marked, one by one, on the pictorial view. It will be seen that several lines coincide. For example, *cd* in III not only represents one edge but includes six others. It should also be noted that a full line may cover a dotted line, as at *e* in II.

Reading a drawing

The language of technical drawing has to be written and it has to be read. The reading is more difficult than the writing, as will be clear from figs. 1 and 2. If an engineer cannot read a drawing he is illiterate in the technical sense.

We cannot hope to read a drawing at a glance. In the same way that we can only read this page word by word, so we have to read drawings line by line, view by view. There is no quick and easy method which can give the ability to read drawings. Practice is essential.

Exercises

1. Cover the drawings in fig. 2. Using fig. 1, draw, freehand, the five views of the object, inserting all lines, full and dotted. Compare your drawings with fig. 2.
2. Cover the drawings in fig. 4. Using fig. 3, draw, freehand, three views of the object from front, top and right. Compare your sketches with fig. 4. Could you make the object from fig. 4? If not, add any additional views, or notes, and make complete dimensioned drawings. (Hint: from fig. 4 it could be assumed that the upper slots were rectangular, whereas they are triangular.)
3. A guide stop is shown in fig. 5 with cutting planes AB and CD which will be referred to later (page 148).
   Draw, scale full size, in 3rd angle projection:
   (*a*) the given front elevation.
   (*b*) the given plan view.
   (*c*) and (*d*) side elevations to the left and right of view (*a*).
4. From fig. 5 draw full size in 1st angle projection:
   (*a*) the given plan view. Re-name this as the new front elevation.
   (*b*) a plan view projected from view (*a*).
   (*c*) a side elevation seen from the left of view (*a*).
   (*d*) an auxiliary view seen in the direction of the arrow P.

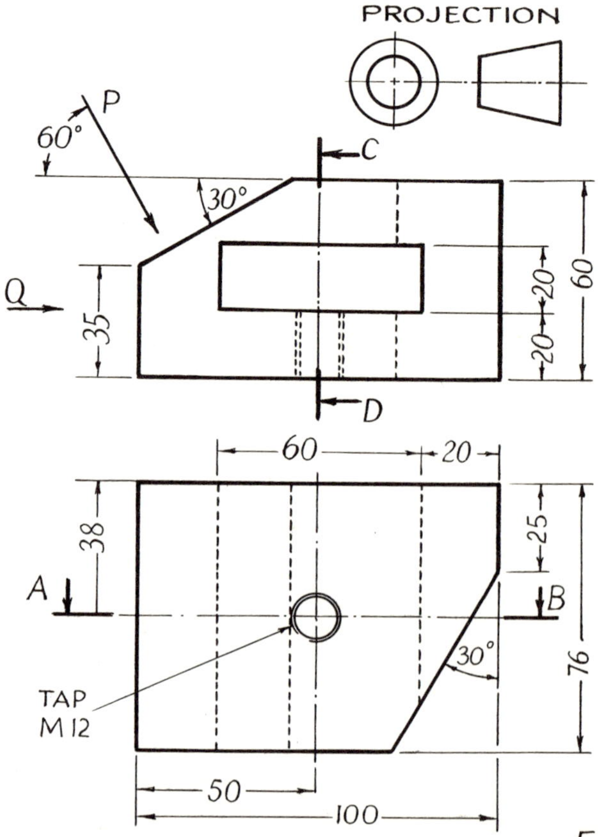

Fig 5

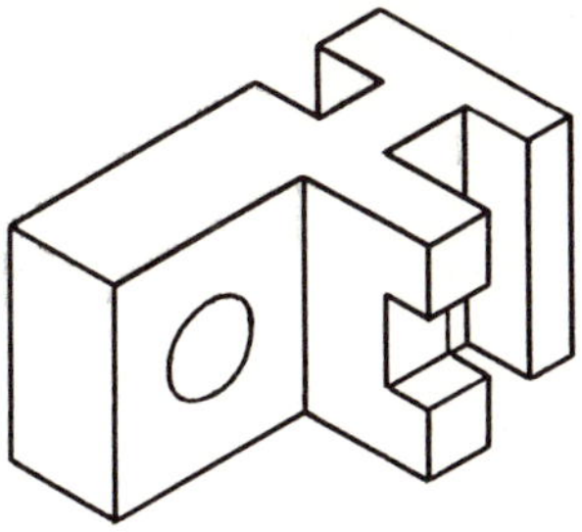

Fig 1

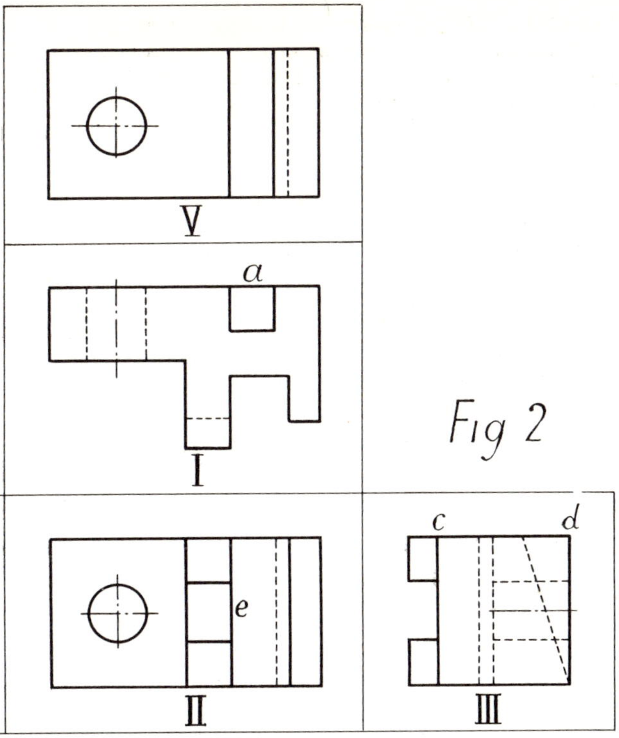

Fig 2

PROJECTION

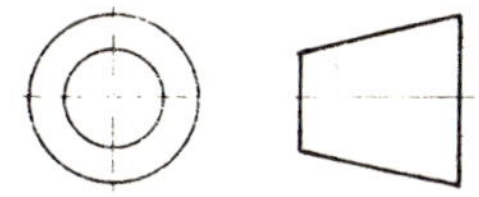

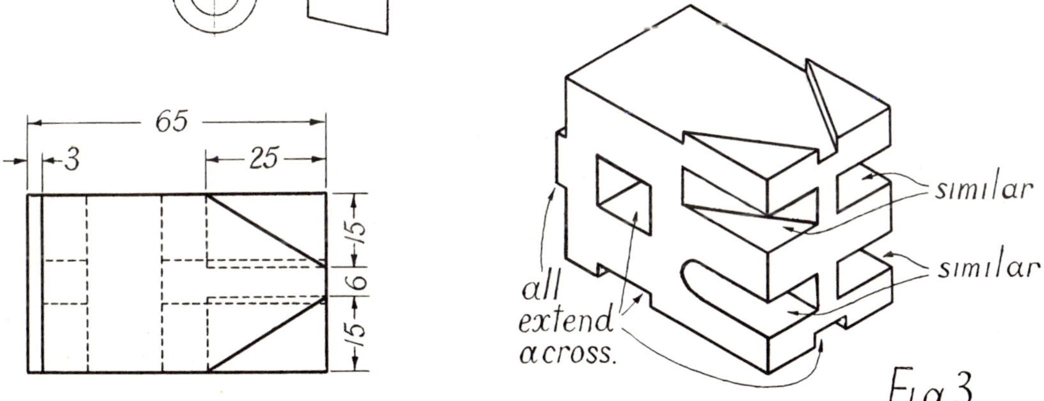

Fig 3

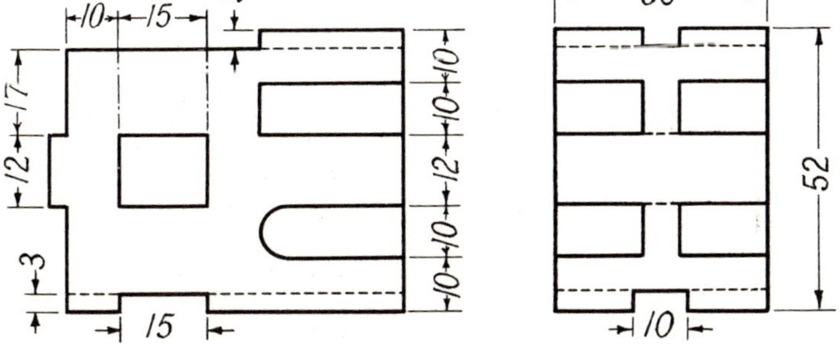

Fig 4

Drawing Procedure

Much time may be saved by using an order for drawing. In reading what follows the student should glance at the step-by-step views opposite, and at the pictorial view below of the object to which they relate.

Scale and spacing

A not uncommon error is the advancing of one view only to find that there is not room for the others. This may be avoided if, after first deciding how many views are necessary, we write down the extreme sizes of the object in each view and then arrange, on the paper provided, the three or more rectangles required for the views.

This may involve us in a choice of scale. If an object is very small, a full-size drawing is unsuitable and a multiple scale must be used. If an object is very large it cannot be drawn full size and a fractional scale must be used.

Scales recommended are : $\frac{10}{1}, \frac{5}{1}, \frac{2}{1}, \frac{1}{1}, \frac{1}{2}, \frac{1}{5}, \frac{1}{10}$.

Arrangement of object and views

The object is usually drawn either in its natural position or, if it is a machined part, in a position it would occupy during machining. In engineering practice separate drawings are often made for each part. Some parts are so simple that one view only is required; others are better conceived from two views; but for many objects, three views are required—sometimes four or even five.

It is customary to place front and side views adjacent horizontally, with a top view above them as in IV (3rd Angle Projection) or below them (1st Angle Projection).

Progress of views

It is best to progress all views together, beginning with the drawing of centre lines (I) and circles (II). We may readily project from circles and draw lines in other views. Successive stages are shown in III, and in IV, the final stage before dimensioning.

Sectional views

These will be incorporated only if they are really necessary for the understanding of the drawing. Section lining is time-wasting and should be limited.

Even in the early stages, speed of draughtsmanship can and should be practised. When marking out the horizontal centre line in the front elevation, keep the tee square in position and mark out the centre line of the side elevations. Vertical features can be carried forward from elevation to plan view. In this way all views will grow together more rapidly than if the views were drawn separately. This will save valuable time, particularly in examinations. The student should take one, two or, if possible, three dimensions and make use of them in all views before again looking for further information.

Exercise

Draw three views of this object, adopting the steps shown opposite. Choose suitable fillet radii. Scale half full size, in 3rd angle projection.

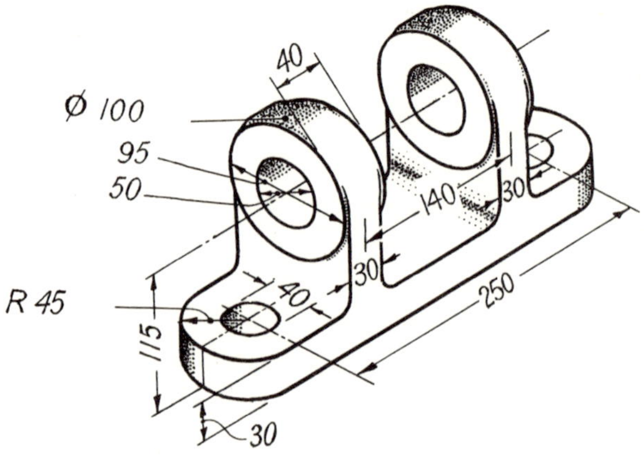

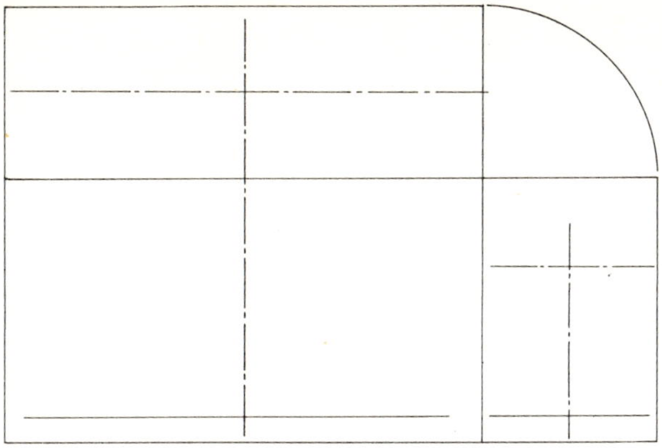

I

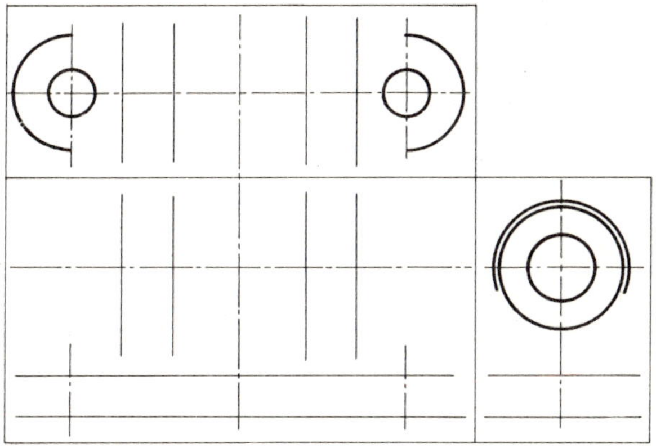

II

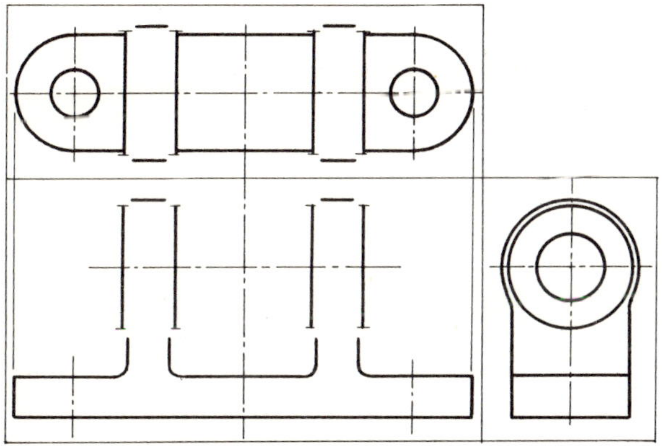

III

PROJECTION

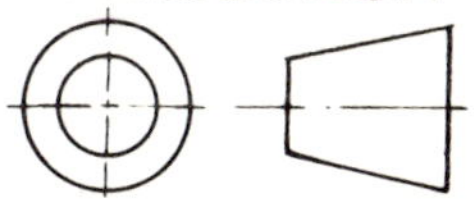

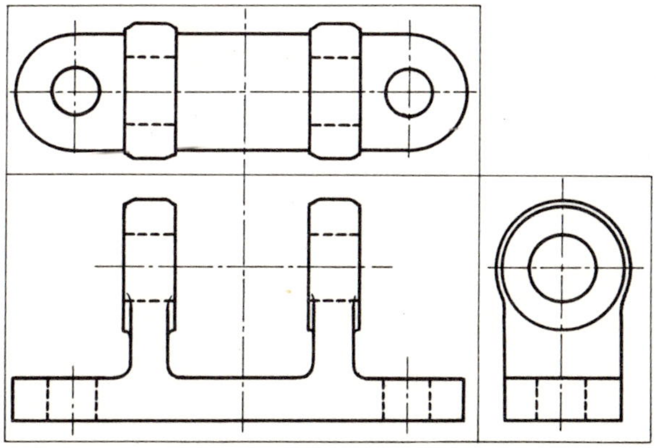

IV

Curves of Intersection

These were dealt with in Part II where, on pages 122 and 124, intersections between cylinders and cones were projected. Where the *development* of surfaces is required, as in Sheet Metal Work, an essential first step is the plotting of the curves of intersection; but where the only purpose of these curves is to show the shape of the part, then the common drawing-office practice is to draw curves which are approximately correct. The time taken to plot the curves accurately would not be justified. Nevertheless, the student should know how to obtain the projections of curves of intersection if only to be certain that his approximate curves are reasonably correct.

Forked end

A common engineering part is the forked end of a rod shown in fig. 1. This is forged and then machined all over. The surface marked T is a solid of revolution formed by turning in a lathe, the rod rotating about its axis; the surfaces marked F are flat and parallel. The round and the flat surfaces intersect in the interesting curve shown.

To obtain any point on this curve, take a plane such as PP and project the curved outline cut by the plane—the circular arcs drawn about *o*. These arcs intersect the parallel sides in points such as *b*. Project from *b* to obtain the required points b_1, which must lie in PP. Note the positions of the extreme points c_1 and d_1.

Approximate curves

Three examples of common engineering articles are given in figs. 2, 3 and 4, in which curves of intersection are drawn without plotting points. Consider the bent link in fig. 2. It would be unhelpful to leave the drawing of an end as shown by the part in the rectangle to the left of fig. 2. Notice also the deliberate use of thin lines at *a*, where surfaces meet smoothly; the lines do not in fact exist, but to show them assists in the reading of the drawing. The intersections in fig. 4 follow closely the true lines—which are straight when the intersecting cylinders are of equal diameter (see page 122, Part II).

Exercises

1. Draw the views in fig. 1, carefully plotting the curve of intersection. Add a third view looking on the broken end of the rod. Insert dimensions—which are given in millimetres.
2. Glance at figs. 2, 3 and 4 and make freehand sketches of similar components, inserting curves of intersection. Compare your sketches with the drawings in the book.
3. More advanced. Make an isometric drawing of the forked end in fig. 1.

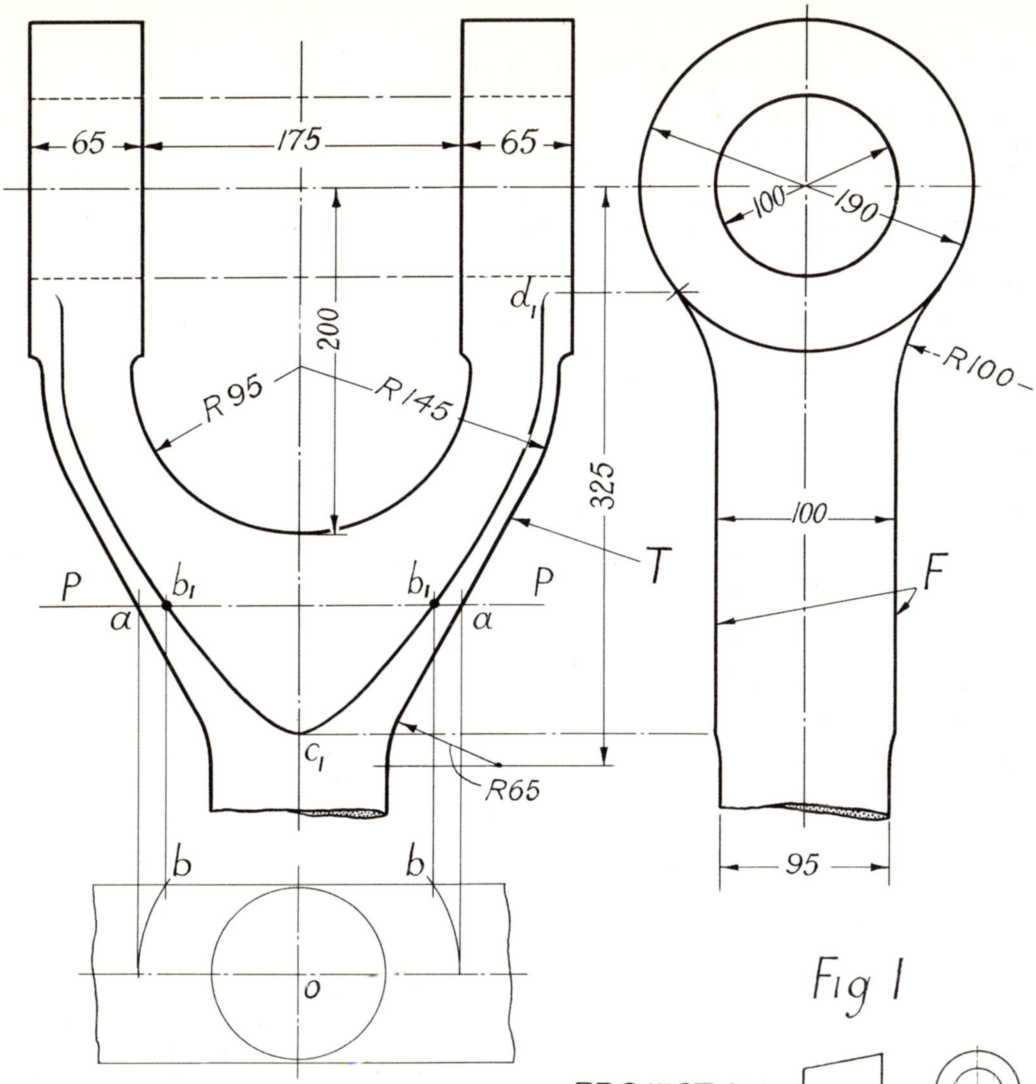

Fig 1

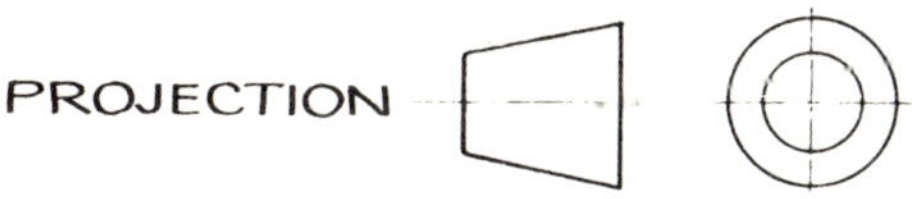

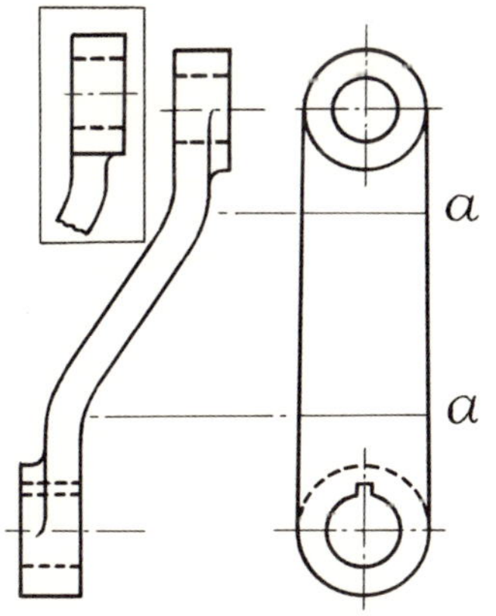

Fig 2

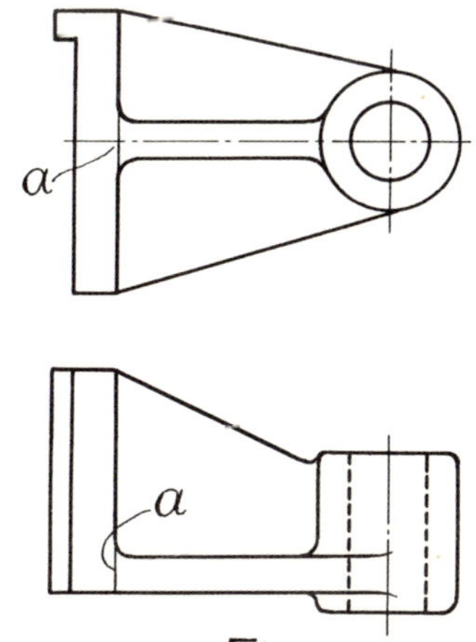

Fig 3

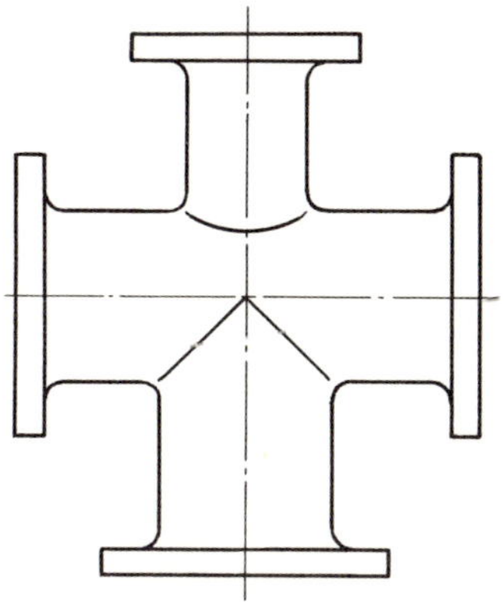

Fig 4

Letters and Numbers, Drawing Headings

Nothing spoils a drawing so much as poor lettering. Skill in forming letters and numbers comes only with practice and the student should try constantly to improve his style.

In the Drawing Office the lettering will be done by the use of Stencils, which give uniformity and clearness. Nevertheless, it is necessary for the student to be able to complete his drawing with all notes, titles and dimensions, and to produce a satisfactory result,

Size and clarity

Lettering and numbering should be done freehand but always with the aid of parallel lines. These lines should normally be horizontal and their spacing depends upon the size of lettering required. Usually the lines need never be more than 5 mm apart except for main titles, and even for these, broadening of the figures is often preferable to heightening. An illustration of the effect of widening letters was given in Part I.

Clarity is the first requirement of lettering on drawings. The student will know that engineering practice is to have *prints* made from drawings which are either traced upon, or made directly on, transparent paper or cloth. The prints may be roughly handled in the workshop and are often creased. Numbers or words which were clear on the original may be blurred on a used print; a 3 may be taken for a 5 or an 8. Every possible care therefore must be taken to use letters and figures which are distinct. No decoration whatever is required.

Alphabet

Two alphabets for Upper Case letters (Capitals) are shown opposite, one vertical and the other inclined. The vertical letters are slightly more difficult to make than the inclined, but they are more impressive. The slope of the inclined letters should be 5 in 2, as shown by the triangle. The letter forms in the script alphabet shown for Lower Case letters (Smalls) have been simplified as much as possible.

Drawing headings

Several kinds of Headings are set out in B.S. 308, but that shown opposite is fairly typical. It contains the information usually required and should be studied and adopted.

Exercise

Draw lines 10 mm apart and, without looking at the book, draw an upright and inclined alphabet, for capitals and smalls, adding figures. Compare your proposals with those shown opposite.

A B C D E F G H I J K L M N O P Q R S

T U V W X Y Z 1 2 3 4 5 6 7 8 9 0

A B C D E F G H I J K L M N O P Q R S

T U V W X Y Z 1 2 3 4 5 6 7 8 9 0

abcdefghijklmnopqrstuvwxyz

DRAWING NO. SYSTEM OF PROJECTION

TYPICAL DRAWING SHEET

DRAWN		MATERIAL	TOLERANCES	NAME OF FIRM	
CHECKED				TITLE	DRAWING NO.
APPROVED		FINISH			
ISSUED					

Sectional Views

General rules

Some consideration has already been given in Parts I and II to sectional views, and we may sum up the conclusions reached there as follows:

(1) The rules of projection apply to sectional views, which therefore take the positions of the external views they replace.

(2) When a sectional view replaces an external view, all outlines visible beyond the section plane must usually be shown; sometimes it is sufficient to show only the part sectioned.

(3) The trace of the imaginary cutting plane, i.e. the line of intersection of the cutting plane and the plane of reference, is shown by a thin chain-dotted line, thickened at each end as shown on page 34. The section plane is lettered and arrows are inserted to give the direction of view.

(4) The section lines should be spaced by eye and should slope at 45°, the exception being where a surface is bounded by lines at 45°, when the section lines may be at 60° or 30°. The distance between the lines is governed by the areas to be covered. If the lines are too closely spaced their drawing is lengthy and tedious. For the examples in this book a spacing of 4 mm is suitable, with a closer spacing of 2 mm for small parts.

Bearing step

The above rules are all illustrated in fig. 1, which shows a simple bearing step for supporting a shaft. The imaginary plane in I reveals the section in II, and three views of the part are shown in III.

Sections of composite objects

The two views in fig. 2 describe a guide for a shaft which slides horizontally. The engineering names of the parts are marked. The *bush* is cylindrical and is forced into a hole in the guide; it is thus replaceable and can be made of a more suitable material than the guide. The bearing surface in the bush is reduced so that the shaft touches the bush only at its ends.

When such a composite object is to be sectioned we distinguish the parts either by sloping the section lines in opposite directions or by changing their spacing. Notice that the shaft is not sectioned; this convention, with others, is referred to on page 152. It is common practice to show only half the sectioned view, as in fig. 3.

The arrangement in fig. 4 differs in an important respect from that in fig. 2 because the guide and the bush are each in halves, bolted together. The section plane now cuts parts *which are not continuous*, and this fact is recorded by sloping the section lines for adjacent parts in opposite directions. The method of securing the halves of the bush is not shown.

The differences between figs. 2, 3 and 4 should be carefully studied because of their general application.

Exercises

1. Draw fig. 2 freehand. Without referring to the book draw the half-sectioned view, and the section when the top and bottom are separate halves. Compare your drawings with figs. 3 and 4.
2. A guide stop is shown on page 140, fig. 5, in 3rd angle projection. Draw the following views full size:
   (*a*) the given elevation.
   (*b*) a plan view in section on AB.
   (*c*) a side elevation seen in the direction of the arrow Q.
   (*d*) a side elevation in section on CD.

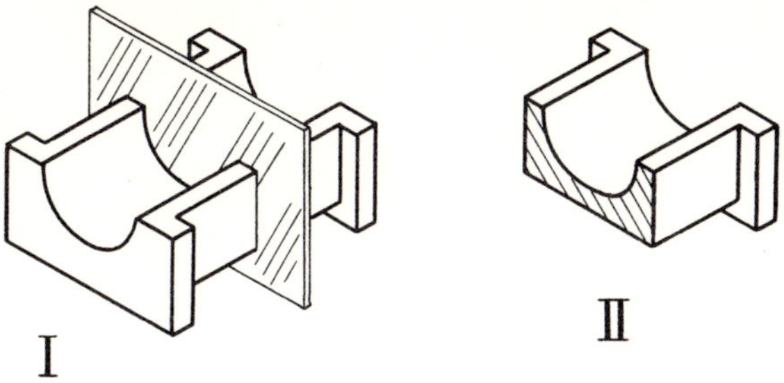

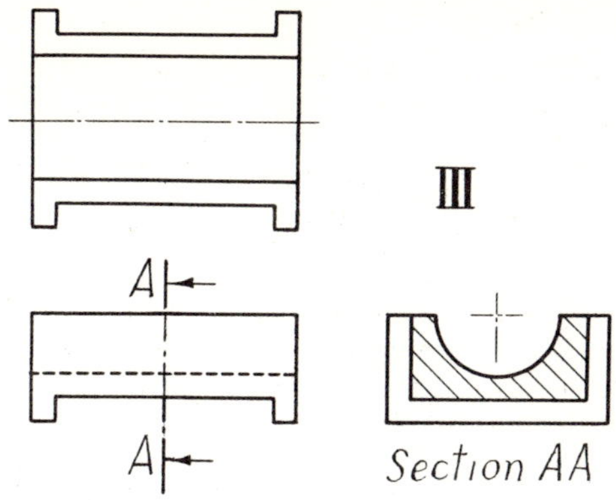

Fig 1

PROJECTION

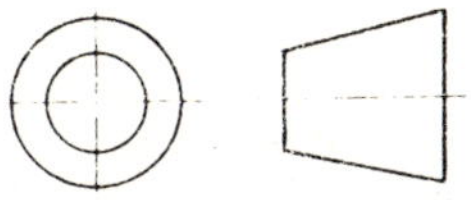

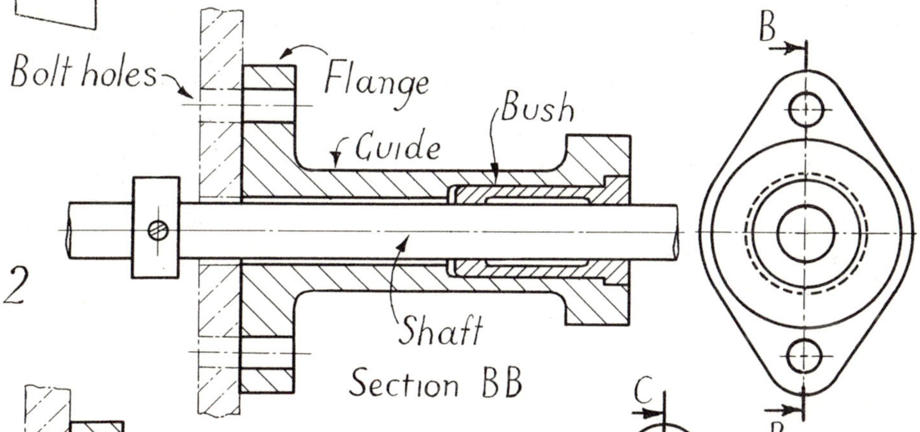

Fig 2

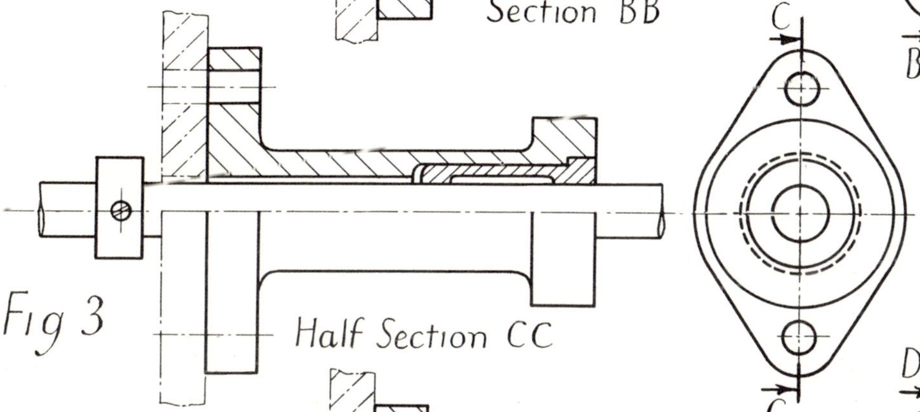

Fig 3

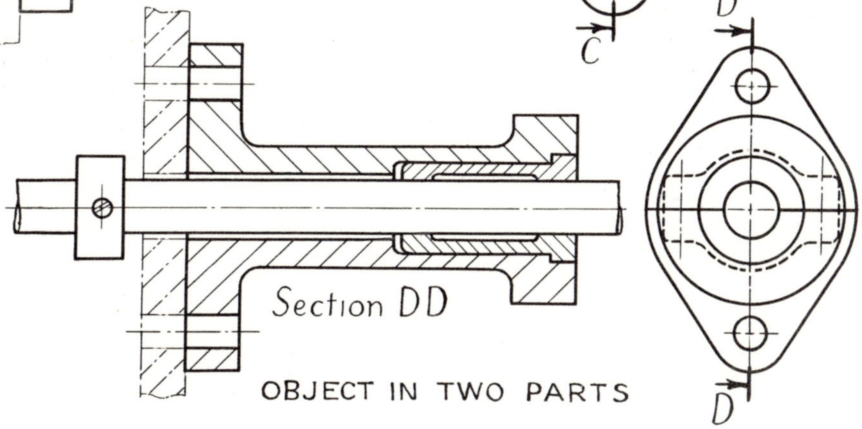

Fig 4

Revolved sections

Three revolved sections are shown in fig. 1, I, II, III. The method is used for bars and cast members to show the shape of the section without drawing the usual view. The object is sometimes shown broken to make room for the revolved section, as in I.
As an alternative, we may remove the section, and then revolve it, as shown in fig. 2 for a crane hook having a varying section.

Departures from true sections

Webs or ribs are commonly used to strengthen castings, as shown for the cylinder on a square base in fig. 3. When a section plane passes through a web *along its length*, the section should be shown as in fig. 3, II, not as in III, in order to avoid the appearance of solidity. If the section plane cuts *across* the web, then the section is shown in the usual way.
A drawing of a pulley with five arms is shown in fig. 4, I. A true central section II is not as satisfactory as the modified sectional view in III, in which the arms are not sectioned and the lower half is slightly incorrect. Fig. 4, III really represents a section on two planes, chain-dotted in I, and this device of a staggered section is commonly employed to give a clearer idea of the object and often to save time. It may be termed a zigzag section.

Exercises

The dimensions for figs, 4, 5 and 6 are in millimetres.

1. Write down the various dimensions for the pulley in fig. 4. Without referring to the book, draw, to a scale of $\frac{1}{5}$ full size, two views of the pulley corresponding to I and III. Compare your drawings with those opposite.
2. Draw, full size, two views of the object in fig. 5, one view looking on the end but half in section, and the other giving a central section through the axis of the bore. The recess at the bottom is rectangular. Insert dimensions.
3. Draw, full size, the left-hand view of the object in fig. 6 and add another view taken centrally through the holes, half in section. Dimension your drawings.

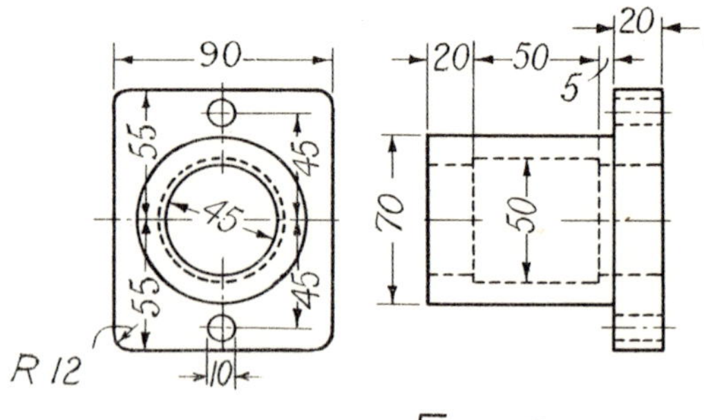

Fig 6

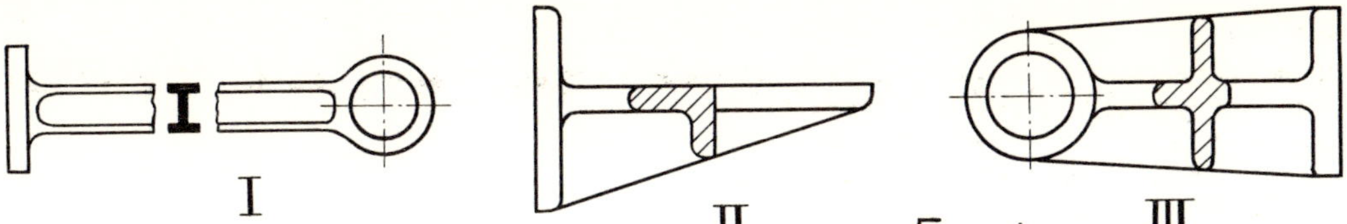

Fig 1

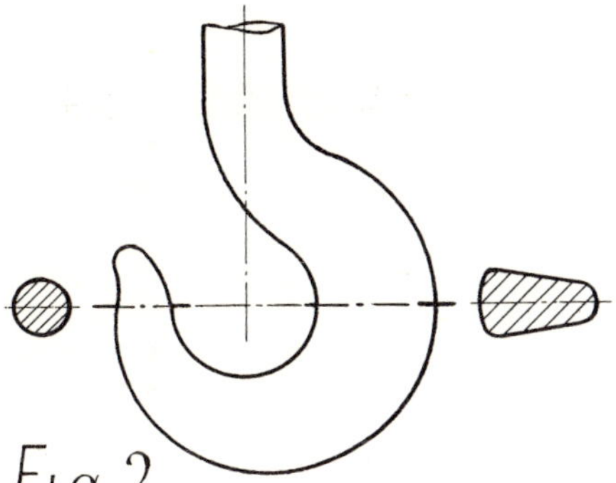

Fig 2

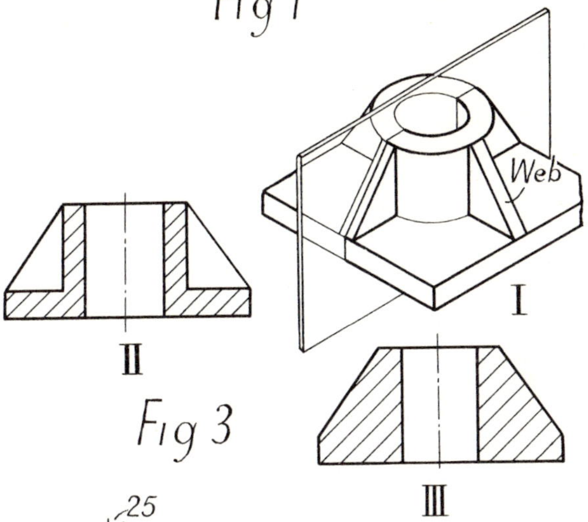

Fig 3

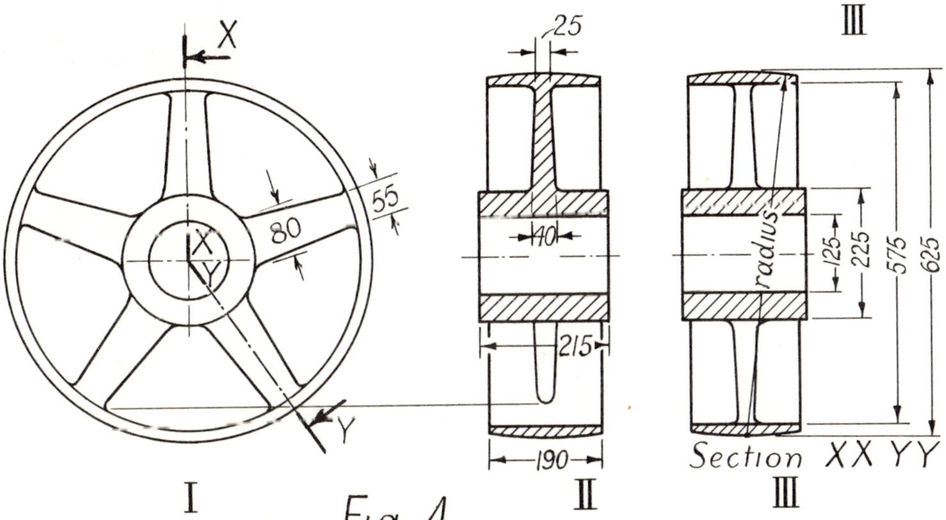

Fig 4

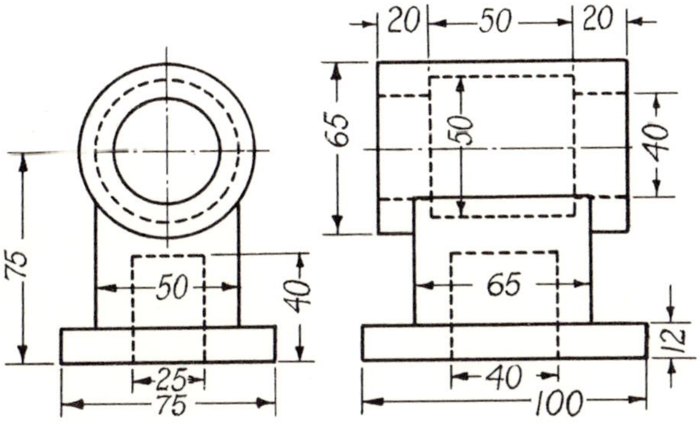

Fig 5

Common engineering terms

From time to time engineering terms are used here which may not be familiar to the student. He may know such names as *shaft, rod, bolt, nut, washer,* and *bracket*; but he may not know *web, rib, spigot, flange, bush, valve, tap bolt, key, cotter, split pin.* He should become familiar with these parts in the workshop, but they are discussed, described or illustrated on the following pages, to which reference should be made.

Ribs and webs: 151, 155, 177, 191, 192, 193.
Tap bolt, stud: 167, 169.
Key: 183.
Cotter: 185.
Flange: 149, 177, 181, 183, 193.
Bush: 149, 162, 177, 188, 189, 192, 203.
Spigot: 181, 183.
Valve: 192, 193.
Split pin: 171.

Some of these parts are shown in the drawing opposite.

Parts not sectioned

There are many exceptions to the rule that parts cut by the imaginary plane should be sectioned. In fig. 1 opposite a composite solid is cut by a section plane. Although this plane passes through the shaft and through rivets, bolts, a key and a cotter, these parts are not shown in section. Many other components are similarly treated, e.g. tapered and split pins, valves, ball bearings.
The student should refer to the sectional drawings on the following pages: 169, 185, 192, 200, 202, 203.

Exceptional sections

Where a part is relatively thin, it may be wholly blacked in. This procedure is commonly adopted for structural shapes, for sheet metal and for packing, etc. Where several thin parts make up a composite body, spaces are often left between the blacked-in sections, as in fig. 2. See also fig. 6 on page 176.

Symbols for welds

It is becoming more and more common to fabricate engineering components by welding together simple shapes and standard parts and sections. The process of welding has brought into being an elaborate Scheme of Symbols for Welding, and these are set out in British Standard 499. Two are shown in fig. 3. Examples of welded practice are given on page 181.

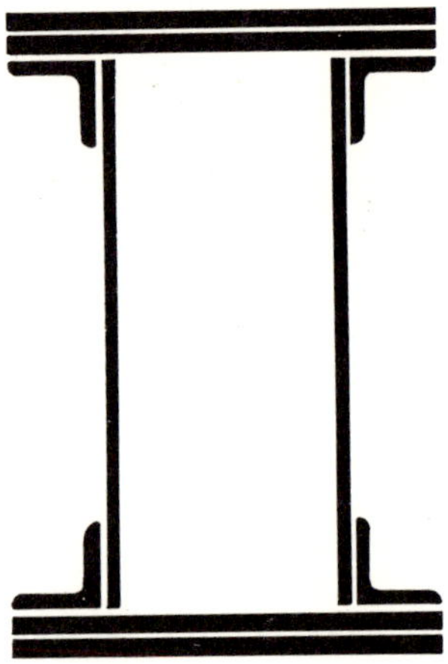

Fig 2

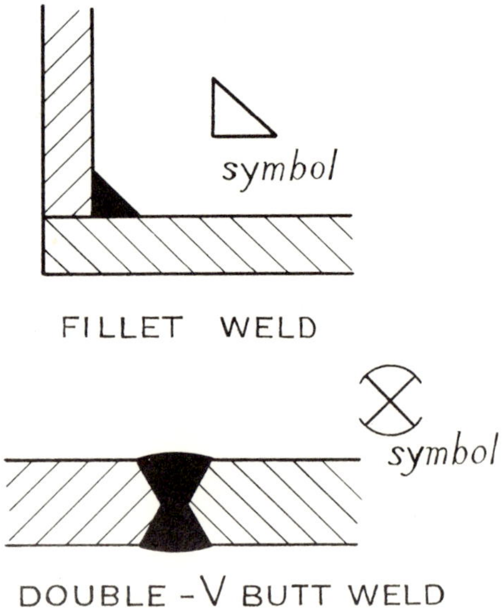

Fig 3

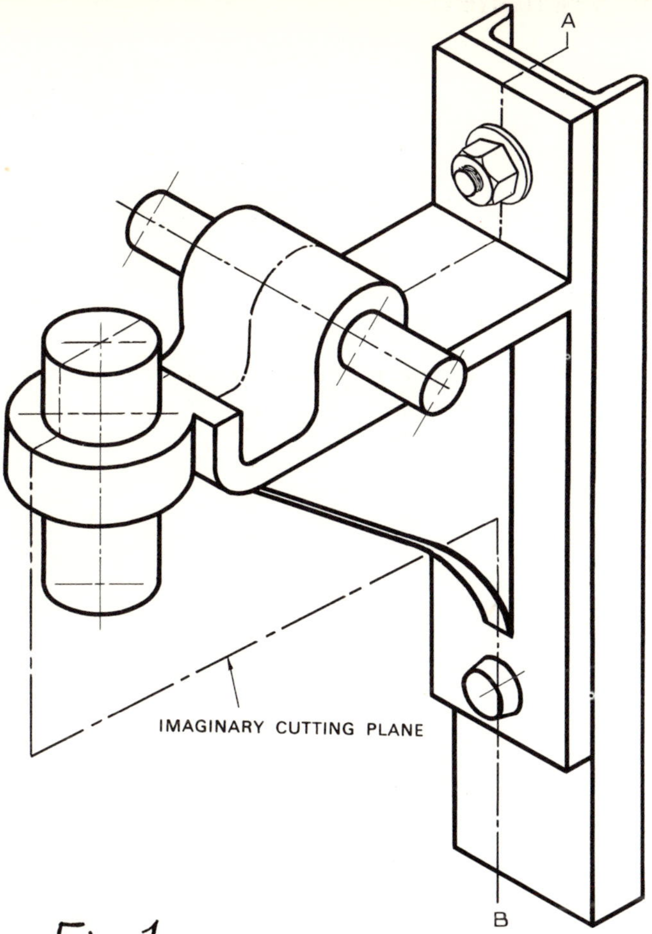

Fig 1 *(i) An isometric view*

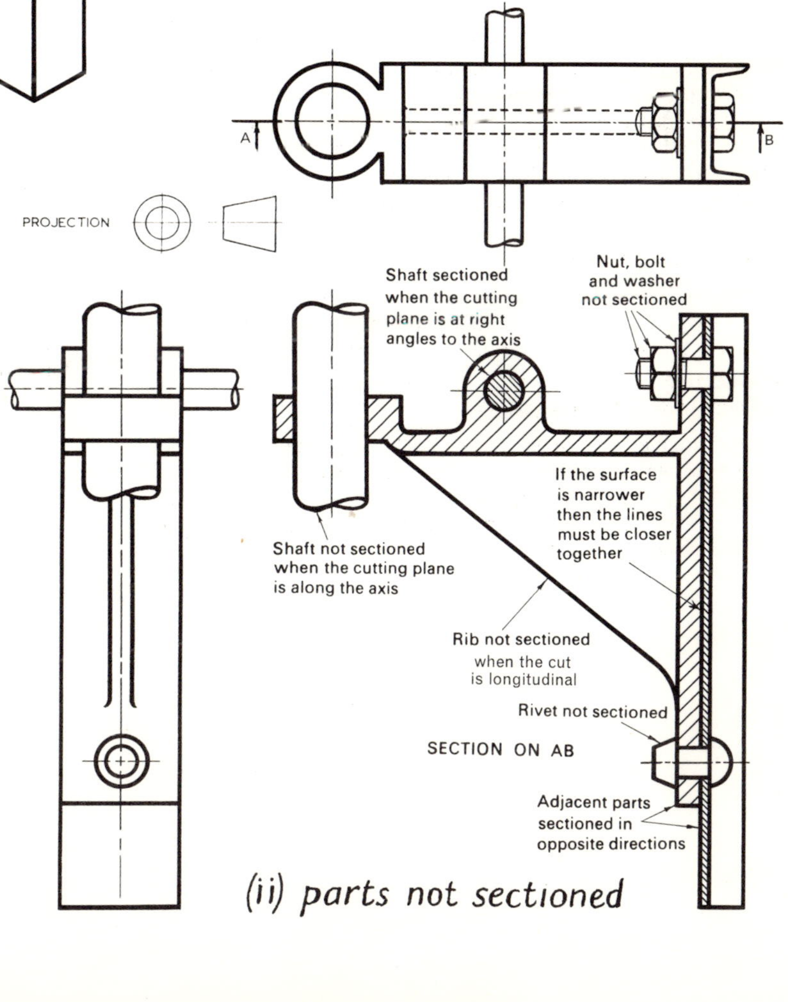

(ii) parts not sectioned

Dimensioning

The insertion of dimensions is essential for the completion of a drawing. Although the drawing may be accurately made to scale, *sizes of distances must not be measured from the drawing.* The dimensions used in manufacture must be those which are actually given in figures.

Rules

The following rules were given in Part I:
Figures should normally be at right angles to the dimension lines and should read either from the bottom or from the right-hand side of the drawing. (Easier reading may result from having all dimensions vertical, and this is the practice of some firms.)
The principal view of an object should carry the main dimensions, which should not generally be repeated. Overall sizes should be given for reference.
Centre lines should not be used as dimension lines. The positions of cylinders should be dimensioned from their centre lines.
Diameters are preferable to radii.
Dimensions should not normally be taken from dotted lines. In general, dimensions should be placed outside tne main outlines.

Position and size dimensions

Most engineering parts consist of combinations of simple geometrical solids: e.g. cubes, prisms, pyramids, cylinders, cones. Each of these can easily be dimensioned for *size*. If we do this first, and if we then link the solids together by dimensions which settle *relative positions*, we shall have dealt logically with the dimensioning of the part The size dimensions alone should be sufficient for us to make each part separately.
Let us do this for the simple bracket shown in fig. 1. It is to carry a shaft and to rest on a horizontal surface with its projecting edge XX against the edge of the support. Before we can insert dimensions for position we have to select surfaces which can be used for reference. Obviously one must be XX, another the lower surface YY; for the third we shall take ZZ.

Size dimensions

These are marked *S* in fig. 2: for the lengths and diameters of the cylinders; for the length, breadth and thickness of the base; and for the thickness of the ribs.

Position dimensions

These are marked *P* in fig. 2. They settle the position of each of the parts in relation to the surfaces we have chosen. The webs are centrally placed and their position has not been fixed exactly.
Note.—Very often the insertion of *P* dimensions makes some of the *S* dimensions unnecessary.

Exercises

1. *Sketch* the two views in fig. 2 but without the letters *S* and *P*. Close the book, and insert size and position dimensions on your sketches. Compare with fig. 2.
2. Measure the sizes of the parts in fig. 2. Draw the two views, twice the size, and draw views in the directions A and B. Dimension your drawings.
3. Insert size and position dimensions on the drawings in fig. 3. See page 156.

PROJECTION

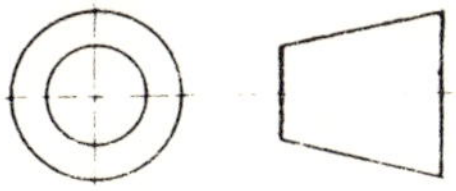

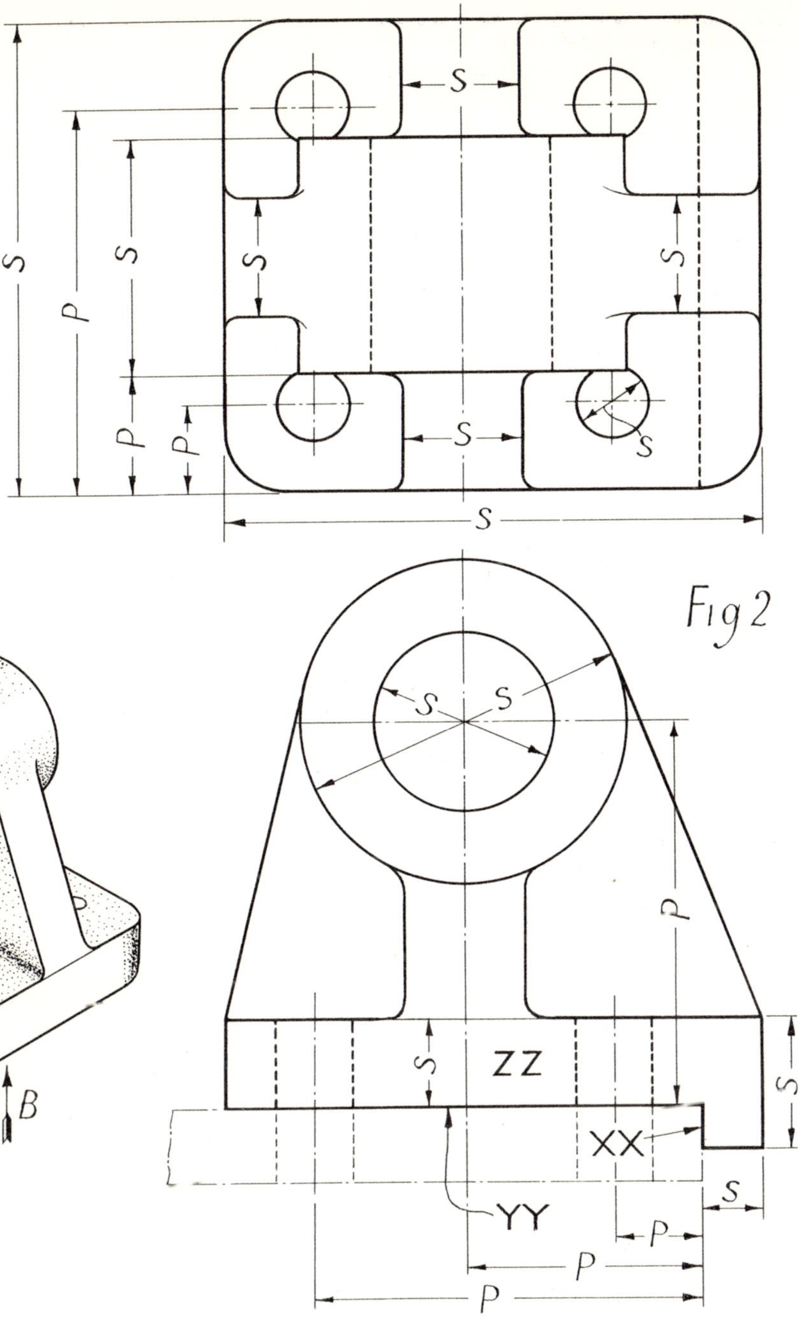

Fig 2

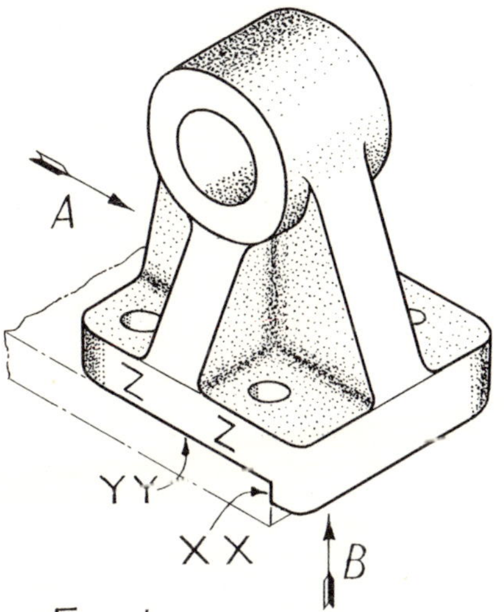

Fig 1

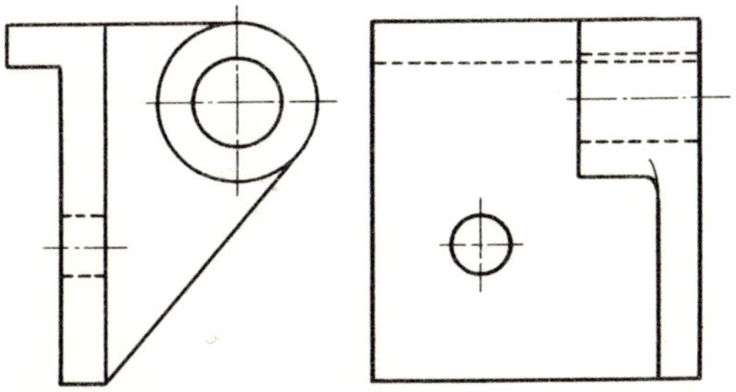

Fig 3

Size and position dimensions (continued)

When an object is to be machined all over, there is often no particular reason for selecting one reference surface rather than another, but many engineering components are not machined all over. For example, articles which are forged or cast may have only selected surfaces machined, and for such objects it is important to avoid using a rough surface for reference.

Consider the simple article shown in fig. 1. This is a casting, and those parts which are to be machined are indicated. The three machined *plane* surfaces will be used as references for the positional dimensions. Edge views of these surfaces are shown by the thick lines in fig. 3. The various *size* dimensions are given in fig. 2 and need no comment. It will be noted that the length of the cylindrical part is not given a size dimension. This is because it is settled by the difference between the *S* and *P* dimensions 'ringed', both of which are important.

Systems of dimensioning

Three ways of dimensioning one view of the object are shown opposite. In fig. 4, I the dimensions are wholly outside; they need not be cramped and may be made prominent. In fig. 4, II the dimensions are wholly within the views; few extension lines are required and the outline of the object is shown without interference. A compromise arrangement is shown in fig. 4, III, and this often combines the advantages of I and II.

Although no definite rule can be given, in general the first system should be adopted. The *shortest* dimension lines should always be arranged *nearest* the outline of the view.

Relative importance of dimensions

The student should realize that a dimension marked 50 mm really means that the size may be a little more or a little less than 50 mm, within the usual limits of manufacture. If this is too loose a definition of the size, then we shall have to say within which extreme limits the actual size must lie. This practice is commonplace in some branches of engineering, and it makes the dimensioning of a drawing a much more difficult task. Some consideration of the difficulties is given on pages 194 and 195.

Exercises (solutions are given on page 206)

1. The object in fig. 5 has three machined surfaces at right angles to each other, with two holes also at right angles to each other. Sketch freehand, about twice the size given, three views and insert size and position dimensions.
2. The machined parts of the bracket in fig. 6 are indicated. The two large holes are parallel. Sketch freehand, about twice the size given, three views of the bracket, and insert size and position dimensions.

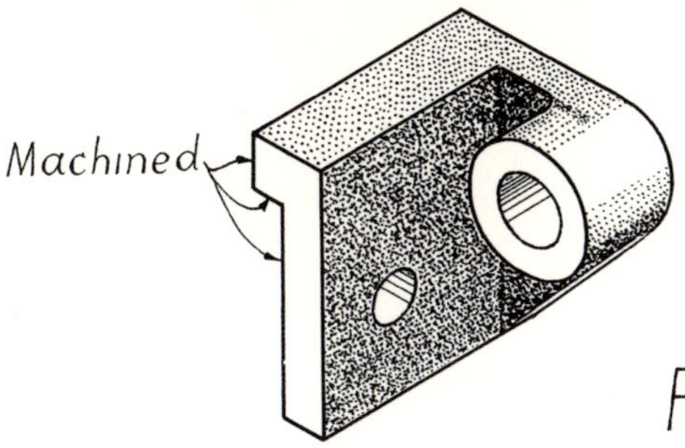

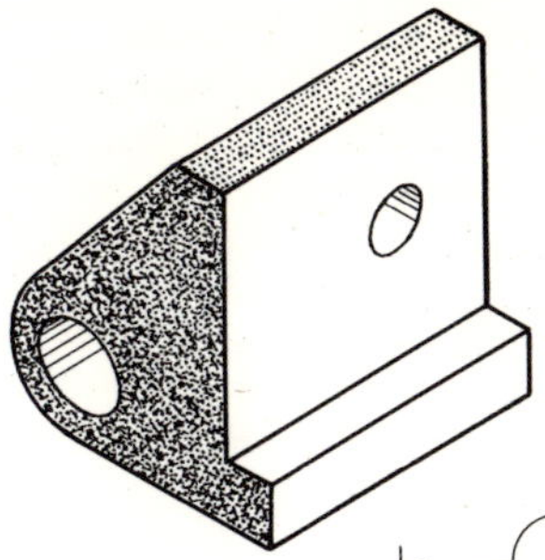

Fig 1

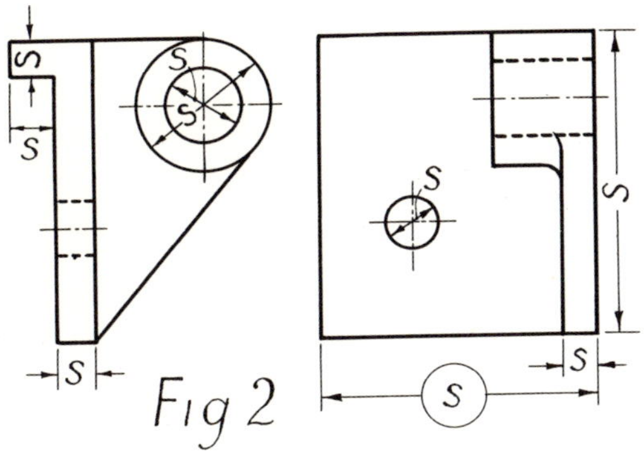

Fig 2

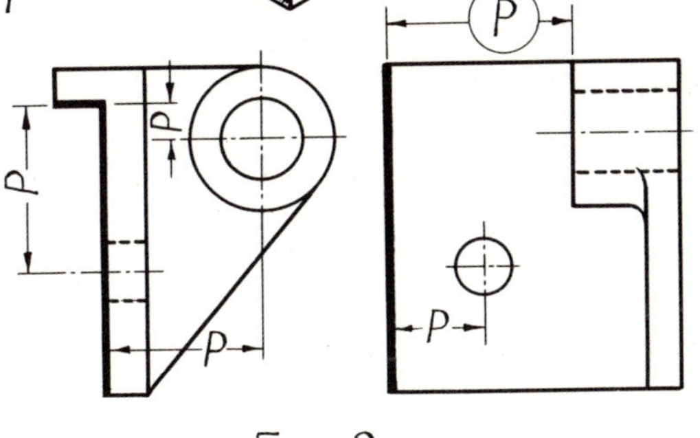

Fig 3

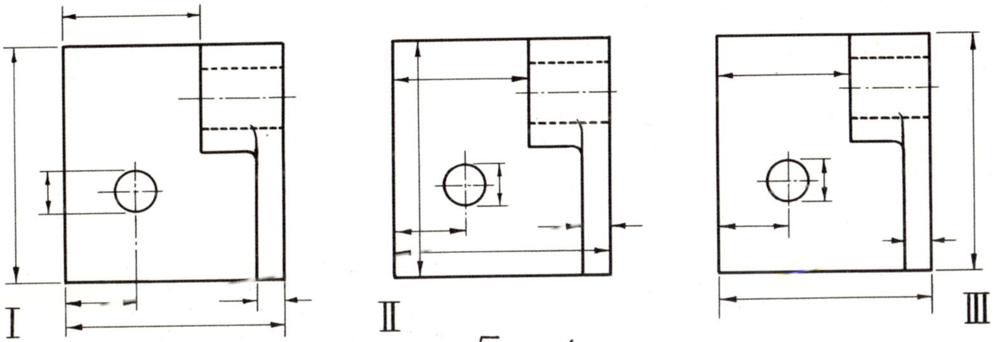

Fig 4

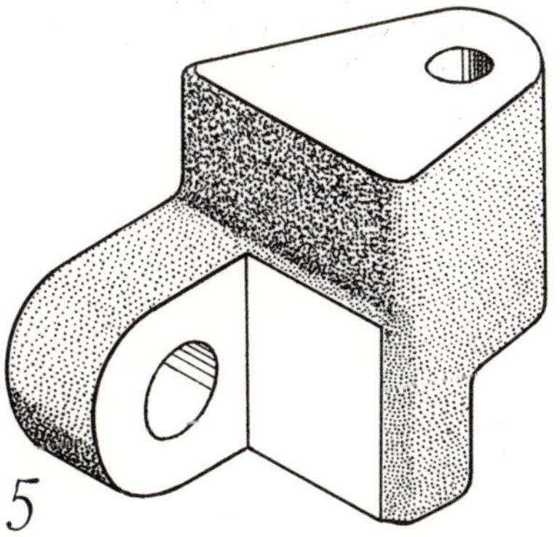

Fig 5

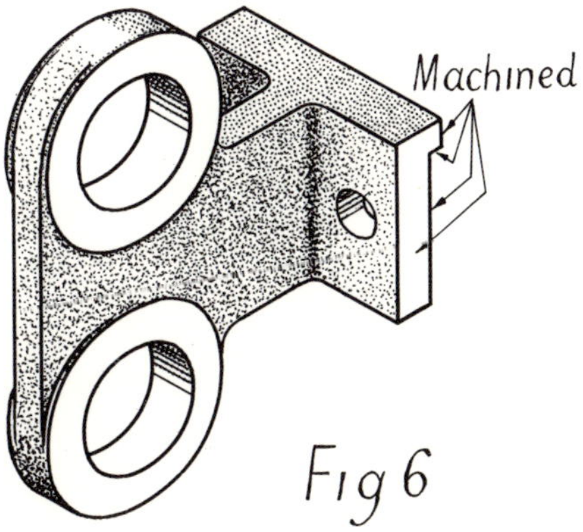

Fig 6

Dimensioning: Special Cases

Restricted space

The example shown in fig. 1 gives an acceptable treatment where many dimensions are involved and space is limited. Sometimes a double arrowhead is used on the unfinished dimension line.
All the dimensions shown are diameters, but it would be tedious to apply ø or Dia to each of the many dimensions. A general note would of course make the matter clear; but if in some other example there is the least doubt about the shape, the abbreviation ø or Dia should be inserted. Fig. 1 could relate to an article in which all parts were *square*, but this would be so unusual as to call for another view.

Remote centres

Centres of arcs of large radius are often too far away from the drawing of the arc to permit their true position to be shown. When this is so, the method given in fig. 2 may be adopted. The centre marked is obviously not the true centre of the arc.

Dimensions from a reference face: Auxiliary dimensions

A grooved drum which is to be very accurately machined is shown in figs. 3 and 4. The drum consists of two cylinders, bolted together, each having a number of narrow grooves cut into the surface. The bolting arrangements are not shown.
The dimensions are given in millimetres, without any indication of the degree of accuracy required. In practice, each dimension would have attached to it another, very small, dimension, called a *tolerance.*
The dimensioning in fig. 3 is neat but not very satisfactory where tolerances are involved, for the reason given on page 156, when the parts are being machined, measurements will be taken from the common reference surface F, and it would be much more convenient to the craftsman to have dimensions given from this surface, as in fig. 4.
There is another and more important reason for dimensioning as in fig. 4 rather than in fig. 3, which will be understood after page 194, dealing with tolerances, has been read. For the same reason, overall dimensions, which ought not in fact to be given when tolerances are involved, must be regarded as provided only to give useful information (for example, when ordering material). These redundant but useful dimensions are called *auxiliary* or *reference* dimensions, and they are enclosed in a bracket or a rectangle.

Exercises

1. The diameters in fig. 1 taken from top to bottom are, in millimetres: 180, 160, 60, 40, 80, 110, 220, 270. There are four equidistant holes, 20 dia. The overall thickness is 80. The recess depths are 25 and 12, and the central part is 60 long. Make a drawing of the part, showing the view given but with only half in section, and add a view from the top. Insert dimensions, using *both* views, and reduce as far as possible the number of uncompleted dimension lines.
2. Draw fig. 4, to a scale of $\frac{1}{5}$, and complete the drawing by giving an outside view of the lower half. Insert all dimensions, which are given in millimetres, regarding this as a test for neatness and method.

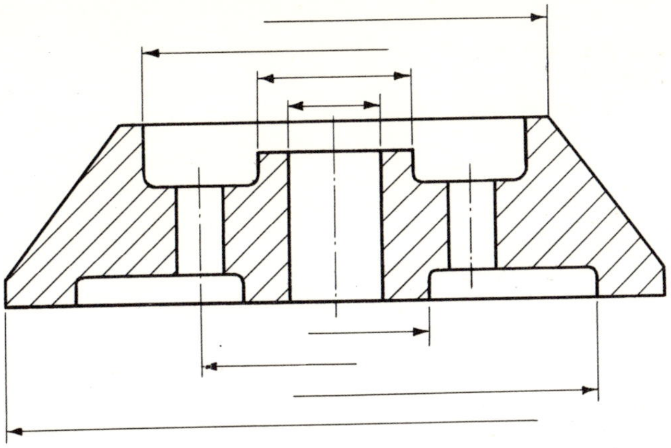

Fig 1

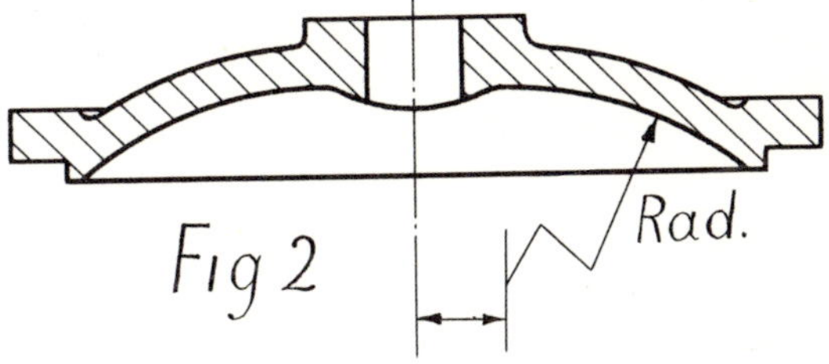

Fig 2

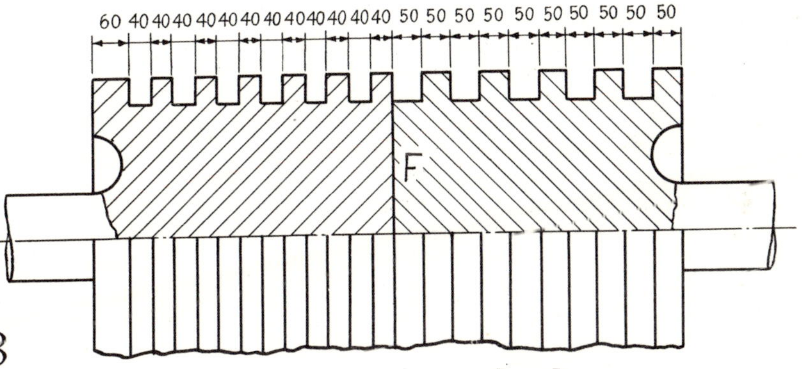

Fig 3

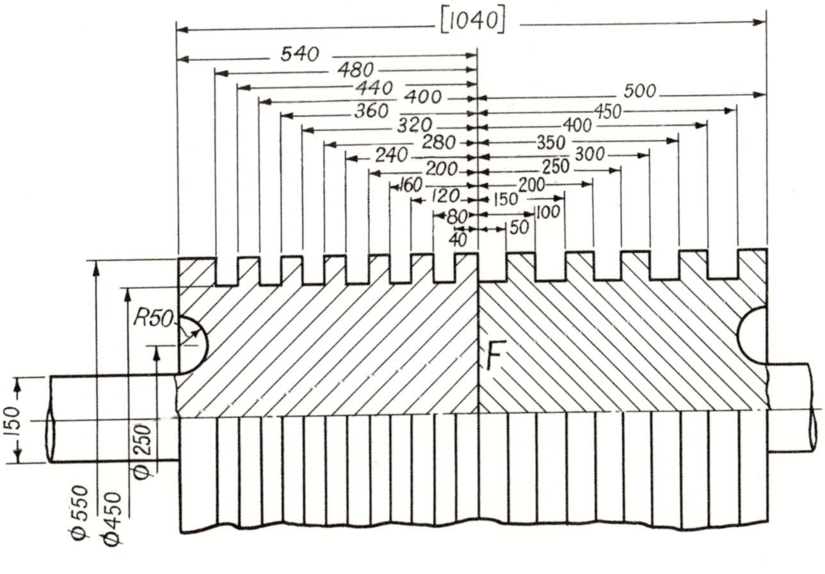

Fig 4

Abbreviations, Machining Symbols

Abbreviations

A large number of abbreviations are used on drawings, and for a full list of these the student should refer to B.S. 308. The following selection includes those most likely to be used in the classroom. All abbreviations must be printed in capital letters.

Centre Line	℄	Minimum	MIN
Chamfered	CHAM	Pitch Circle Diameter	PCD
Cylinder	CYL	Radius (before a dimension)	R
Diameter (in notes)	DIA	Screwed	SCR
Diameter(before dimension)	Ø	Square (in notes)	SQ
Hexagon	HEX	Square (before dimension)	□
Maximum	MAX		

Machining symbols

It will be clear from what has been said about dimensioning, and notably from the drawings on pages 155 and 157, that it is essential to mark on drawings the surfaces which are to be machined. These are the surfaces to which reference is made for Position Dimensions, rather than to unmachined surfaces.

The symbol formerly used in this country was the letter *f*, meaning *finish*, applied as in fig. 1. In this example it is clear that the part is to be machined only on the top, bottom and bore.

The *f* has now been replaced by the tick, applied as in fig. 2. The angles of the tick are 60° and the point must touch the line representing the surface. Note the positions for the ticks at the bottom and for the bore. The presence of the tick means that some material must be allowed on the design size for machining. A well-established practice on the Continent of Europe is to use small equilateral triangles as in fig. 3.

A more complete illustration of the application of the machine symbol is shown in fig. 4. The finish of a surface is often an extremely important factor where loads and speeds are high. Surfaces are now described by numbers which clearly indicate their quality, the numbers being added to the tick or triangle in important work. This matter is however outside the scope of this book.

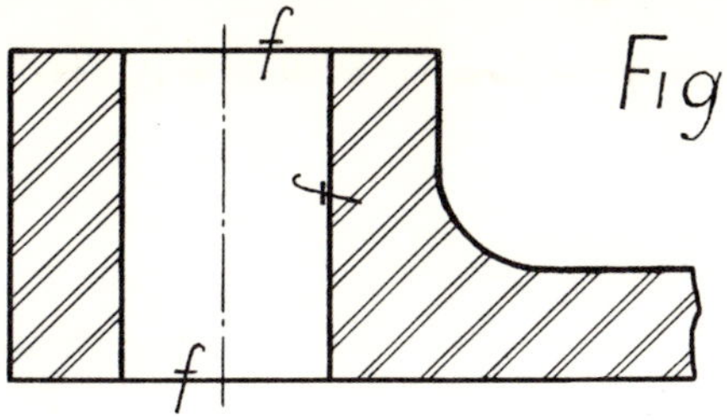

Fig 1

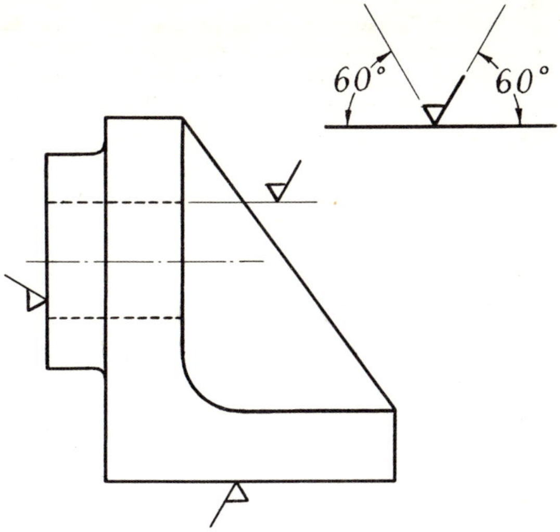

Fig 2

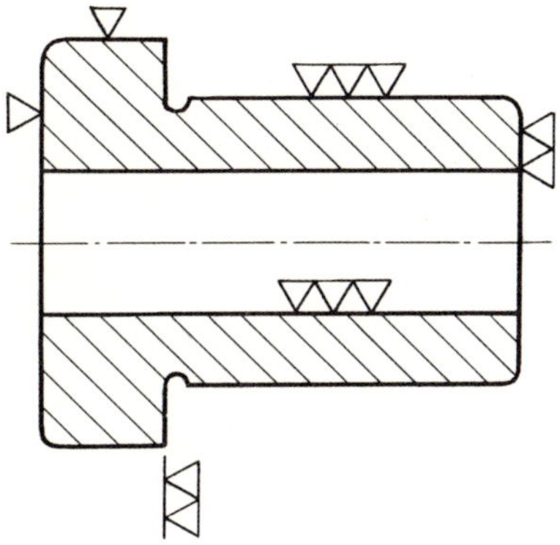

Fig 3

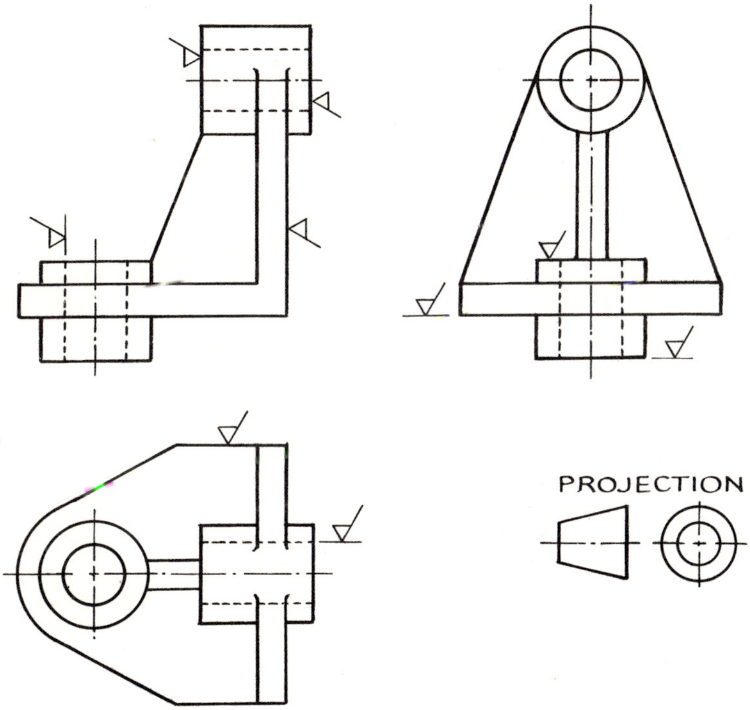

Fig 4

Dimensioned Drawing for Reference

The dimensioned drawings opposite show recommended line thicknesses. The object, shown pictorially below, consists of five parts. When assembled, the end of the rocking arm lies between the bushes and is secured to the shaft when this is inserted. The arm and the shaft are to move only through 90°.

Line thicknesses

Outlines, bold; *extension and dimension lines*, thin and continuous; *centre and 'ghost' lines*, thin chain; *section line* (AB), thin chain, thickened at the ends.

Dimensions

Because the casting is to be machined only on the base, the two stops and the bores, *position* dimensions (underlined) are taken from these and from centre lines. Less-important dimensions have been omitted.

Exercises

1. (*a*) Draw full size the two views given, without dimensions, using lines of the right kind. (*b*) Insert dimensions, omitting underlining. Without referring to the book, mark all *position* dimensions. Check. (*c*) Mark with a tick other dimensions assumed to be covered by those given. (e.g. All four holes are covered by the dimensions for one.) (*d*) Mark with a dot other *minor* dimensions. Do not repeat unnecessarily.
2. *More advanced.*
   Draw a section through AB with the rocking arm in position and against the lower stop. Propose a way of (*a*) securing the arm to the shaft; (*b*) securing the two bushes.

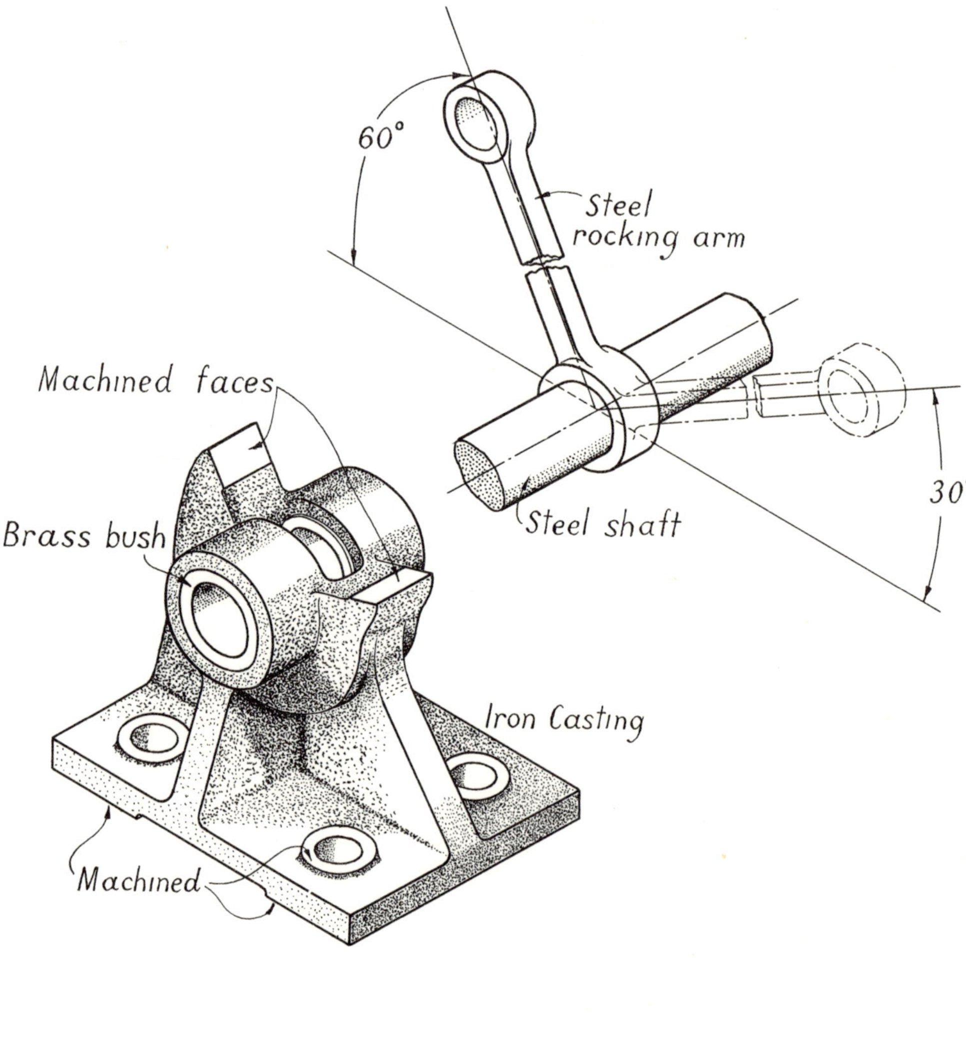

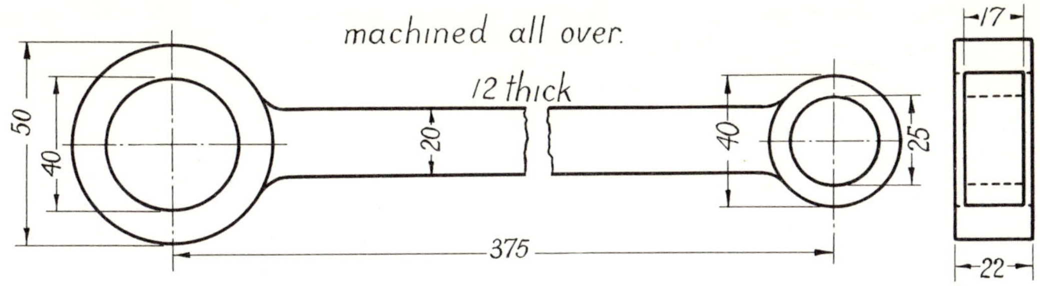
machined all over.
12 thick
50
40
20
40
25
375
17
22

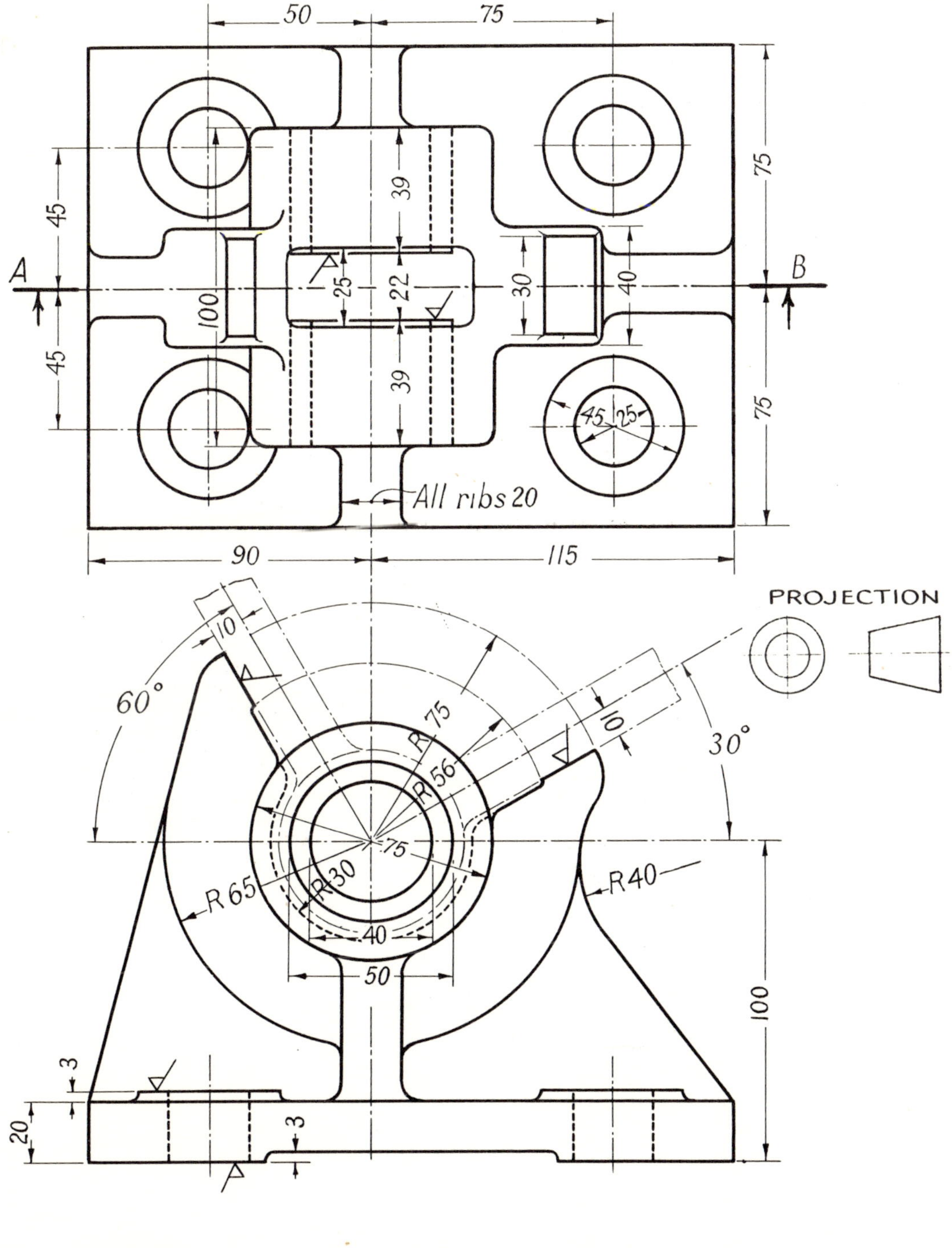
50
75
75
75
45
45
A
B
100
25
22
30
40
39
39
45
25
All ribs 20
90
115
PROJECTION
10
60°
R 75
R 56
10
30°
R 65
R 30
75
R40
40
50
100
3
20
3

Rivets and Riveted Joints

Rivets

Rivets are used to connect together permanently two or more pieces of metal; if the joint is to be a temporary one, bolts and nuts, or screws, are used. The rivet is a cylinder with a head forged at one end. It is inserted in a hole drilled through the parts which are to be connected and projects sufficiently for a head to be formed at the other end. This is done either by hammering by hand or by pressing with a machine—the rivet being either hot or cold when inserted

Rivet heads

These have been standardized and only three of the many forms available are shown opposite, fig. 1: I, Snap; II, Pan; III, Countersunk. The proportions given are very simple to apply on a drawing. The countersunk head is forced into a conical hole and is used when a smooth, or flush, surface is required.

Lap and butt joints

In Lap Riveting, fig. 2, one plate overlaps another and the rivets pass through both plates. In Butt Riveting, fig. 3, the plates are kept in alignment and a cover plate or strap is placed over the joint between them and riveted to both plates; frequently two straps are used, placed one on each side of the plates, as in fig. 4.

Single-riveted lap joint

This is shown in fig. 2. There is only a single row of rivets. The distance between the centres of the rivets is called the *pitch*, which must be at least twice the rivet diameter. We may calculate a suitable rivet diameter from the formula $d = 1{\cdot}2\sqrt{t}$, where t is the thickness of the plate.

Single- and double-riveted butt joint

The joint in fig. 3 is a Single-riveted Butt Joint with a single strap. The joint in fig. 4 is a Double-riveted Butt Joint with two straps; notice that the rivets in fig. 4 are placed zig-zag, or are staggered. Butt Joints are sometimes Triple Riveted (not shown here), then having three rows of rivets in each plate.

The *margin* should not be less than the rivet diameter.

Exercises

1. Two plates 10 mm thick overlap a distance of 70 mm. The plates are secured by snap-headed rivets 20 mm diameter having a pitch of 50 mm. Make a drawing of the joint as in fig. 2 and insert dimensions.
2. Make drawings (arranged as in fig. 2) of a single-riveted *butt* joint for plates 20 mm thick as shown in fig. 3. Take rivets 30 mm diameter and a pitch of 80 mm. The single strap is 25 mm thick and the margin is 30 mm. Show pan heads.
3. Make drawings of a *double*-riveted butt joint as in fig. 4. Plates 25 mm thick, two cover plates each 20 mm thick, rivets 30 mm diameter snap head, pitch 80 mm, distance between rows 75 mm, margin 33 mm.

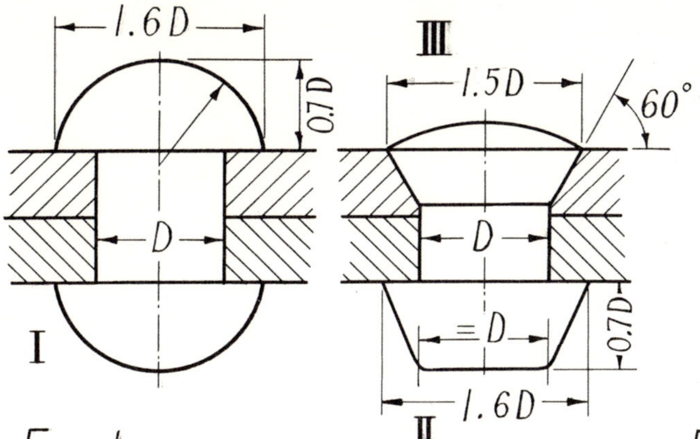

Fig 1

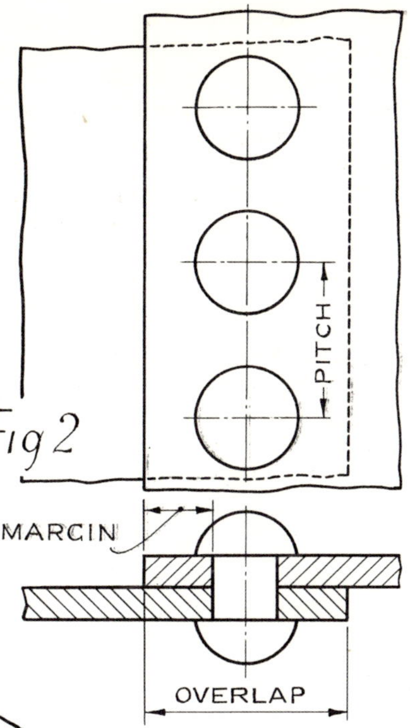

Fig 2

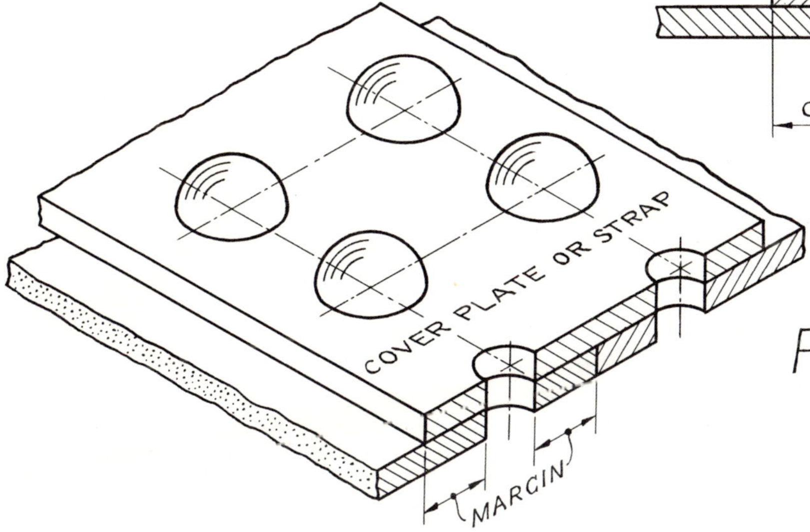

Fig 3

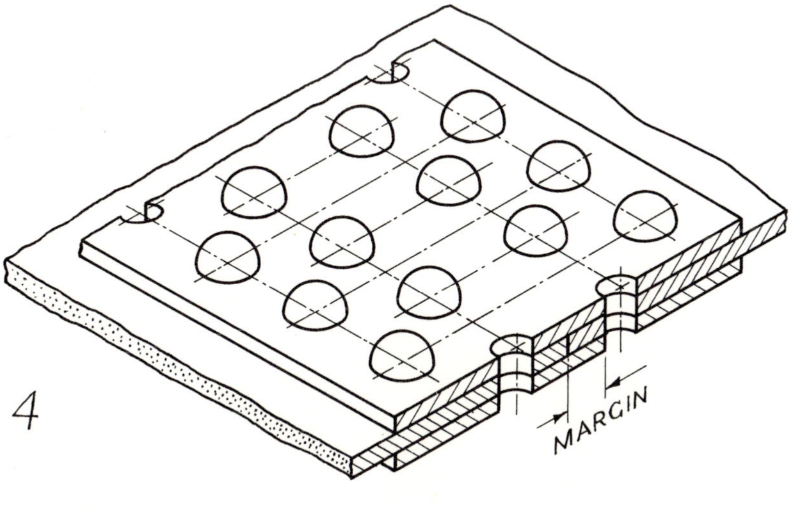

Fig 4

Screw Threads and Screws

The use of screws as fastenings is very old, and such things as bolts and nuts, if not familiar to the student, should be examined by him. We have to consider here, how to show screw threads on a drawing in such a way as to save time and yet convey the desired information.

ISO metric thread

This thread form is shown at the top right of the diagram, along with the Whitworth and U.S. standard threads.

Conventions

The screwed part in fig. 1 has a diameter of 42 mm and pitch 4.5 mm. Strictly speaking, each crest and root should be shown by a helix, drawn as on page 101. This would be a long job and it is never done in practice. In fig. 1 the helices are represented by straight lines, inclined to the axis, but even this method is rarely adopted in the Drawing Office. Fig. 2 shows a simplification which has been and still is widely used. British Standard 308 permits either of the presentations shown in fig. 3 and fig. 4. The vees in fig. 3 are to be drawn freehand, roughly to scale; the parallel lines in fig. 4 are drawn of course to represent the major and minor diameters.
The method of fig. 4 is shown applied to a *stud* in fig. 5. The stud is threaded at each end and is screwed into a part A, remaining fixed to it. Parts B and C fit over the stud and are secured by a nut screwed on the stud. Note that the hole in A into which the stud is screwed is shown a little deeper than is in fact necessary. Fig. 5 would be well understood by any engineer and no further lines are required on the drawing. (See also fig. 6 on page 169.)

Right- and left-hand threads

If the drawing of a screwed part is to be shown as in fig. 1 or fig. 2 we must incline the thread lines in the right direction. That shown opposite is correct for a right-handed thread; for a left-handed thread, the lines must slope downwards from right to left.

Exercises

See page 168

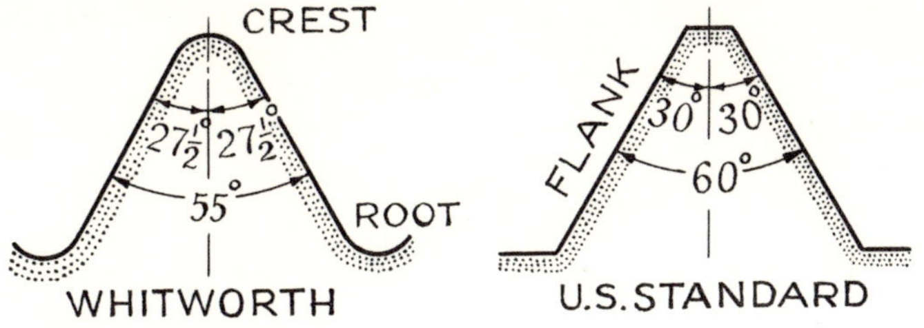

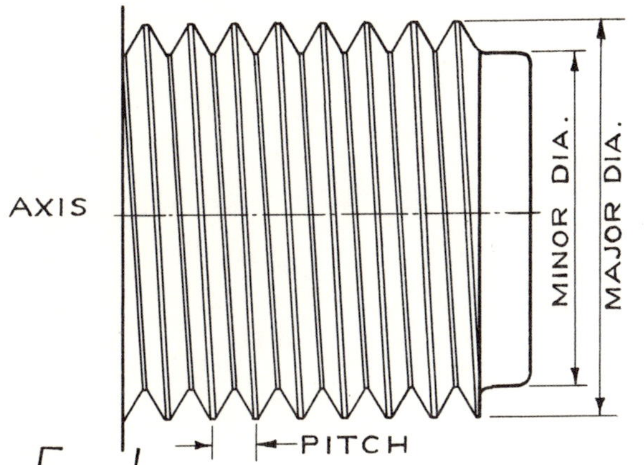

Fig 1

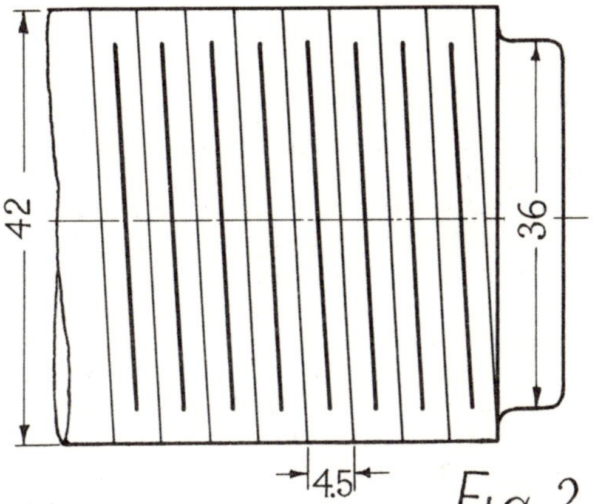

Fig 2

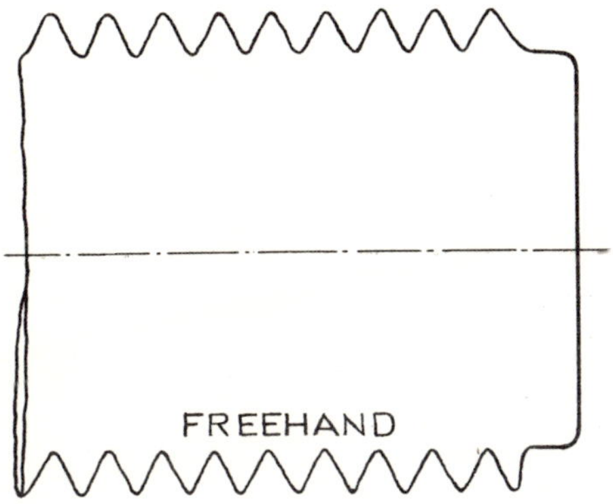

Fig 3

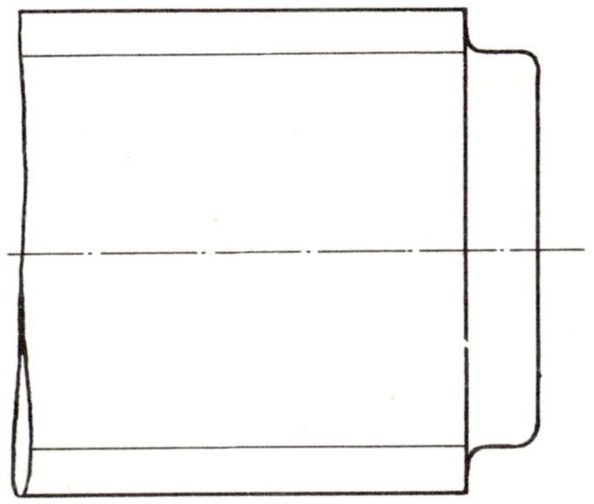

Fig 4

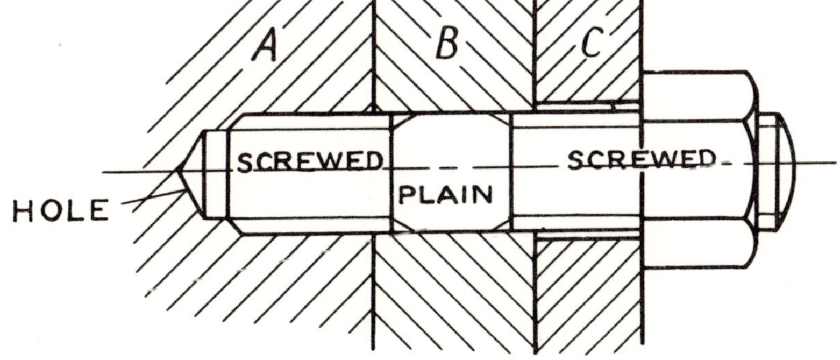

Fig 5

Nuts and Bolt Heads

Nuts and Bolt Heads are usually made in the form of hexagonal or square prisms with the corners of one end rounded off. For bolts of the same diameter, the width across the flats is the same for the square as for the hexagonal nuts.

Standard proportions

The sizes of nuts and bolt heads have been standardized and they are such that they cannot all be given exactly in terms of the diameter.
Thickness of nut is approximately $0.8 \times$ diameter of bolt.
It will be noted that the thickness of the Bolt Head is very nearly $0.6d$.

Drawing nuts and bolt heads

It is necessary to use simple proportions when drawing nuts and bolt heads in order to save time, but the proportions used should not make the article look too small or too large. For example if we took C equal to $2 \times d$ we should be nearly right when d was 5 or 15 mm but much too large when d was 24 mm and upwards. We must therefore take into account the size of work. In the drawings opposite, C has been made $1.75d$.
The corners of the head and nut are *chamfered* at an angle of 30° to the end. As will be clear from the conic sections on page 120, the curves of intersection between the flat sides and the conical surface are hyperbolas, but we shall show them as circular arcs. The positions of the centres of these arcs are given by the simple construction shown in figs. 2 and 3, using the radius $0.5A$—which must be taken from a hexagon. Note the difference between figs. 2 and 3.

Bolts, tap bolts and studs

The tap bolt, fig. 5, is used when we cannot have the bolt head as in fig. 4. The stud, fig. 6, is not withdrawn when the nut is removed so that the thread in the hole is preserved. Three ways of showing screw heads are shown in figs. 4, 5 and 6. The generally accepted methods are as shown in figs. 5 and 6. The tapped hole is shown in section in fig. 6. Notice that here the section lines go across the threads, but not when a stud or set bolt is screwed into the hole.

Exercises

1. Draw three views of a hexagonal bolt and nut, diameter 30 mm, length of shank 130, length of screw 60. Show the nut half-way along the thread. Use the method of fig. 2 on page 167.
2. A stud 36 mm diameter passes through a flange 35 mm thick and is screwed into a tapped hole 42 mm deep, as in fig. 6. The flange is held by a hexagonal nut bearing on a washer, 3 mm thick. Show the arrangement in section, using for the threads the method of fig. 3, page 167.

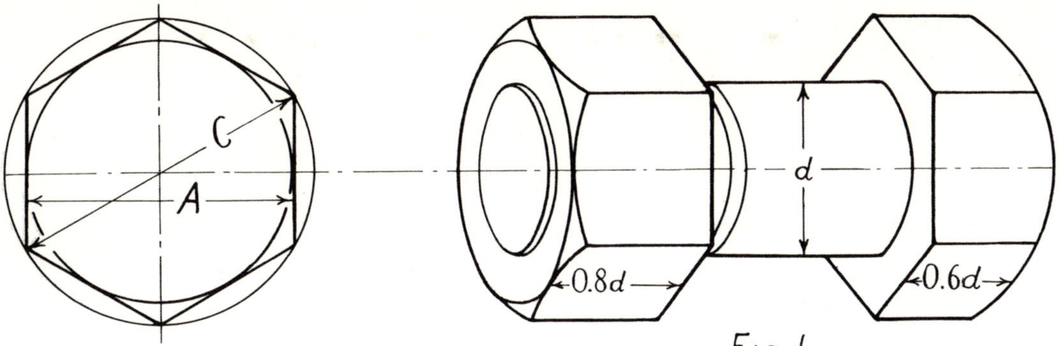

Fig 1

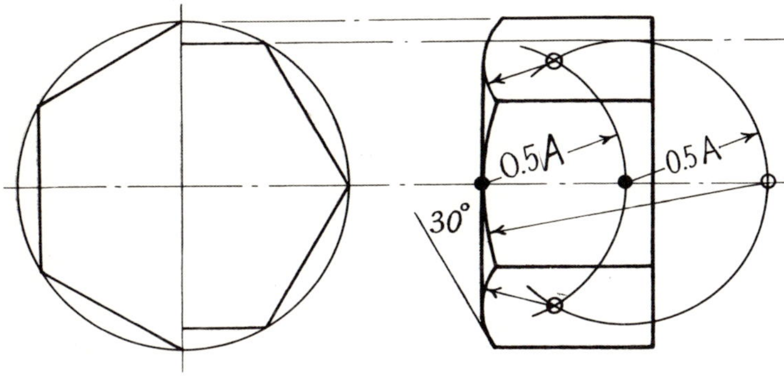

Fig 2

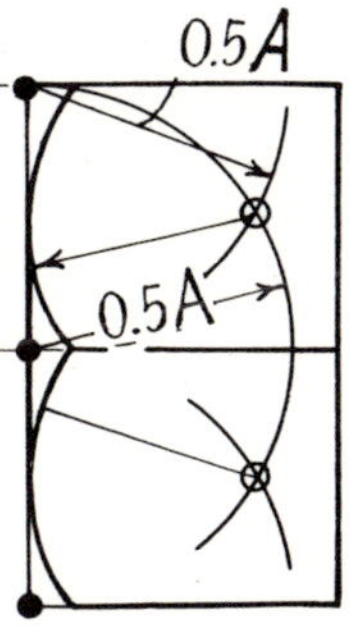

Fig 3

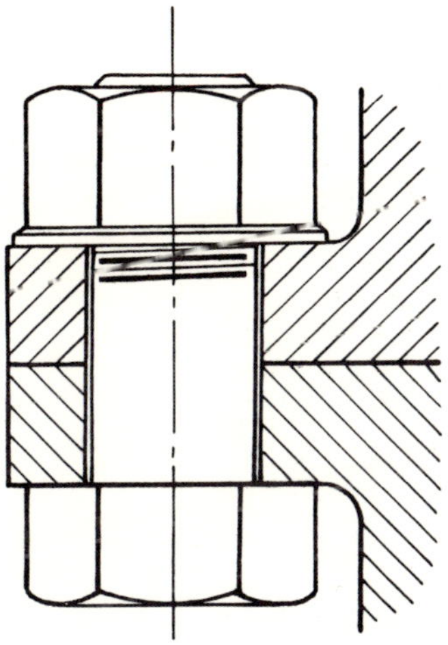

Fig 4

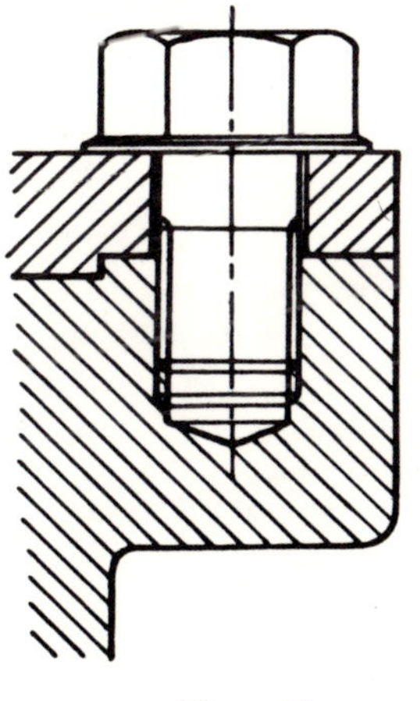

Fig 5

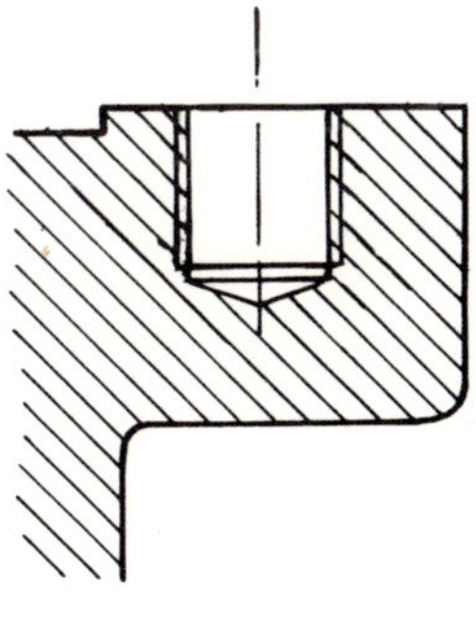

Fig 6

Screw heads

The forms and proportions of screw heads shown in fig. 1, I to V, represent British standards for screws from 3 to 30 mm diameter. The thread for the countersunk-headed screw is taken to the head; for the round and cheese-headed screws it stops at a distance from the head not exceeding twice the pitch of the thread. The dimensions of the slots in the heads have been omitted; the student may use his own judgment as to these sizes.

Set screws

Set screws are used chiefly to prevent relative movement between two machine parts. They are invariably made of steel and are usually case-hardened. In the standard form, fig. 1, VI, the heads are square and the points are flat, both being chamfered. Headless screws, or *grub screws*, are used on moving parts where a projecting head would be dangerous. In the example shown in fig. 1, VII, a special wrench fits into a hexagonal socket. The point of this grub screw is recessed to form an annular bearing surface.

Locking device for nuts

When a bolt and nut which carry a varying axial load are subjected to vibration, there is a tendency for the nut to work loose. For example, the bearing for a connecting rod for the engine of a motor cycle or car is secured by bolts which are pulled repeatedly and vibrated continuously. The consequences of any slackening back would be serious, and many devices are used to keep such nuts in position on the bolt. Only one or two can be considered here.

Castle and slotted nuts

These are shown in figs. 2 and 3 and provide a positive lock. The bolt end is drilled to take a split pin, and the nut is tightened and then slightly turned so that the pin will pass through two opposite slots. The proportions of these nuts have been standardized but *for drawing purposes*, and for bolt sizes not far removed from 25 mm, the following ratios may be used:

d	A	L and B	N and M	K
1	1.5	1.5	0.6	0.8

Collar nut

For large nuts the arrangement shown in fig. 4 is commonly used. The lower part of the nut is made circular and is grooved. It fits into a recess, and a set screw bears on the bottom of the groove and prevents any rotation. A split pin, fig. 5, is also used, in case the set screw slackens, but the split pin will not prevent *slight* motion, although it will stop the nut from slacking back. Probably the most frequently used device for preventing bolts, nuts, set-bolts, etc., from working slack is the thin nut (lock nut) shown in fig. 6.

Exercises

1. A bolt M24 dia. passes through two plates each 15 mm thick. The bolt has a castle nut, of the type shown in fig. 2. Make a drawing of the arrangement, including a view on the nut, showing a split pin 5 mm diameter in position.
2. Draw the arrangements in fig. 4 for a bolt 80 mm diameter. Take width across hexagon = 120 mm, overall thickness of nut 80 mm, diameter of collar 118 mm, thickness of collar 25 mm, diameter of recess 115 mm, diameter of set screw 12 mm, diameter of split pin 10 mm. Use your own judgment for other dimensions.

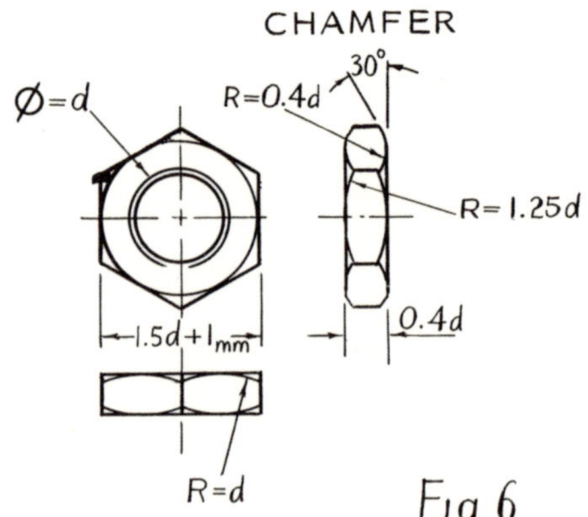

Fig 6

Fig 1

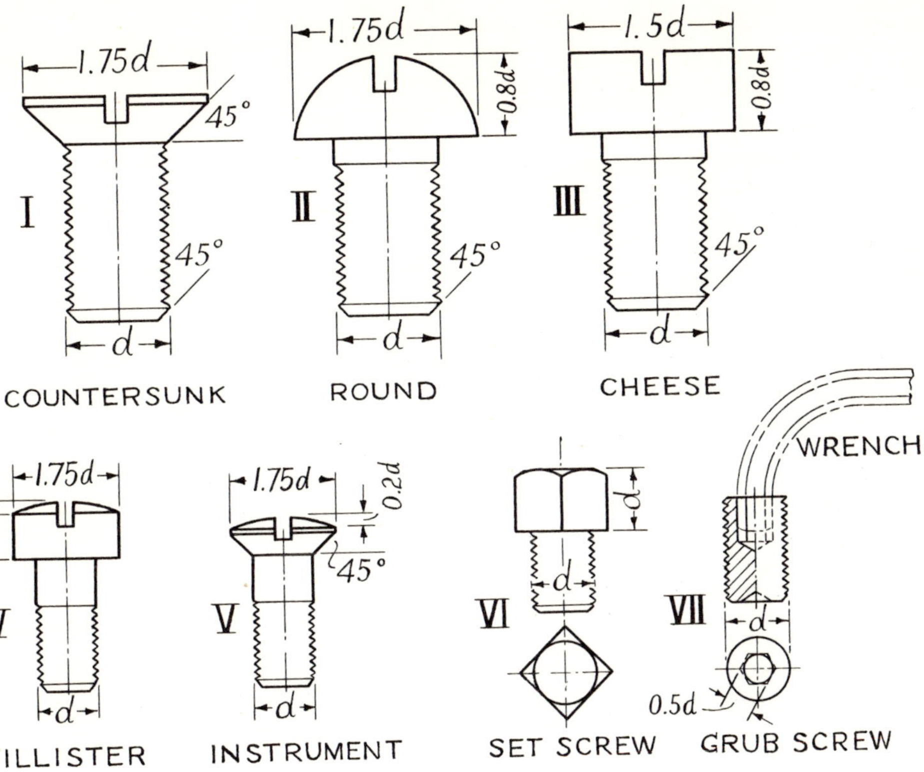

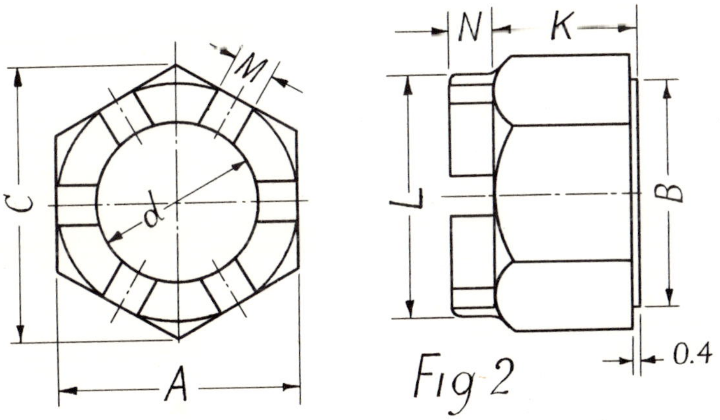

Fig 2

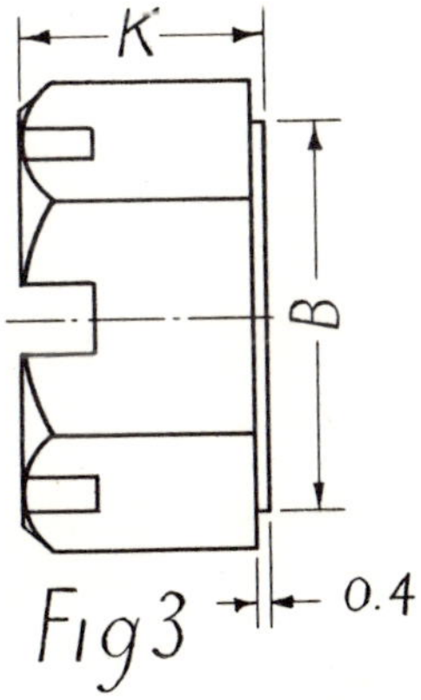

Fig 3

Fig 4

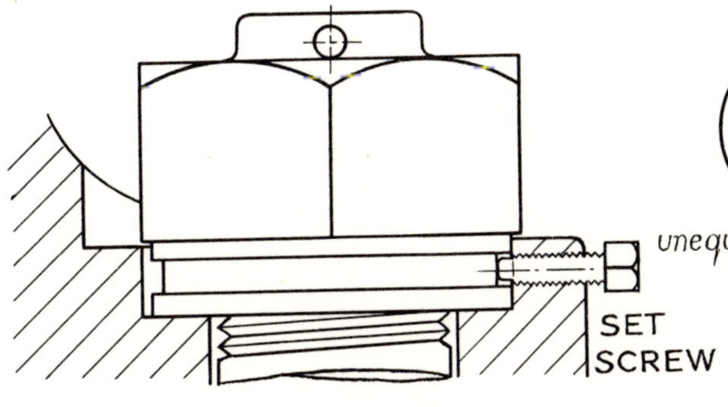

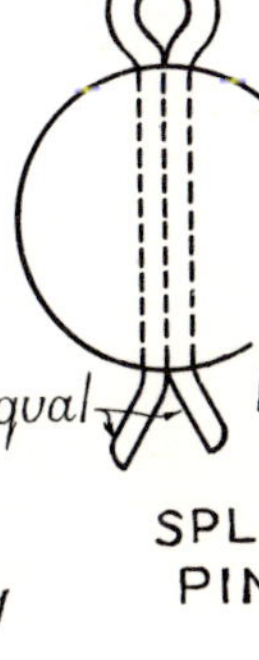

Fig 5

Freehand Sketching

The importance of being able to make freehand sketches was stressed in Part I and it is emphasized again here. The art can be acquired by practice, and the student should make sketches of engineering components and arrangements at every opportunity. Pictorial views are much more informative than the normal views, as will be clear if the drawing of the vice opposite is compared with those on pages 174 and 175.

Use of squared or ruled paper

Although the use of squared paper, or of paper ruled with lines parallel to the three isometric axes, is a great aid to freehand sketching, the temptation is then to make the drawing accurately to scale; and this is not required. All that is necessary is a sketch in which the parts are in fair proportion with each other, and are sufficiently dimensioned.

Kinds of sketch

There are two main kinds of freehand sketch: (*a*) there is the sketch in which a designer 'feels' towards the expression of an idea in his mind and which has yet to take shape; and (*b*) there is the sketch made of an existing article for the purpose of record or manufacture. Type (*a*) is shown in fig. 1, and some of the best examples of this kind of sketch are to be found in the note-books of Leonardo da Vinci—an example has been given in the Frontispiece. Type (*b*) is exemplified in fig. 2, which shows a small vice—fully described on pages 174 and 175.

Procedure

It is usually best to draw the centre lines of any cylindrical parts or to sketch their outlines. If an isometric or pictorial sketch is intended, then circles will usually appear as ellipses. It will be recalled from the work in Parts I and II that the major axis of the ellipse will be either horizontal, or sloping at 60° to the horizontal. The drawing of the major axis in its approximately correct position provides an excellent guide and helps towards the production of a drawing which looks right.

Dimensioning

The importance of including all necessary dimensions cannot be stressed too much. The student should imagine, when he is measuring and sketching an object, that it will not be possible for him to return to it in order to obtain forgotten dimensions—as is so often the case. He will then check and recheck his dimensioned sketch to ensure that the article can in fact be made from the information he has provided.

Exercise

1. Sketch, freehand, the vice shown in fig. 2 and dimension your drawing, taking all sizes from pages 174 and 175. The sketch as shown will not permit the vice to be fully described; for example, the lengths of the screw and the nut could not be given. Add any further pictorial views, or show any parts of fig. 2 in section, so that all necessary dimensions may be inserted.

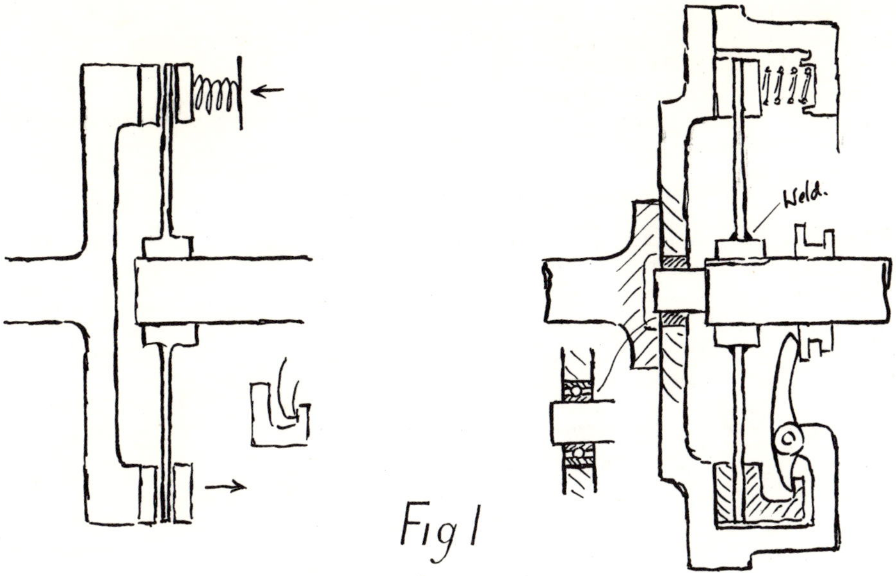

Fig 1

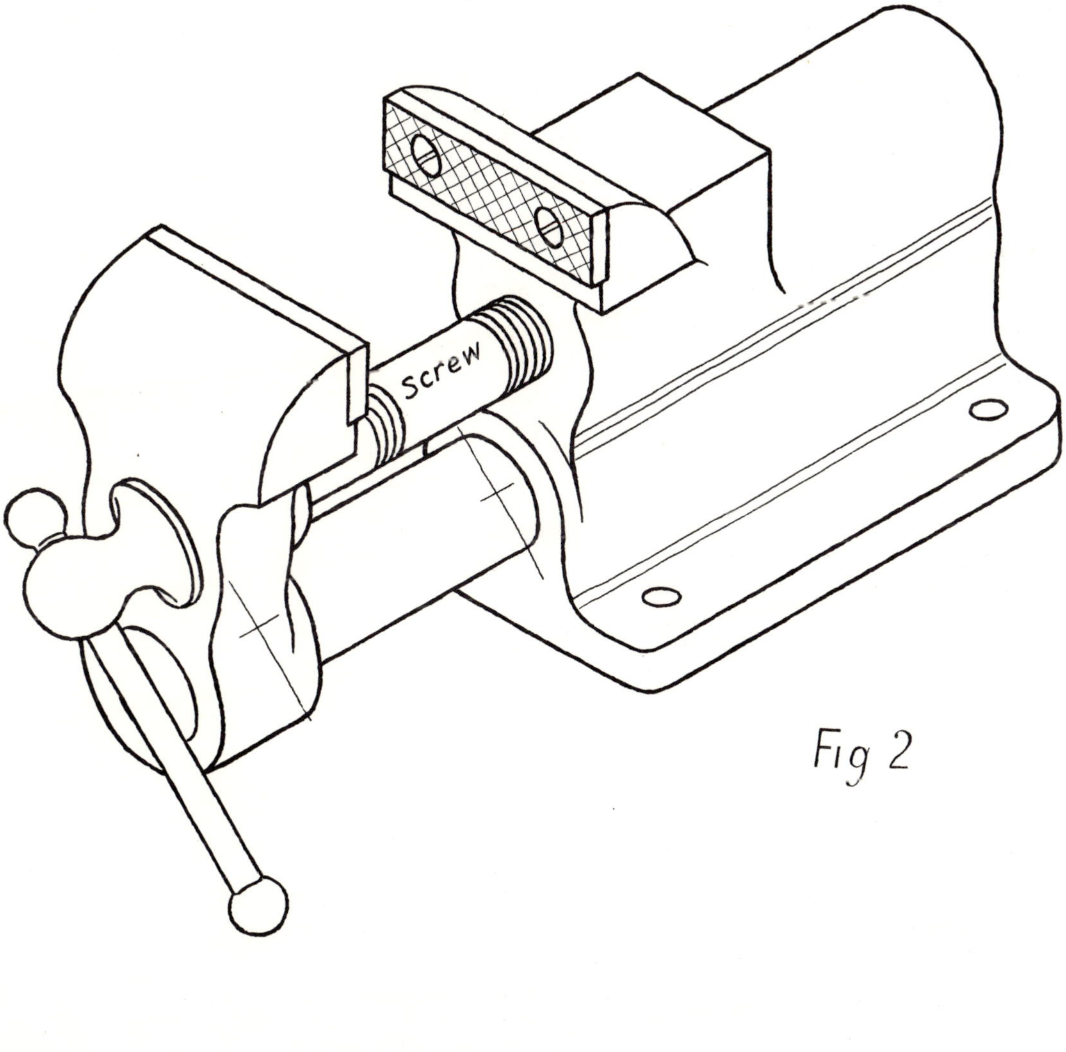

Fig 2

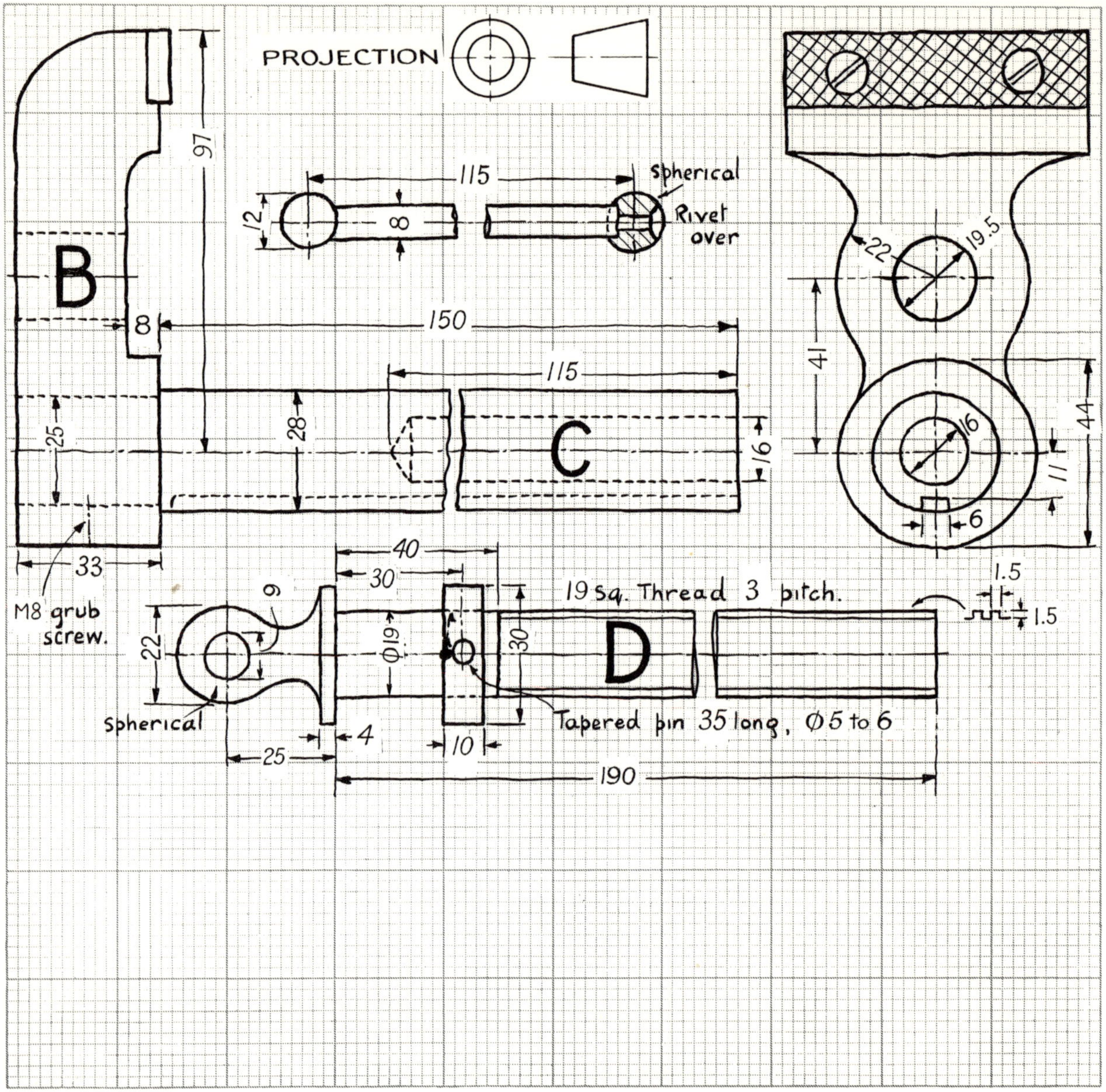

Sketching on squared paper

The sketches give details of the engineers' vice* shown pictorially on p. 173.

The main parts A and B are castings which are machined only where necessary. The actual jaws are steel strips secured by 6 mm diameter screws. The working face of each strip is chisel-marked to provide a grip.

The thread on the steel screw D is square. The steel guide C slides freely in the bore in A, but is a press fit in B, where it is fixed by a grub screw. The groove on the underside of C slides over a small peg key. The screw D revolves freely in B and is located by a collar secured by a tapered pin. The screw is turned by a steel handle bar with ends riveted on.

*** The vice was made by students in a Technical Secondary School; it is sound in design and provides a good workshop exercise.**

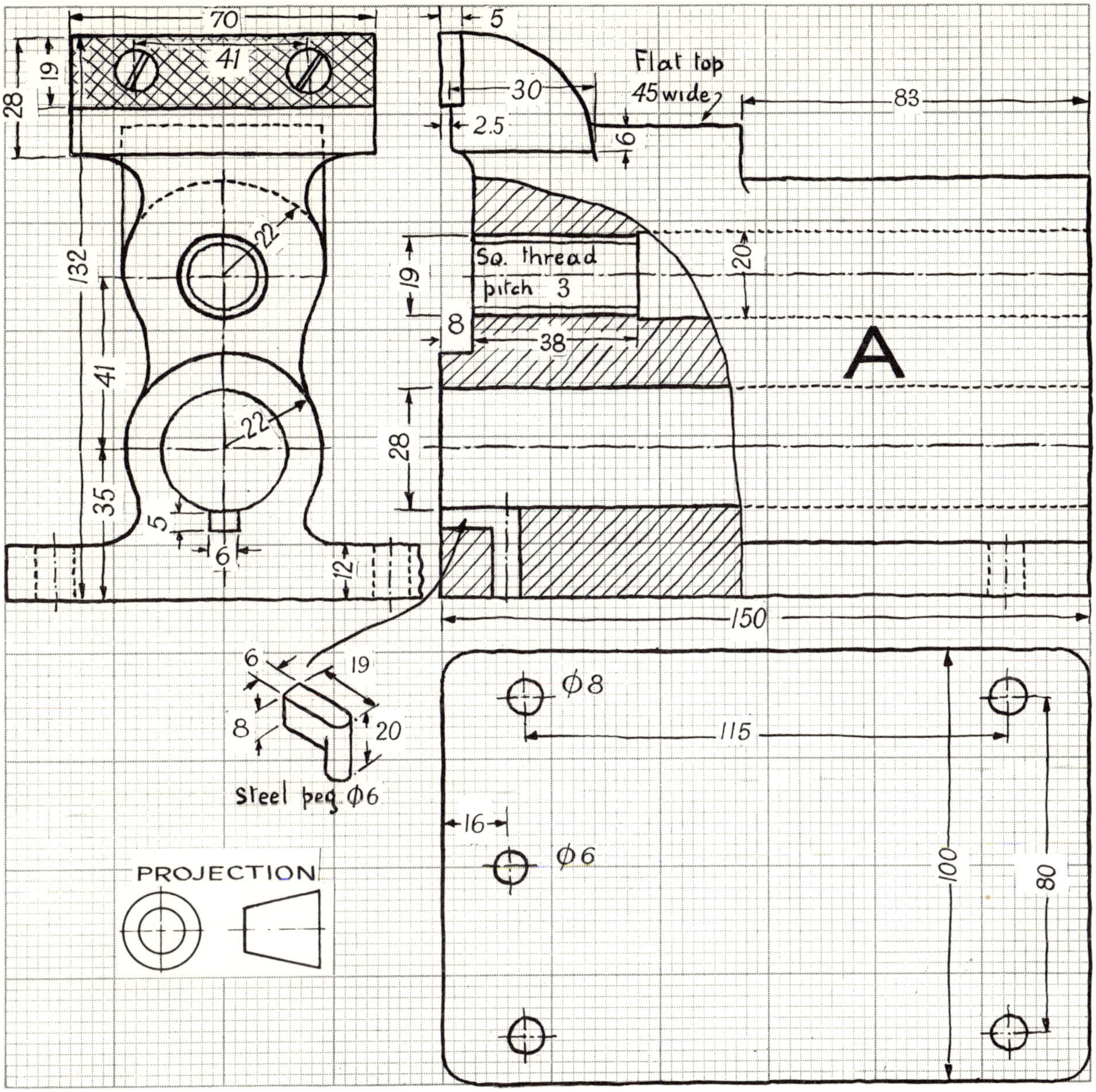

Exercises

1. Draw, full size, three views (front, side and top) of the completely assembled vice, giving any necessary sections. Insert only the main dimensions.
2. Make measured drawings freehand on squared paper of other items of workshop equipment.

Rolled-steel sections are largely used in structures of all kinds and they must be given brief consideration here. The sizes of the sections have been standardized and three of the commoner types are shown opposite: the Angle Bar, fig. 1, is extremely useful in combination with plates and other sections (see below); the Channel Bar, fig. 2 and the Joist, fig. 3, have flanges which taper in thickness.

Built-up beam and column base

A typical built-up Beam or Girder is shown in fig. 6. *Stiffeners* (dotted) are used to prevent the plate web from buckling. The stiffeners are angle sections with the ends 'joggled' to fit the beam angles; they are placed at intervals on *both sides* of the web. The *Column Base,* fig. 7, is built up from channels, angles and plates. It distributes the column load over a greater area.

Exercises (dimensions are in millimetres)

1. The arrangement in fig. 4 is one of a number which support a steam pipe from a channel bar. A strap passes over the bar and carries another strap. From this hangs a 20 mm rod, carried on a spherical block to permit a small movement. A clip around the pipe is bolted to the lower end of this rod. The Channel Bar is 180 × 90 mm; web t_1 10 mm, flange t_2 13 mm. The strap and clip material is 80 × 8 mm. The bolts are all M16 diameter, and the separating bush is 45 mm diameter, with a 20 mm hole. Prepare dimensioned drawings of the complete arrangement as in fig. 5, allowing 25 mm space between the bush and the flange. Use your own judgment for parts not dimensioned.
2. Draw three views of the Column Base in fig. 7. Channels, web 8 mm, flange 10 mm; angles 10 mm. The plates are 12 mm thick and all rivets 20 mm diameter, countersunk on the underside but elsewhere with snap heads. Fully dimension your drawing.

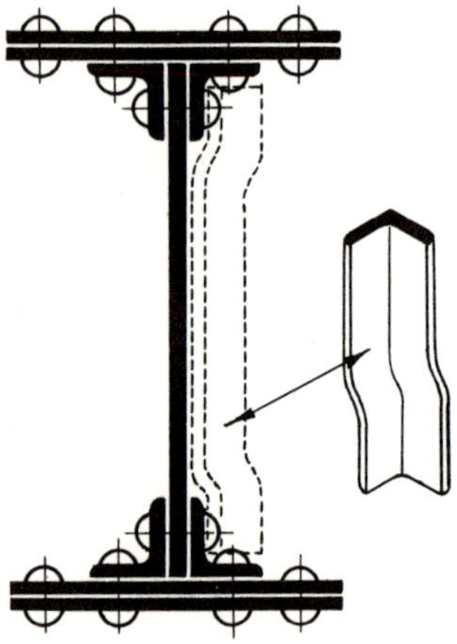

Fig 6

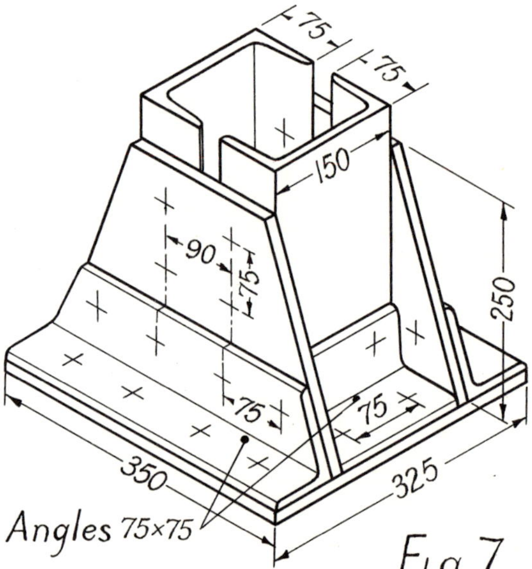

Fig 7

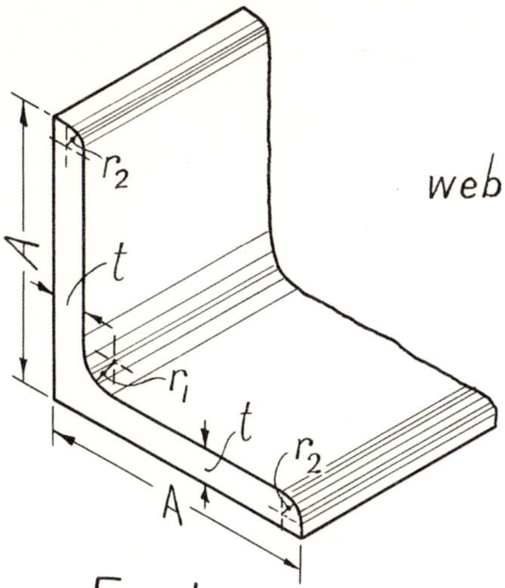

Fig 1

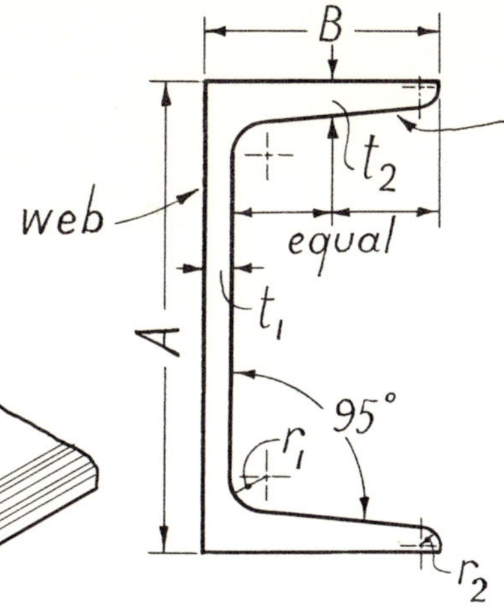

Fig 2

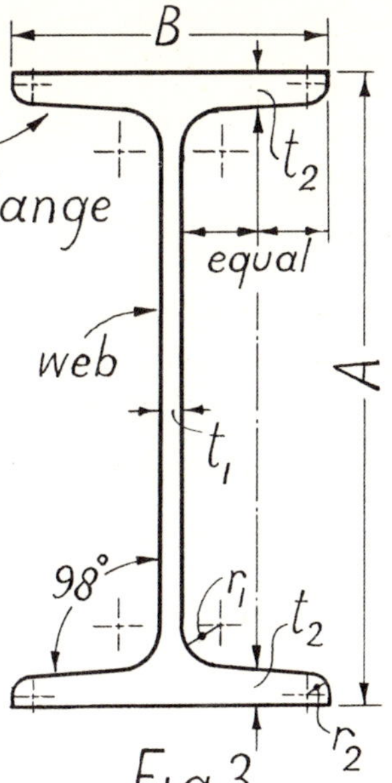

Fig 3

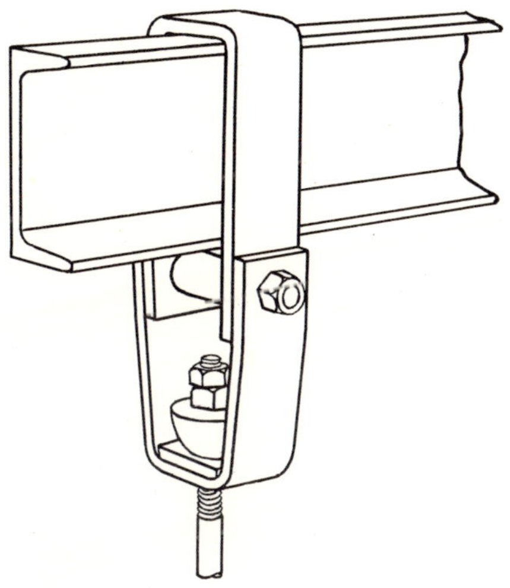

Fig 4

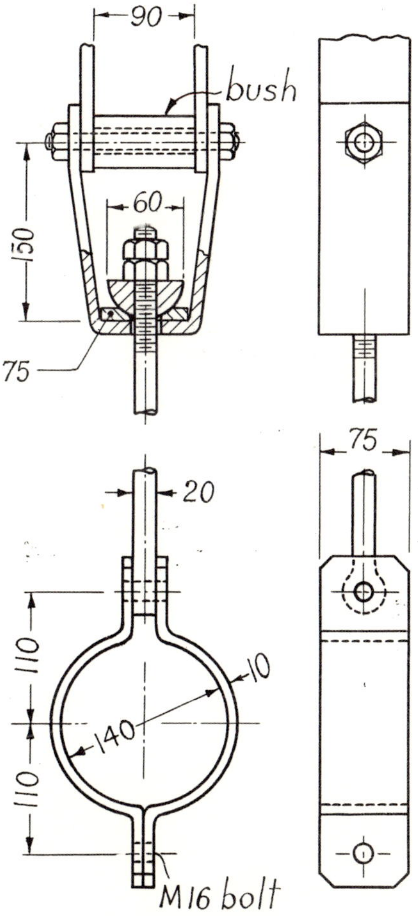

Fig 5

The developments of the cube, prism, pyramid, cylinder and cone have been dealt with earlier. These developments have their application in sheet metal work but usually a margin is required for jointing, as shown dotted in figs. 1, 2 and 3. The amount of margin will be governed by the kind of joint used. We shall now consider a method of dealing with developments, or *patterns* as they are usually called, which is widely used in the industry, and we shall apply it to a conical surface of small taper—where the apex of the cone is too remote to be used as it has been in fig. 3.

Development by triangulation

By this method we locate points on the outline of the development by using a succession of small triangles. Although simple, it calls for most accurate working.

Fig. 4 shows part of a cone of small taper. Some of 12 equidistant lines passing through the distant apex are shown on the surface. If the cone were to roll 180 degrees each way and if the 12 lines marked their positions on contact, we should get the result shown. We have to find the relative positions of these lines on the development.

We cannot readily draw the two boundary arcs because their centre is distant, but we can build up the outlines step by step using triangles such as ADB and ADC.

An elevation and plan of the part cone are shown in fig. 5, and the projections of the two lines AB and CD are shown by *ab*, a_1b_1 and *cd*, c_1d_1. Consider the triangle ABD. We know the length of AB (= slant height *h*), and we can take off the length of BD (= *bd* very nearly). What we now require is the length of AD, and this is the problem dealt with on page 108—given the projections of a line (*ad*, a_1d_1) to find its true length, which is given by a_2d_2. For the triangle ABC we know that BC = AD and we shall require in addition the length of AC (= *ac* very nearly). To set out the succession of triangles we should be helped if we had three pairs of compasses, to avoid resetting; one set to AC, one to AD (= BC), and one to BD. Begin by drawing AB, fig. 6, equal to *h*. From A, with radius AC, describe an arc; from B, with radius BC, describe an arc to intersect it in the required point C. Similarly obtain D, using radii BD and AD. Obtain in sequence all the 24 points and draw fair curves through them to give the solution. In the figure some points on the right fall outside the paper margin.

Exercises

1. In figs. 4, 5 and 6 take the diameters as 910 mm and 1040 mm, and take the slant height as 1520 mm. Using as large a scale as your paper permits plot the pattern as in fig. 6 and draw the shape. Trace your solution on transparent paper and, by folding it about AB, check the agreement of the halves.
2. The pattern given in Ex. 1 is to be extended at one end by 25 mm to permit overlap when shaped and the use of a riveted joint. The material is 1·5 mm thick and the rivets are 10 mm diameter with snap heads spaced at 100 mm. Make a drawing of the joint and dimension it.
3. Measure the area of the pattern, taking it as 12 times the area of ABDC and allowing for the overlap. What is its weight if 1 m^3 weighs 7140 kg?

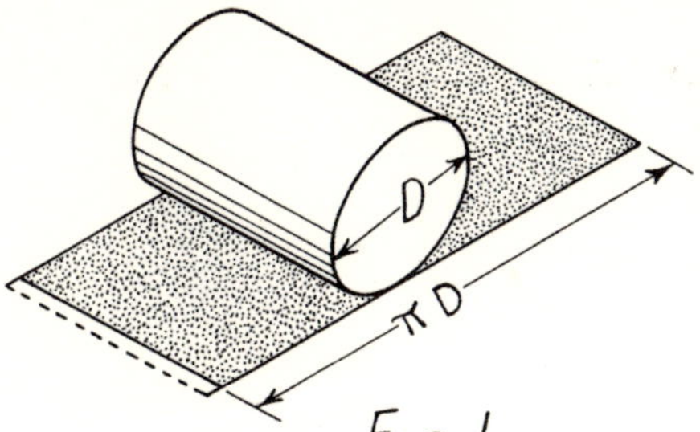

Fig 1

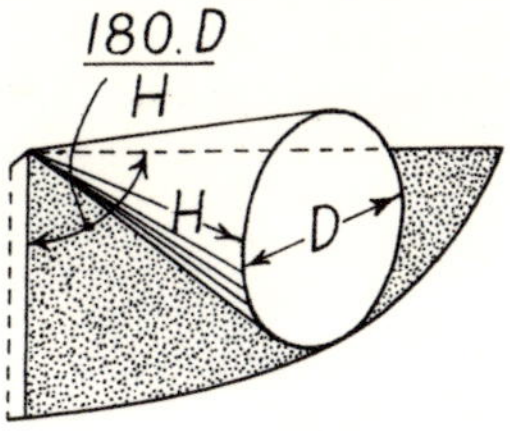

Fig 2

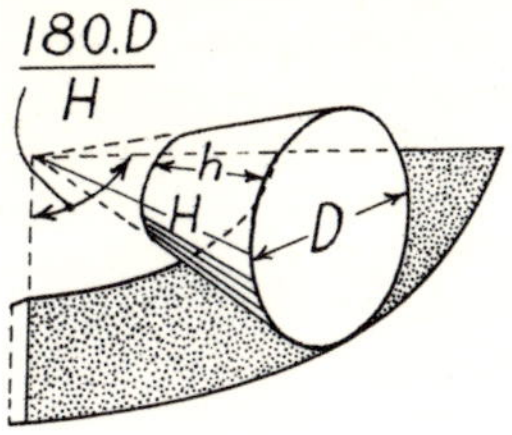

Fig 3

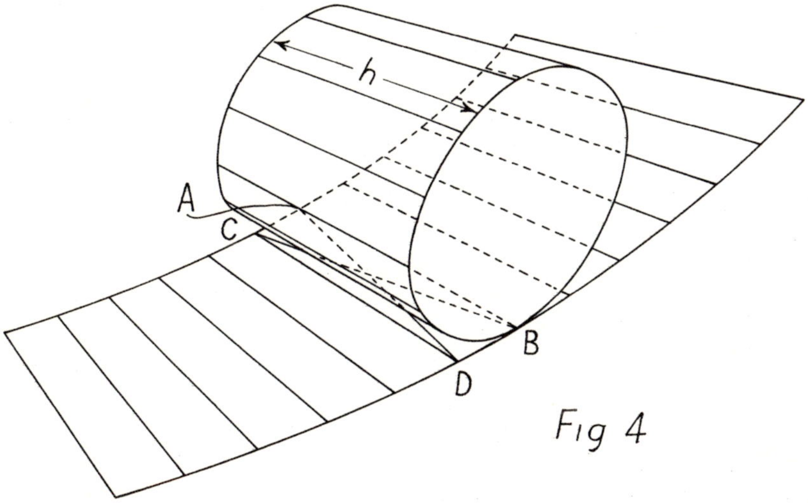

Fig 4

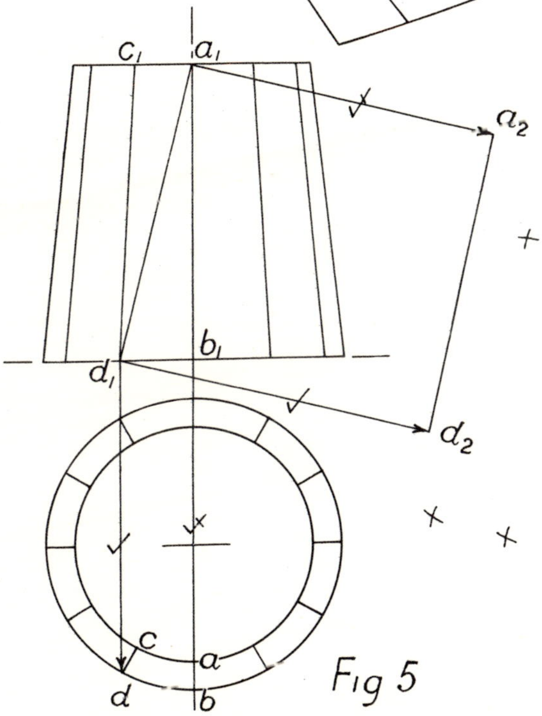

Fig 5

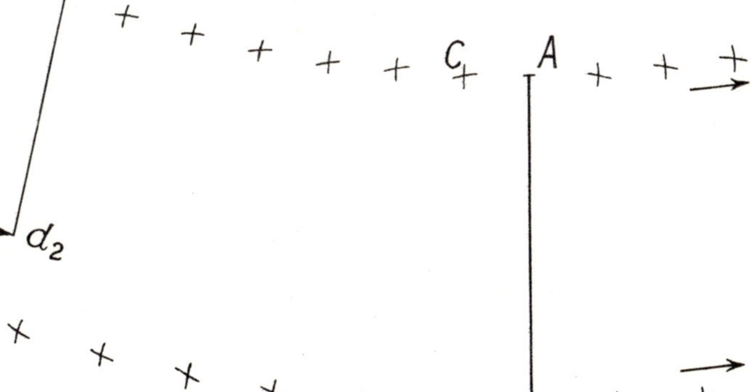

Fig 6

Fabrication by Welding

The introduction of welding, by gas or electric arc, has greatly changed engineering manufacturing processes. Originally the method was viewed with misgiving; but with the improvement in equipment, techniques and, particularly, inspection methods, fabrication by welding is now commonplace. For some parts, e.g. the joining of very-high-pressure steam pipes, it is the only acceptable process.

There are many kinds of welds and two have been shown on page 153. To prevent confusion each type has been given a standard symbol, and these are set out in British Standard 499. Only one, the commonest, is used opposite: the triangle, for a fillet weld. All symbols have to be used with a line and an arrow, as in fig. 1. The rules are: for welds on the arrow side, fig. 1*a*, the symbol hangs on the reference line; for welds on the other side, fig. 1*b*, the reference line supports the symbol; for welds on both sides, fig. 1*c*, the symbols are shown on both sides. In addition to the symbols, weld dimensions are required with any supplementary notes.

Welded flange

A slip-on circular steel flange for a steam pipe is shown half in section in fig. 2. The arrangement is suitable for pressures up to 15 atmospheres. The pipe is fillet-welded to the flange both at the face and at the back.

An alternative superior design used for higher pressures is shown in fig. 3. In both designs it is essential that the flange should be held square to the pipe while being welded on.

Welded link

The link or lever in fig. 4 is a casting, in one piece. It can be fabricated by welding using the three parts shown in fig. 5. The section drawing in fig. 6 shows the welded assembly; the holes would be drilled after the fillet welds had been completed.

Where such a part is to be made in large numbers, 'projection welding would give a good solution; then the bosses would be spigoted into the plate as in fig. 7 and the annular surfaces (shown by the thickened lines) would be prepared for arc welding under pressure.

Exercises

1. Make a full-size dimensioned drawing of a pipe-end and flange as in fig. 2 showing the half section and adding an end view on the front of the flange. Dimensions are: Pipe diameters 80 mm and 90 mm, flange 200 mm diameter and 20 mm thick, bolt holes eight 18 mm diameter on a circle 165 mm diameter, fillet welds 6 mm.
2. Make a two-view dimensioned drawing, full size, of the link in fig. 6 taking the following dimensions: Boss at large end: bore 25 mm, diameter 50 mm, thickness 25 mm; boss at small end: bore 16 mm, diameter 32 mm, thickness 12 mm; plate 6 mm thick, pitch of holes 150 mm; fillet welds 6 mm.
3. Changing the design in Ex. 2 for projection welding as in fig. 7 make dimensioned drawings of a suitable arrangement using your discretion for the dimensions of the fillets.

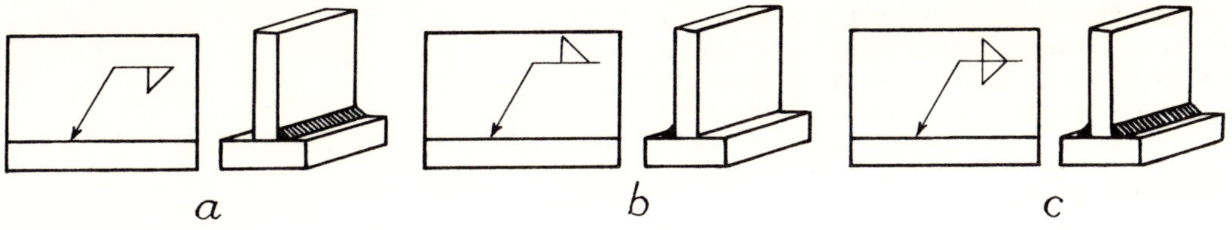

Fig 1

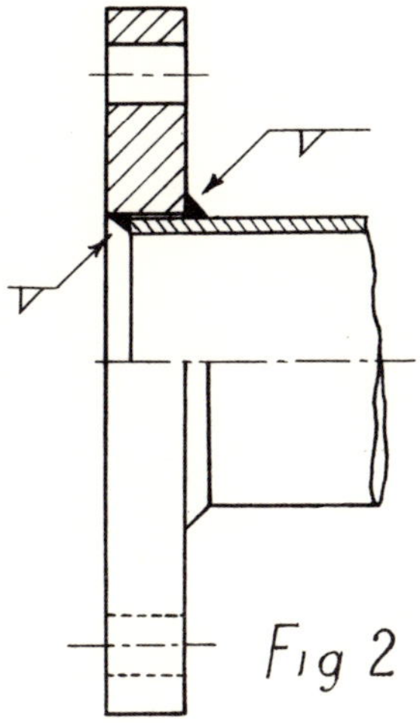

Fig 2

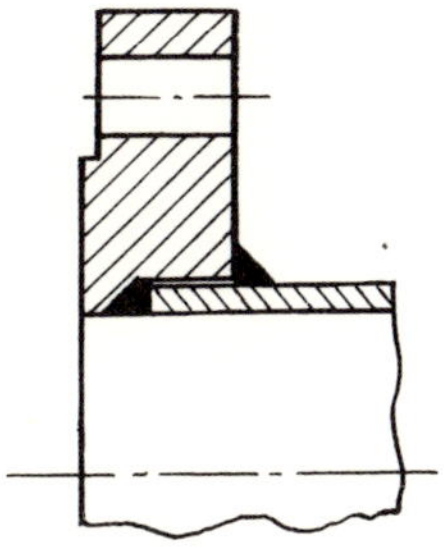

Fig 3

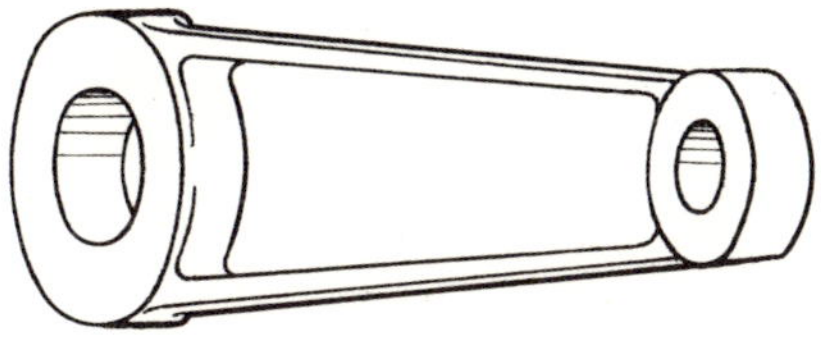

Fig 4

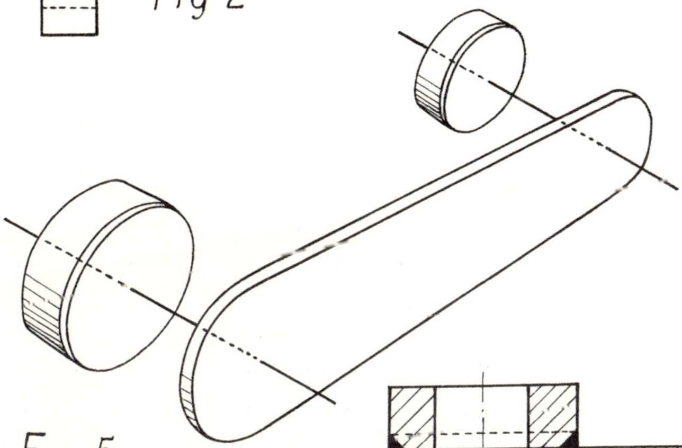

Fig 5

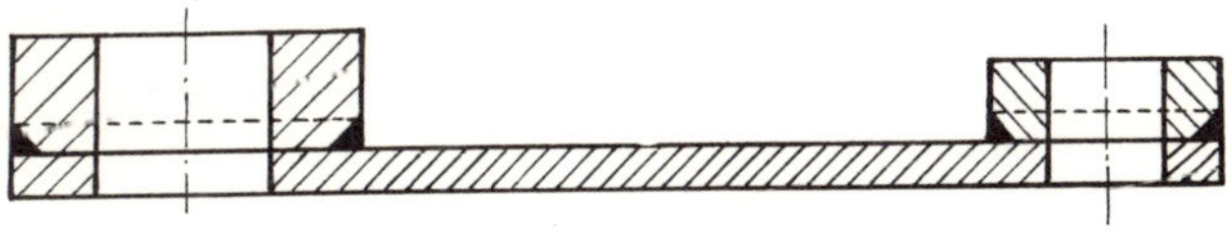

Fig 6

Fig 7

Shafts and Couplings

Cylindrical shafts or spindles are very common in all branches of engineering. We shall consider here some of the more simple ways in which shafts may be secured together, end to end. Before doing so, we must deal briefly with keys and keyways.

Keys and keyways

A key is a steel piece inserted between a shaft and hub to prevent relative rotary movement between the shaft and the hub. The keyway is usually a rectangular slot cut along the shaft, as in fig. 1, and also along the hub. The key is driven between them, the key being a tight fit at the sides only, but not tight between the top and bottom of the keyway. Often the key has a projection at one end, called a *gib head*, fig. 2, to assist in its removal. Another type of key is one which fits in a recess in the shaft, as in fig. 3, and is secured by screws so that the hub may slide over it along the shaft; this key is called a *feather* key.

The proportions of keys have been standardized, but for our purpose we may assume the following:

Width of Key $W = 0.25D$, where D is the shaft diameter.

Thickness of Key $T = 0.7W$.

Flanged coupling

This coupling transmits turning only and is shown in fig. 4. Each flange and shaft has a keyway cut in it. The flanges, usually of cast iron, are pressed on the shafts, and keys are driven in from the face of the flange. The flanges are held together by steel bolts which are a good fit in the holes. The flanges are also kept in alignment by a *spigot* arrangement, a projection machined on one flange fitting into a corresponding recess on the other. To guard against accident, the flanges are extended at the rim to form a shield covering the nuts and bolt heads.

The proportions in fig. 4 are given in terms of the shaft diameter D. They should be 'rounded off' when calculated; for example, for a 50 mm shaft, the outside diameter is $2{\cdot}6D + 75$, i.e. $130 + 75 = 205$.

The numbers of bolts and their diameters may be taken as:

Shaft Diameter (mm)	50	60	70	80	90	100
Bolt Diameter (mm)	12	16	16	20	20	24
No. of Bolts	4	4	5	5	6	6

Exercises

1. **Draw full size the following views of a flanged coupling for a 50 mm diameter shaft, showing the flanges bolted together: side view, upper half in section through a bolt; end view looking on the nuts.**
2. **Draw full size views of *one flange* (the right-hand one in fig. 4) of a coupling suitable for a shaft 75 mm diameter. Dimension the flange completely and mark those surfaces which are to be machined. (Hint: only the face and projecting ring, the bore, the holes, and the back strip on which the bolts press, need machining.)**

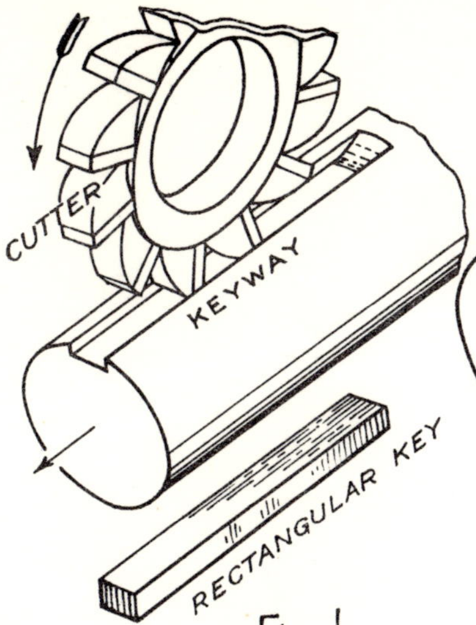

Fig. 1.

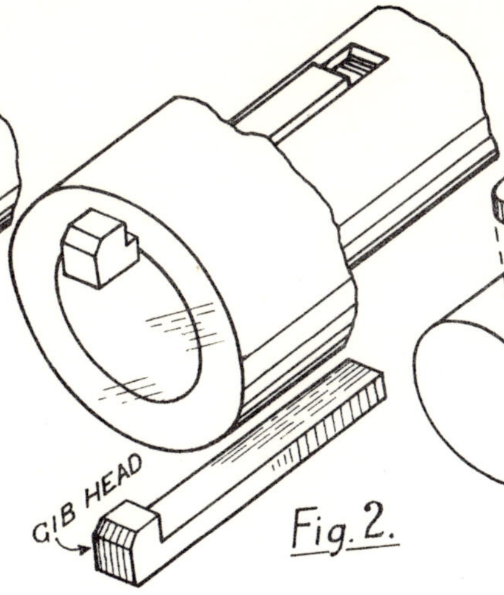

Fig. 2.

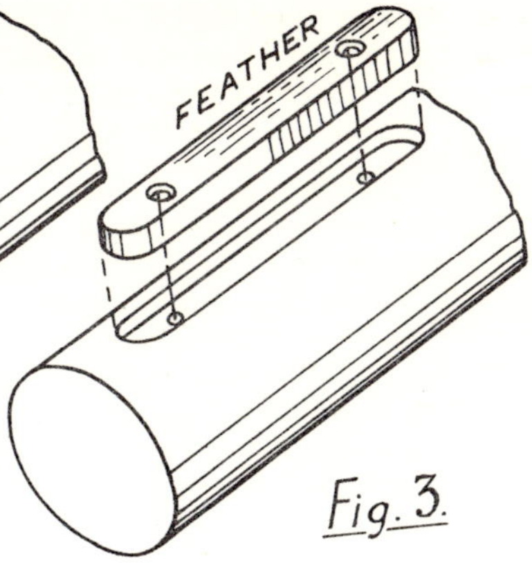

Fig. 3.

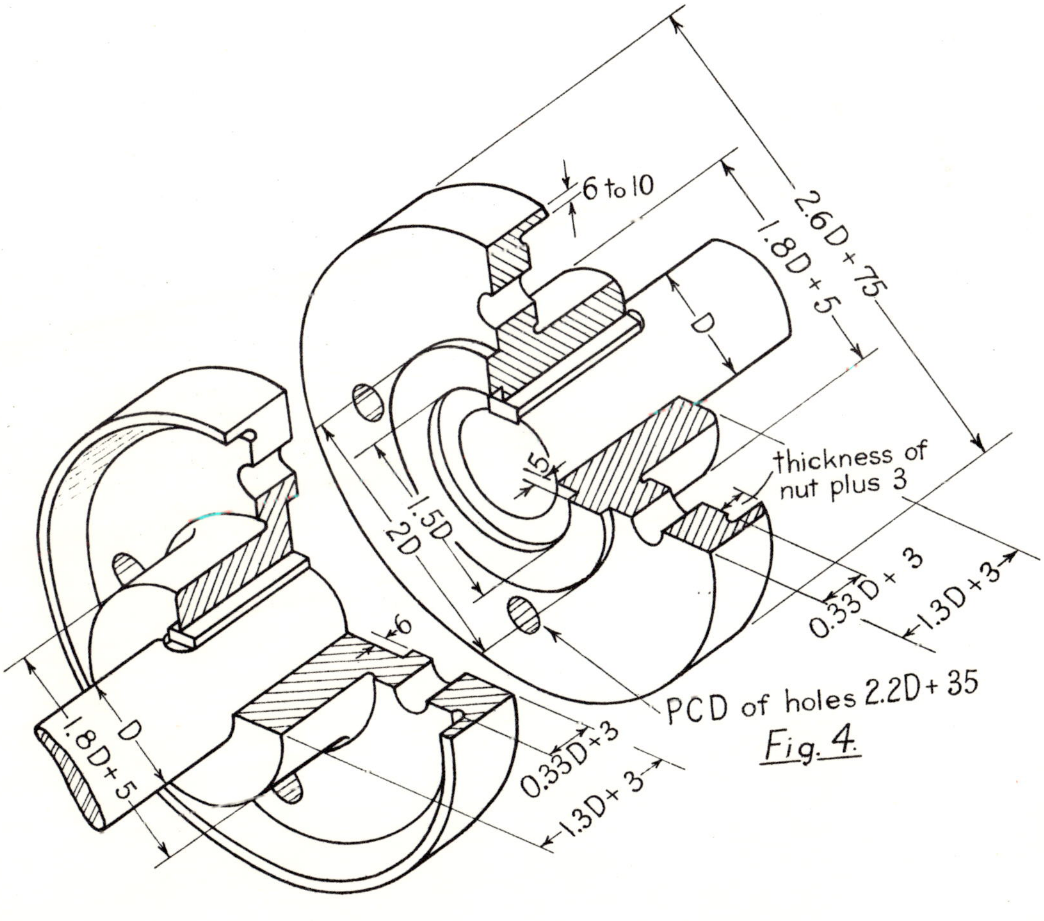

Fig. 4.

Cottered Joints

The cottered joint is a temporary fastening, temporary in that it can readily be disconnected, as distinct from riveted or welded joints which are classified as permanent. The purpose of the cottered joint is to fasten together two rods or shafts which are to transmit a push-and-pull force only. It uses a cotter, which is a tapered piece of metal with rectangular or circular cross-section. When driving the cotter into position, the taper will cause the two parts to be pulled together. Figs. (*a*), (*b*) and (*c*) are nearly self-explanatory. Referring to fig. 1 (*b*), the gaps shown on either side of the cotter are to allow for the two parts to be fully tightened together by the cotter.

At fig. 1 (*c*) a U-shaped strap is fixed around a two-piece gun-metal bearing, the end of the rod is inserted, and the cotter driven into position to pull the parts tightly together. Should the cotter be driven too tightly, the lower side of the strap will be distorted as shown in fig. 1 (*d*). This can be avoided by the use of a gib as shown in fig. 1 (*e*). The gib also provides for the slot in the strap and rod end to be made parallel, while the gib and cotter are tapered (1 in 25 max).

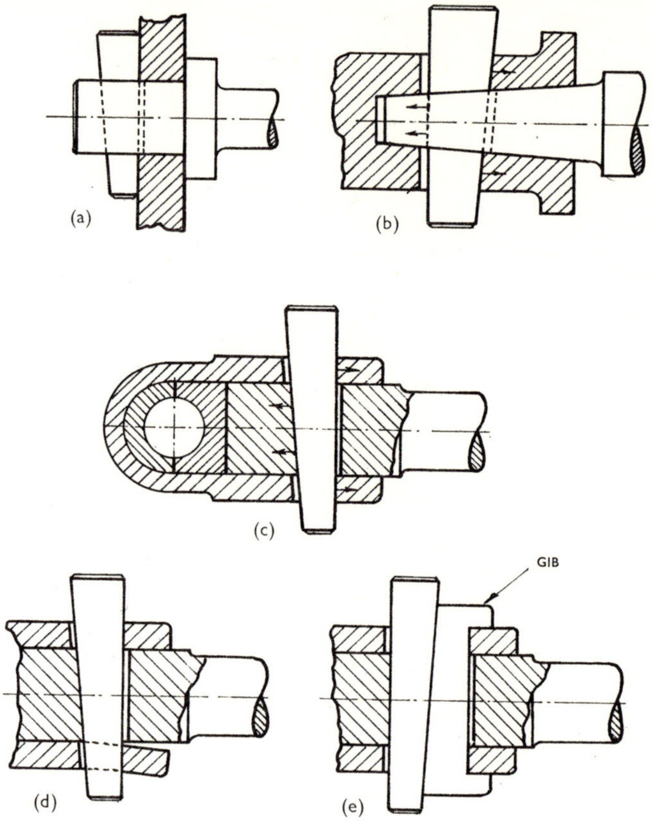

Fig 1

Exercises

1. Make a detailed drawing of the parts required to construct a cotter joint as show in fig. 1 (*c*). The joint should connect a pin 25 mm in diameter to a rod 30 mm in diameter. The cross-section of the cotter should be approximately 0.5 of the rod cross-section.
2. The connecting rod end, fig. 2, connects a pin 76 mm in diameter and a rod of the same diameter. Make a working drawing showing all the separate details of fig. 2, except the nuts, giving whatever views are necessary to manufacture the parts. All dimensions should be added to each part drawing so that each may be made without referring to other details..
3. Having made the detail drawings in Ex. 2, make an assembly drawing using the parts drawn. The assembly should be as fig. 2 with an end view added. In this way a check will be made on the parts shown in the solution to Ex. 2.

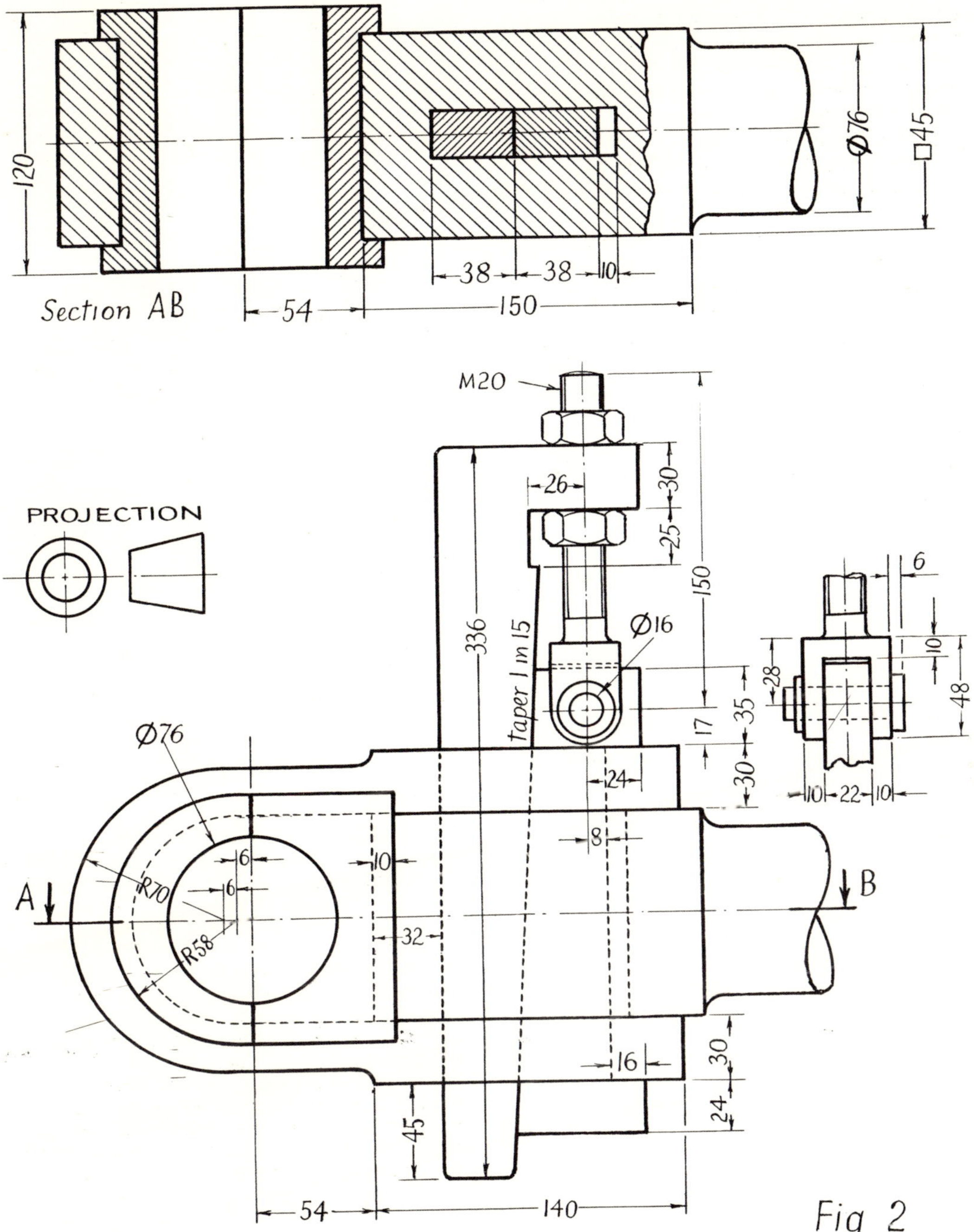
Section AB
120
38
38
10
54
150
Ø76
□45
M20
PROJECTION
26
30
25
150
336
taper 1 in 15
Ø16
Ø76
17
35
30
24
8
10
6
6
R70
R58
32
A
B
16
30
24
45
54
140
6
10
28
48
10
22
10

Fig 2

Knuckle Joints

Knuckle or Pin Joints are used to connect together two or more rods whose axes meet in a common point, this point being on the axis of the pin. The joint has many forms, one of the most common being that shown in fig. 1. The ends of the rods are forged to shape, one being forked, and the other provided with an eye to fit between the jaws of the fork. The pin of the joint passes through both eyepiece and fork, and acts as a hinge. The joint will transmit both rotary and push-and-pull motion and it is widely used, in one form or another, not only for machine parts but for linking together parts of steel structures where the movement is small.

The pin and its fittings

The pin usually has a head at one end and is provided with a collar at the other. The pin is often prevented from rotating in the fork by means of a small stop pin or peg. This is driven into a hole drilled into the pin where it joins the head, and a small part of the peg is allowed to project; this projection enters a recess formed in the fork. The collar is secured by a tapered pin which passes through the collar and the pin. In the alternative a thread is provided at the end of the pin and a thin nut is used, this being fixed in position by a split pin.
In small knuckle joints, one eye of the forked end is often screwed to take a thread provided on the end of the pin. The head of the pin may be of the countersunk type.

Improvements

In good-class work the parts in contact are machined—that is, the sides of the fork and the eyepiece, the holes and the pin. It is often desirable to lubricate the pin to ensure free motion; one method is to drill a hole centrally through one half of the pin to meet another drilled radially and to provide for a grease lubricator at the end of the pin. It would also be good practice to provide a gunmetal bush in the bore of the part within the fork.

Proportions

Suitable proportions are given in the drawings where the diameter of the rods is taken as 1. If the diameter is 20, then all the figures given should be multiplied by 20 and rounded off.

Joint for three rods

This is shown in fig. 2. The proportions given in fig. 1 will not apply here as some changes are involved: e.g. the diameter of the pin would have to be increased.

Exercises

1. Draw, full size, the following views of a knuckle joint of the type shown in fig. 1 for rods 25 mm diameter: side view, top view, and a sectional view on the axis of the pin.

2. Draw separate dimensioned working drawings for each of the parts of a knuckle joint suitable for rods 35 mm diameter. Use a collar and tapered pin. Mark with a tick surfaces which are to be machined.

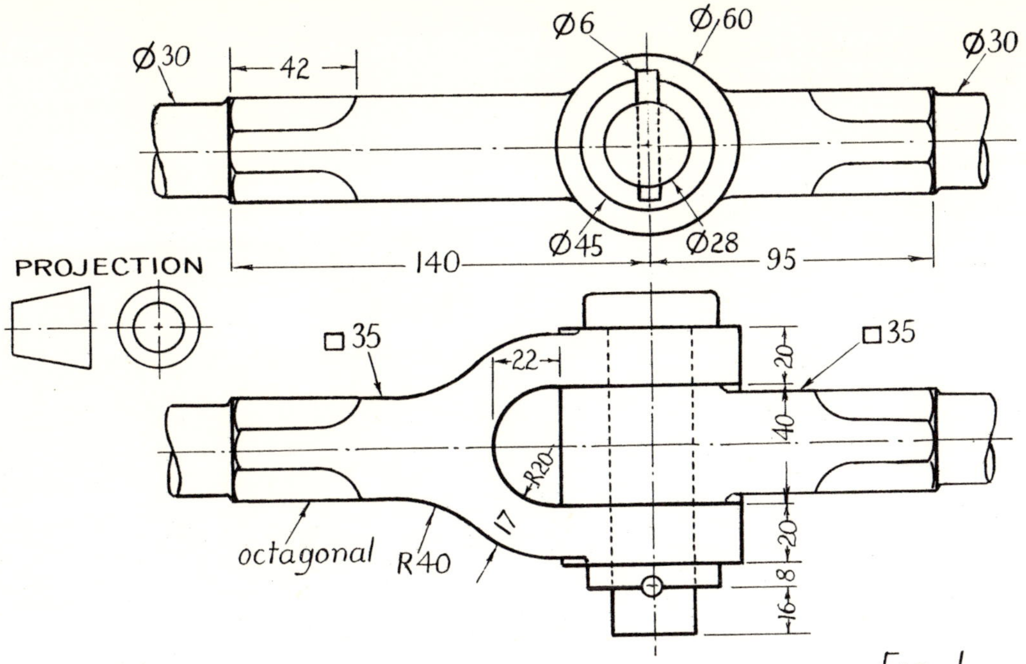
Ø30
42
Ø6
Ø60
Ø30
Ø45
Ø28
140
95
PROJECTION
□35
22
□35
20
40
R20
20
17
octagonal
R40
8
16

Fig 1

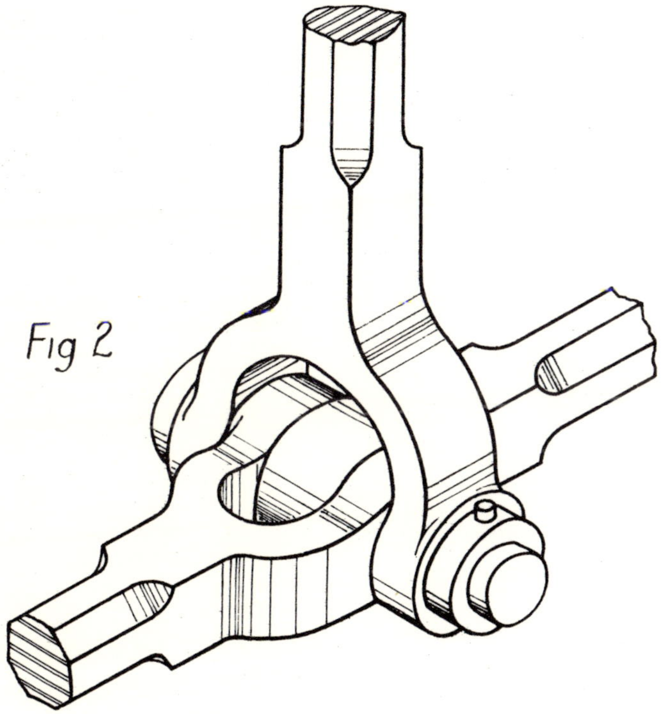

Fig 2

Simple Bearings

Bearings which support revolving shafts are very common in engineering. They will be seen by the student in the College Workshop.

If one examines an old clock it will be found that the spindles revolve in holes drilled in brass side pieces. Because loads and speeds are small, wear is also small and its effects may be neglected. But if a spindle or shaft carries a large load and revolves rapidly, then bearing wear must be provided for. Arrangements are usually made so that a part of the bearing may be renewed. In fig. 1, a brass bush is forced into a bore in a cast-iron base. A hole is drilled for oil. The bush wears oval after a time and can be replaced.

Split bearings

A more convenient bearing is shown in a simple split form in fig. 2. When wear occurs, metal can be removed from A to permit some adjustment.

A much superior bearing, fig. 3, embodies a split bush. One half-bearing fits into the Base and the other into the Cap. On assembly the halves are held together when the Cap is bolted to the Base but the parts marked A and B do not touch the parts marked C and D. A projection on the upper half-bearing fits into a hole in the cap and stops the bearing from turning with the shaft.

Many improvements are in fact made to the type of bearing shown. For example, the Base and Cap are lightened considerably by removing surplus metal and by recessing the underside of the Base; the bearing is lubricated automatically and so on. These refinements cannot be considered here.

The parts are held together by means of M16 bolts passing through the 18 mm diameter holes, the bolt heads being recessed into the underside of the cast-iron base. The bearing would be mounted on a wall bracket similar to the one shown on page 191, fig. 2, and secured by M20 bolts.

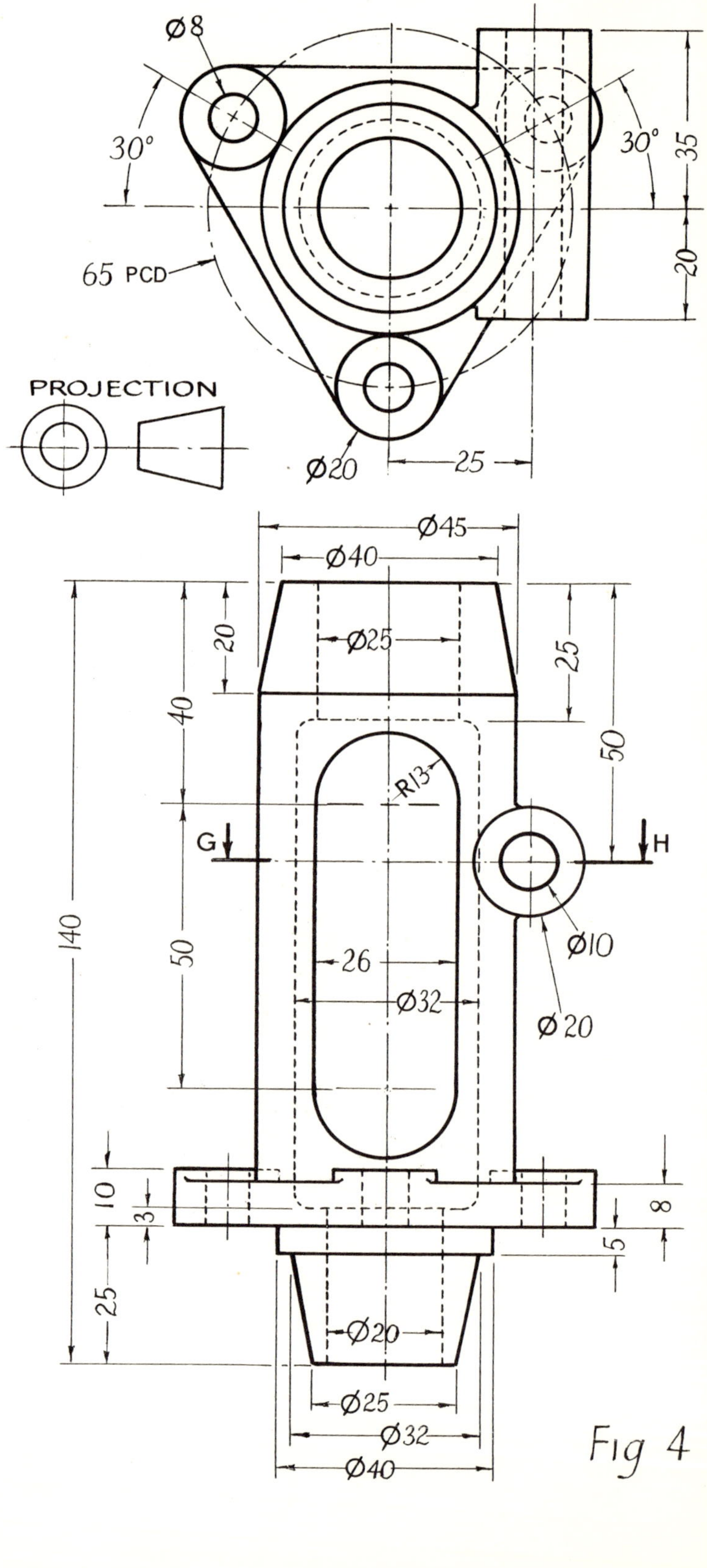

Exercises

1. Assemble the parts of the bearing in fig. 3 and draw views in the directions P and Q.
2. More advanced. Add a view in the direction R showing the bearing in section. Dimension the complete drawing.
3. A governor bearing is shown in fig. 4 in 3rd angle projection. Draw, scale full size, in 3rd angle projection:
   (*a*) the given elevation.
   (*b*) a plan view in section on the plane GH.
   (*c*) a side elevation in section on the vertical centre line of view (*a*) and looking from the left of view (*a*).

Fig 4

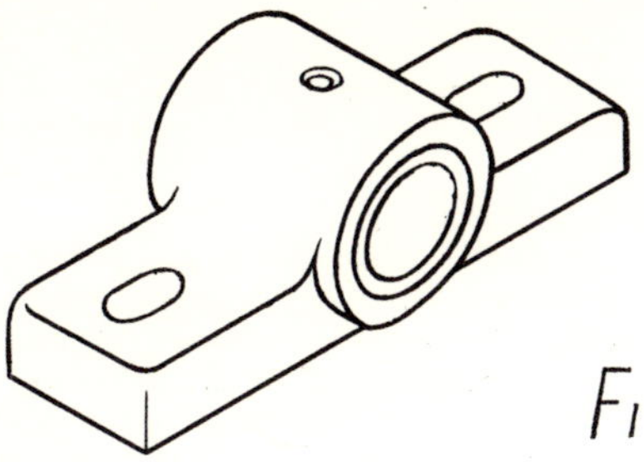

Fig 1

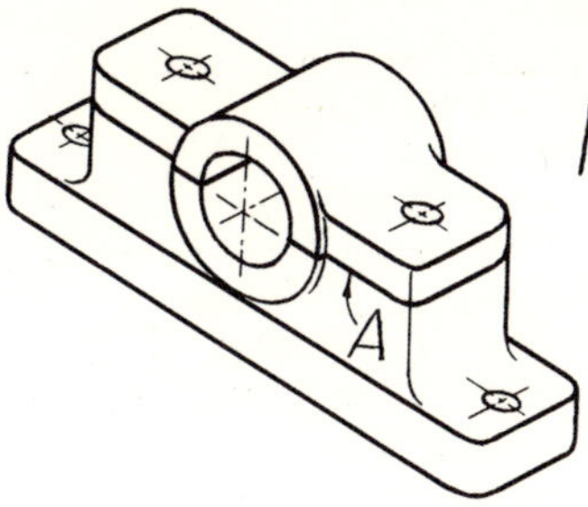

Fig 2

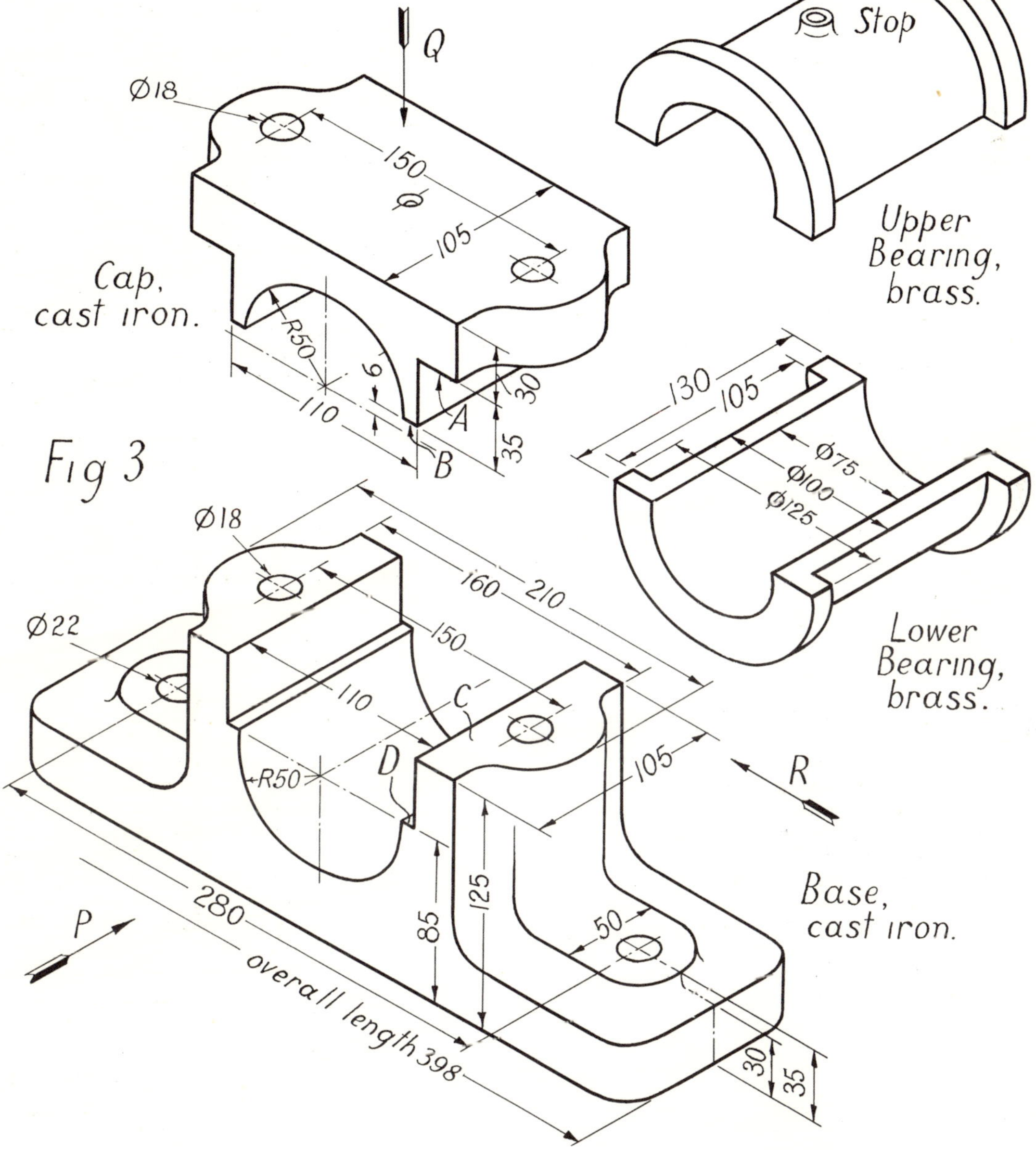

Fig 3

A great many engineering articles fall under the title of Brackets. These are usually castings in iron and their purpose is often to support firmly some shaft or rod which rotates or slides. With all cast brackets it is desirable to reduce both the amount of metal in the article and the extent of machining operations. Weight may be reduced by the use of ribs or webs or by making parts hollow. Castings are usually only machined on such surfaces as bearing faces or position strips—castings are not machined to improve their appearance.

Two examples of brackets for horizontal shafts are shown opposite and used for the exercises below. All dimensions are in millimetres.

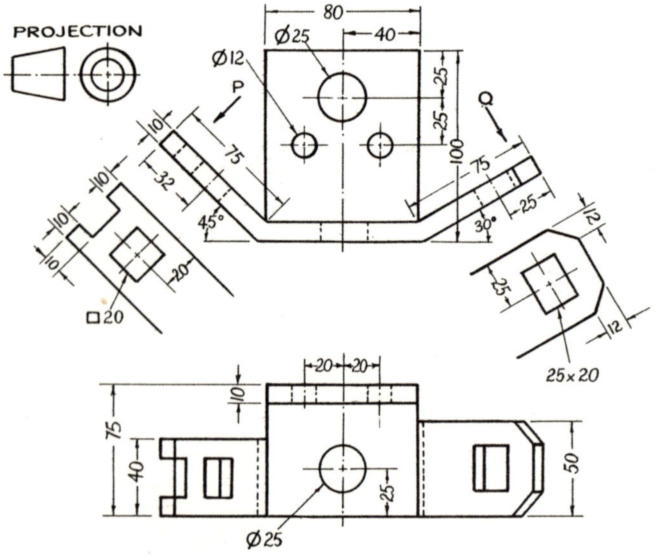

Exercises

Note.—The student is expected to use his own judgment for sizes which are not dimensioned: for example, various radii.

1. *Wall Bracket*
The cast-iron bracket in the upper drawing is in the form of a cantilever. The three members forming the top, the back, and the strut (the inclined member) all have central ribs. Together they provide for the support of the lower half of a bearing for a shaft. The bearing is completed by the addition of a cap, which is not shown, but which would restrain the shaft from any movement upwards.
The face of the half-bearing slopes at 30° to the horizontal, and the plane of the top surface of the horizontal member passes through the centre line of the bearing.
The only machined surfaces are: the back, the two holes, and the bore and ends of the bearing.
Draw, full size, three views of the bracket in the directions P, Q and R, showing by revolved sections the shape of the members. Insert the essential dimensions and add finish marks.

2. *Addition to Ex.* 1
Draw a suitable cap for the bearing, in place. It should match the lower half and should have two holes for securing it, 10 mm diameter 60 mm apart. Modify the design by including split gunmetal bearings, of the type shown in fig. 3, page 189, to take a shaft reduced to 30 mm diameter.

3. Draw views of the wall bracket shown in the lower drawing opposite seen in the direction of arrows A and B. Add a plan view. The system of projection is left for the student to decide.

4. *Wall Bracket and Bearing*
The bracket in the lower drawing carries the shaft bearing shown in fig. 3 on page 189. The bearing is secured by two 20 mm bolts with square necks.
Draw two views of the bracket with its bearing, in the directions A and B. Show the bearing half in section through a bolt. Insert only the essential dimensions and add finish marks.

5. A generator bracket is shown below, projected in 1st angle. Draw the following views, scale full size, in 1st angle projection:
   (*a*) the given elevation.
   (*b*) the given plan view.
   (*c*) an auxiliary view seen from direction of arrow P.
   (*d*) an auxiliary view seen from direction of arrow Q.
   (*e*) a side elevation to the left of view (*a*).
   (*f*) a side elevation to the right of view (*a*).

Note.—(i) It is advisable to draw view (*a*) near the top centre of the drawing paper (A2 size) and the plan view near the bottom centre. By doing so, the two auxiliary views can be drawn between the elevation and plan view.
(ii) The side elevations must be complete for, when looking either from left or right, the more distant details will be seen through the holes.

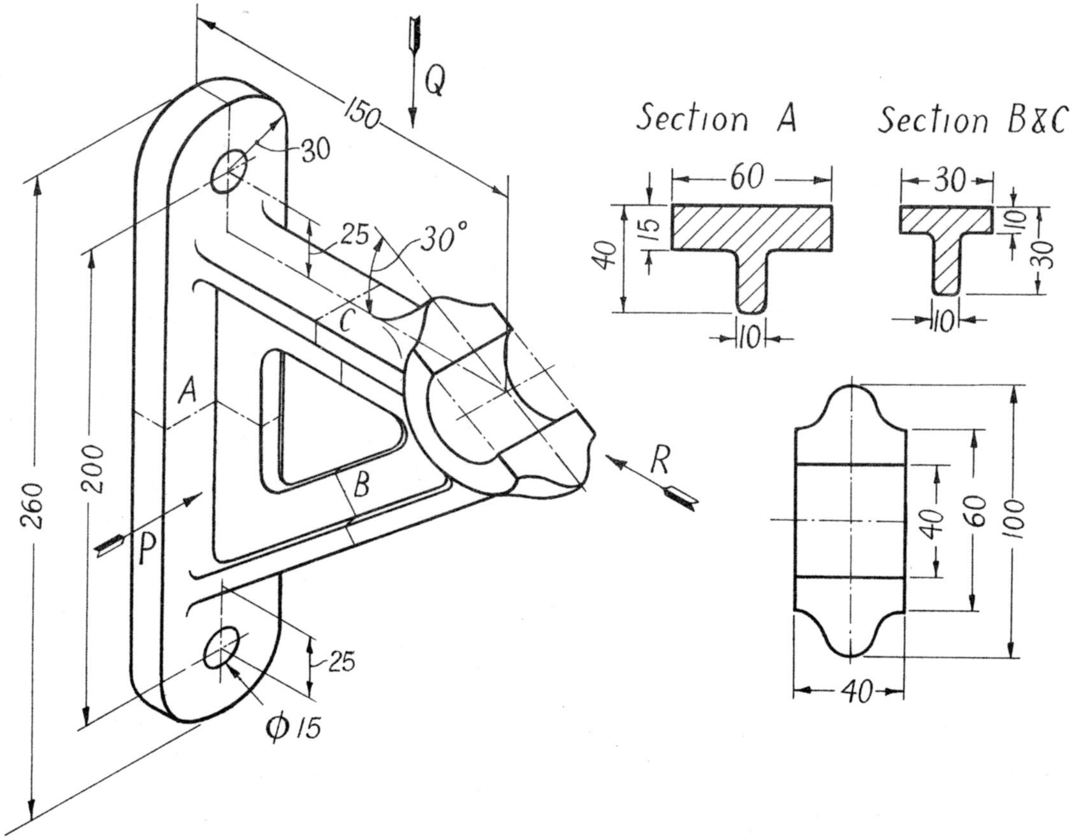
150
30
Q
25
30°
C
A
B
R
P
200
260
25
Ø15
Section A
Section B&C
60
15
40
10
30
10
30
10
100
60
40
40

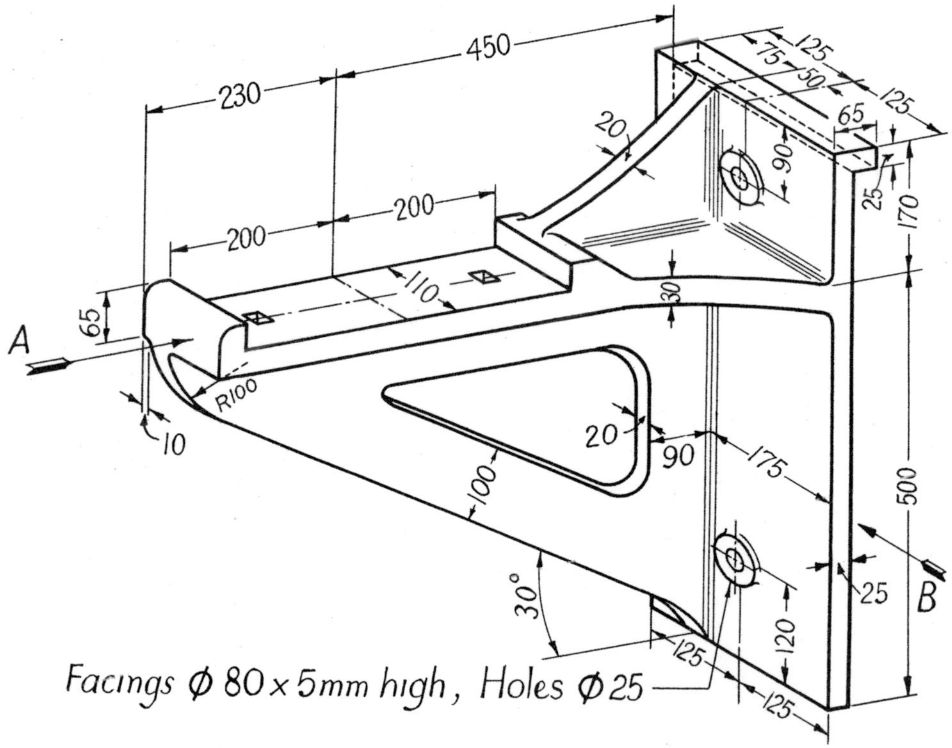
450
230
200
200
110
65
A
R100
10
100
20
90
175
30°
125
125
500
25
B
120
170
25
65
125
75
50
125
20
90
30
Facings Ø 80×5mm high, Holes Ø25

Simple Valves

Valves are used to control the flow of fluids through pipes. Two examples are shown here: in one the valve rotates in opening and closing; in the other the valve rises and falls.

Flap valve—opposite

This simple valve permits flow only in one direction—from left to right: a reversal closes the valve. The flap is of cast iron with a leather disc secured to it by four bolts. Two 'horns' drop into sockets cast on the sides of the cast-iron valve box. The parts machined are the flanges, the cover joint, and the face on which the valve bears.

Lift valve—below

The valve box is of cast iron, and the valve, together with the bush into which it fits, is of gunmetal. The valve is guided in the bush by three ribs, and its lift is limited by a projection from the cover. The face of the valve is inclined at 45° which bears on a narrow ring formed on the end of the bush. The valve permits fluid to surge upwards but not downwards.

Exercises

Note.—In both examples minor dimensions have been left for the student to settle.

1. Draw the following views of the flap valve with the cover bolted in position: side view, with the left-hand part in section to show the valve; view on the top with the front half of the cover removed; sectional end view on the valve face with the valve removed. Insert dimensions and machining marks.
2. Draw dimensioned views of the lift valve: sectional view, as shown; view from the top with the cover and valve removed. Take the flanges as 200 mm diameter, 20 mm thick; use 6 studs for the cover.

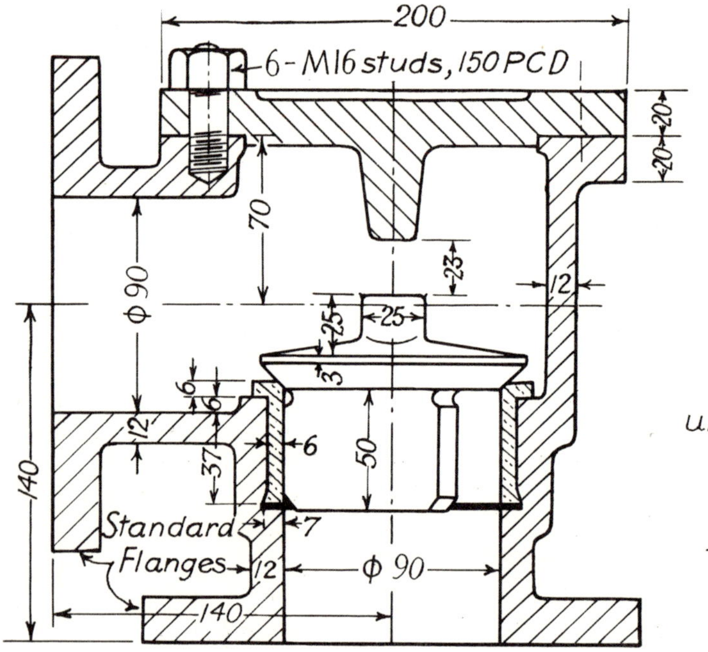

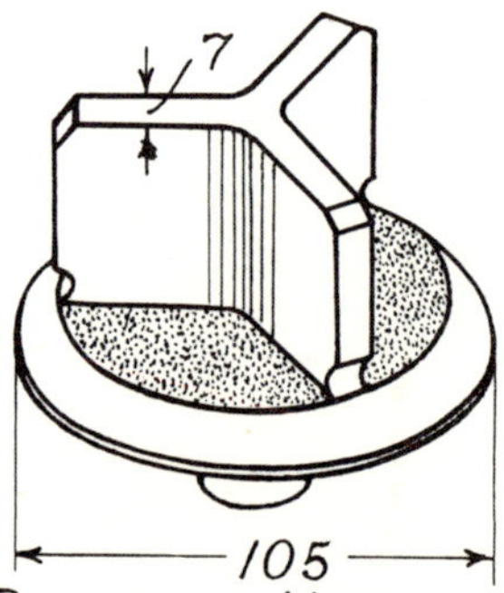

Pictorial View on underneath of Valve

VALVE AND BOX
NON-RETURN TYPE
WORKING PRESSURE
7 atmospheres

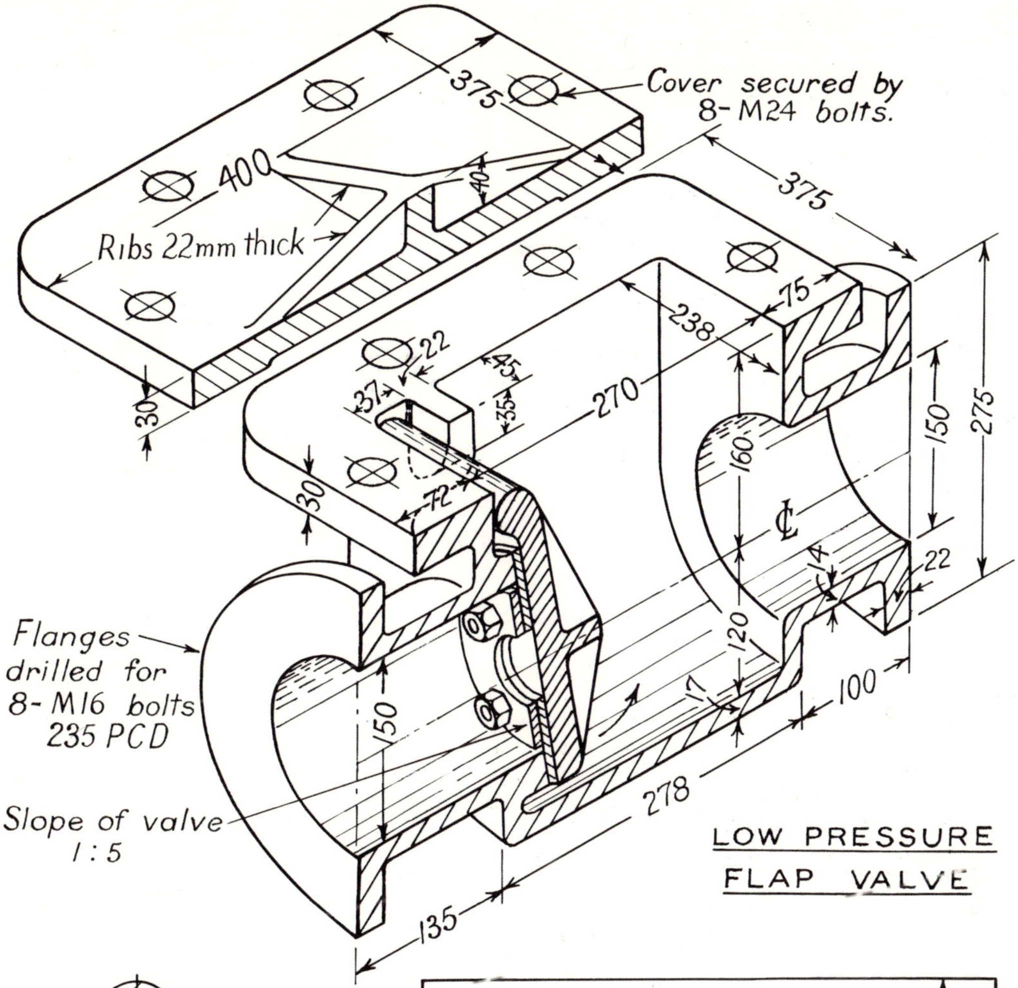

LOW PRESSURE FLAP VALVE

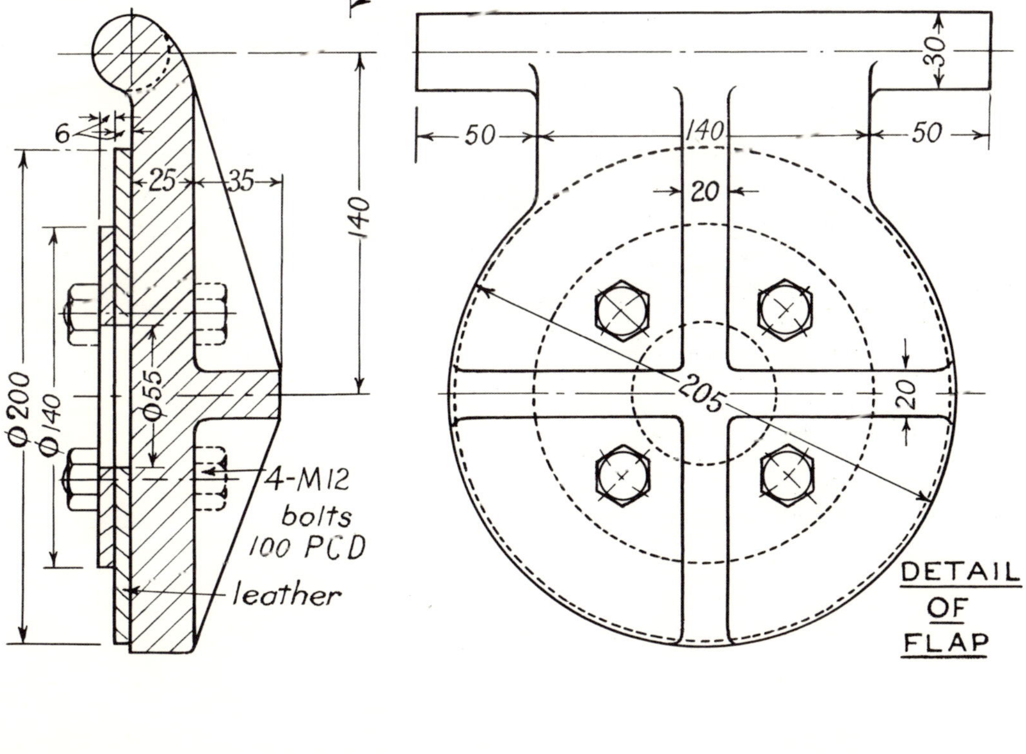

DETAIL OF FLAP

Toleranced Dimensions

Even an elementary book on technical drawing must have some mention of tolerances if only because of their effect on the method of applying dimensions. But tolerances are now so widely used that the student should learn about them as soon as possible.

Limits of size—Tolerance

If we write the dimension 130 mm on a drawing, we are in effect saying that the size is to be between 130·5 and 129·5 mm. If we write 130·0 we imply that the size will be between 130·05 and 129·95 mm.

When components have to be interchangeable we cannot usually leave these extreme dimensions to chance: we have to say what their values may be. The extreme values are the *limiting dimensions* and the difference between them is the *tolerance*, usually expressed as a decimal. We can write the dimension in four ways:

(1)	(2)	(3)	(4)
130·05 129·95	130·05 —·05	129·95 +·05	130 ±·05
(limits of size)	(unilateral)	(unilateral)	(bilateral)

Combinations of dimensions

Two parts, A and B, are shown in fig. 1. Measuring from D, the length of A is to be 75 ±·25 and the length of B is to be 50 ± ·25. The dotted lines show to a very much enlarged scale, the extreme positions of the end surfaces. It will be clear that the extreme overall dimensions will be 125 — ·50 and 125 + ·50; they may be written 125 ± ·50. The same two parts are shown in fig. 2 but, we now give measurements from one end D_1. The length of A is given as 75 ± ·25 and the overall length as 125 ± ·25; the dotted lines again show, to a much enlarged scale, the tolerances. We see that the length of B is given by 50 ± ·50. We cannot possibly have 125 ± ·25 minus 75 ± ·25 = 50.

Two important Rules follow:

(1) *Whether we add or subtract dimensions we must add the tolerances.*

(2) *It follows that we cannot have more than one dimension in the same direction controlling the size of a part;* for suppose we dimension the part as in fig. 3, using three dimensions *a*, *b* and *c*.

Then, if we work from surface D_1, using *a* and *b*, we get for *c* 50 ± ·75, not 50 ± ·25.

Again, if we work from surface D_2, using *a* and *c*, we get for *b* 75 ± ·75, not 75 ± ·25.

One dimension in fig. 3 *is unnecessary.*

Cumulative effect of tolerances—Holes in a plate

If we use 'chain' dimensions with tolerances ± ·25, as in fig. 4, then we get the ever-increasing tolerances shown in fig. 5 if we refer our measurements to the left-hand edge. If we dimension from the left-hand edge, as in fig. 6, again using tolerances of ± ·25, then the tolerances on the distances between the holes are all ± ·50, as shown in fig. 7; for 100 ± ·25 minus 50 ± ·25 = 50 ± ·50; 150 ± ·25 minus 100 ± ·25 = 50 ± ·50; and so on.

The method of dimensioning in fig. 6 is to be preferred to that in fig. 4.

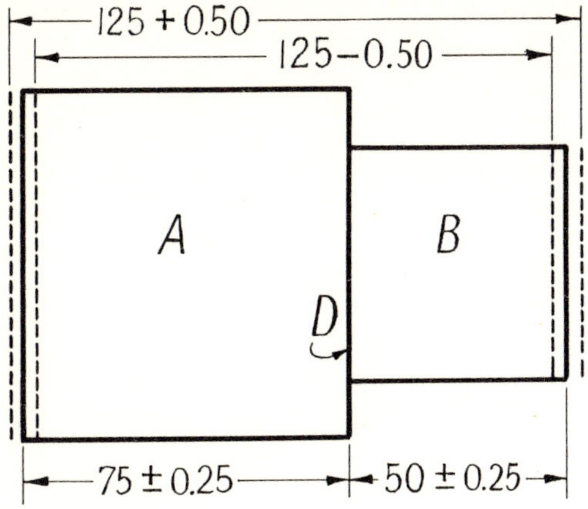

Fig 1

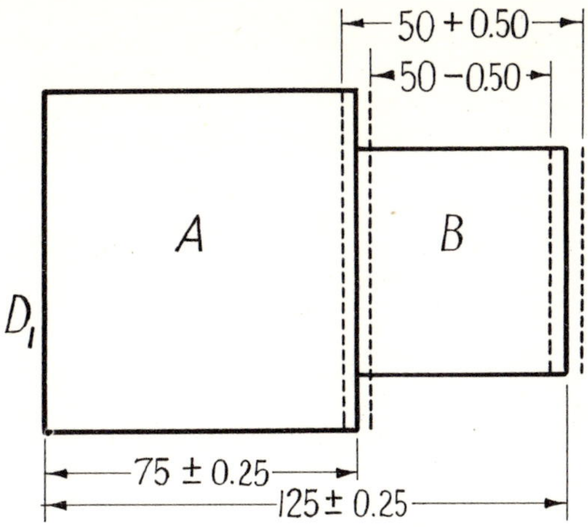

Fig 2

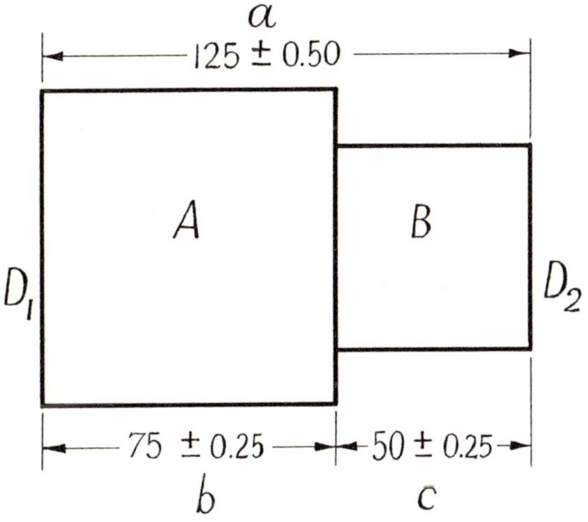

Fig 3

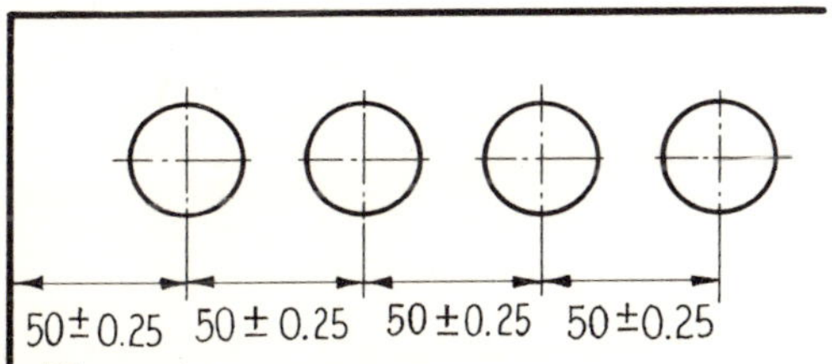

Fig 4

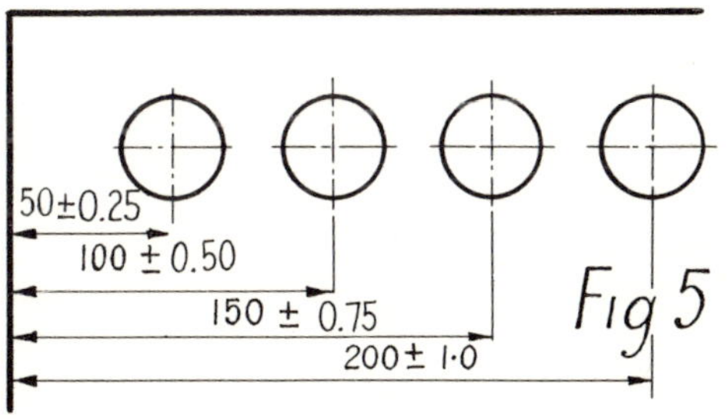

Fig 5

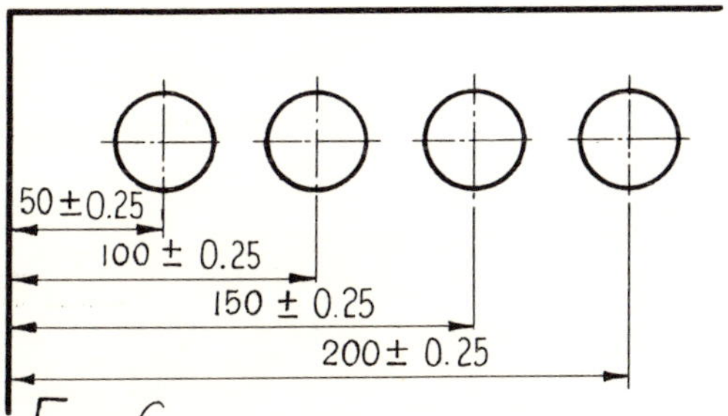

Fig 6

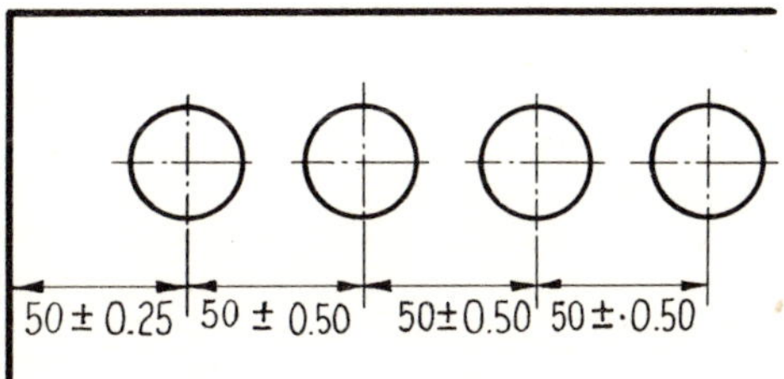

Fig 7

Graphical Symbols

Symbols generally

We have already considered some of the conventions used in the making of engineering drawings; for example, those for screw threads set out on page 167. These conventions have been devised to save drawing time. For the same reason various *symbols* have been standardized, particularly for use in diagrams showing the arrangement rather than the details of components. These symbols represent a kind of shorthand and they are listed in several publications of the British Standards Institution. One of these, B.S. 1553(1) gives symbols for pipes and joints, valves and cocks, drains, strainers, and so on. Another, B.S. 1553(2), deals with engines and boilers, pumps and condensers, and all plant which would be found in a power station. When arrangement drawings of such power plants are made, the actual shapes of the units are not shown; instead, simple outlines are used. For example: a square represents a boiler; a shortened cone, a turbine; a circle, a pump; two triangles, a valve.

Electrical engineering symbols

Symbols in commoner general use than those for power plants are symbols descriptive of components in electrical engineering. Some of these are shown opposite. The student may be familiar with many of those for radio and television, because he can hardly avoid meeting circuit diagrams in which they are used. These symbols are the alphabet of the electronics engineer.

Those shown opposite have been selected from the large number, several hundreds, given in B.S. 108 and B.S. 530. The student should refer to these if he can, if only to become acquainted with the scope and detail of these two publications.

The symbols have been reproduced to the size which would normally be used in circuit diagrams. Note that many of the symbols explain themselves and that all are simple in form. Some of the symbols have variants; compare for example the transformers shown with transformer T1 in fig. 5 on the next page. Stencils are available for drawing the commoner symbols.

Exercises

1. Memorize, a few at a time, the symbols given, by covering the descriptions and writing the name of the component.
2. Write out the descriptions and, without referring to the book, supply the symbols, freehand. Compare your effort with the symbols shown opposite.
3. Without referring to the diagrams, make drawings to about the right size for the following: battery, variable resistance, variable capacitor, transformer with ferromagnetic core, transistor, triode, microphone.
4. Refer to any of the current radio or television magazines and identify the various symbols used in the circuit diagrams shown.

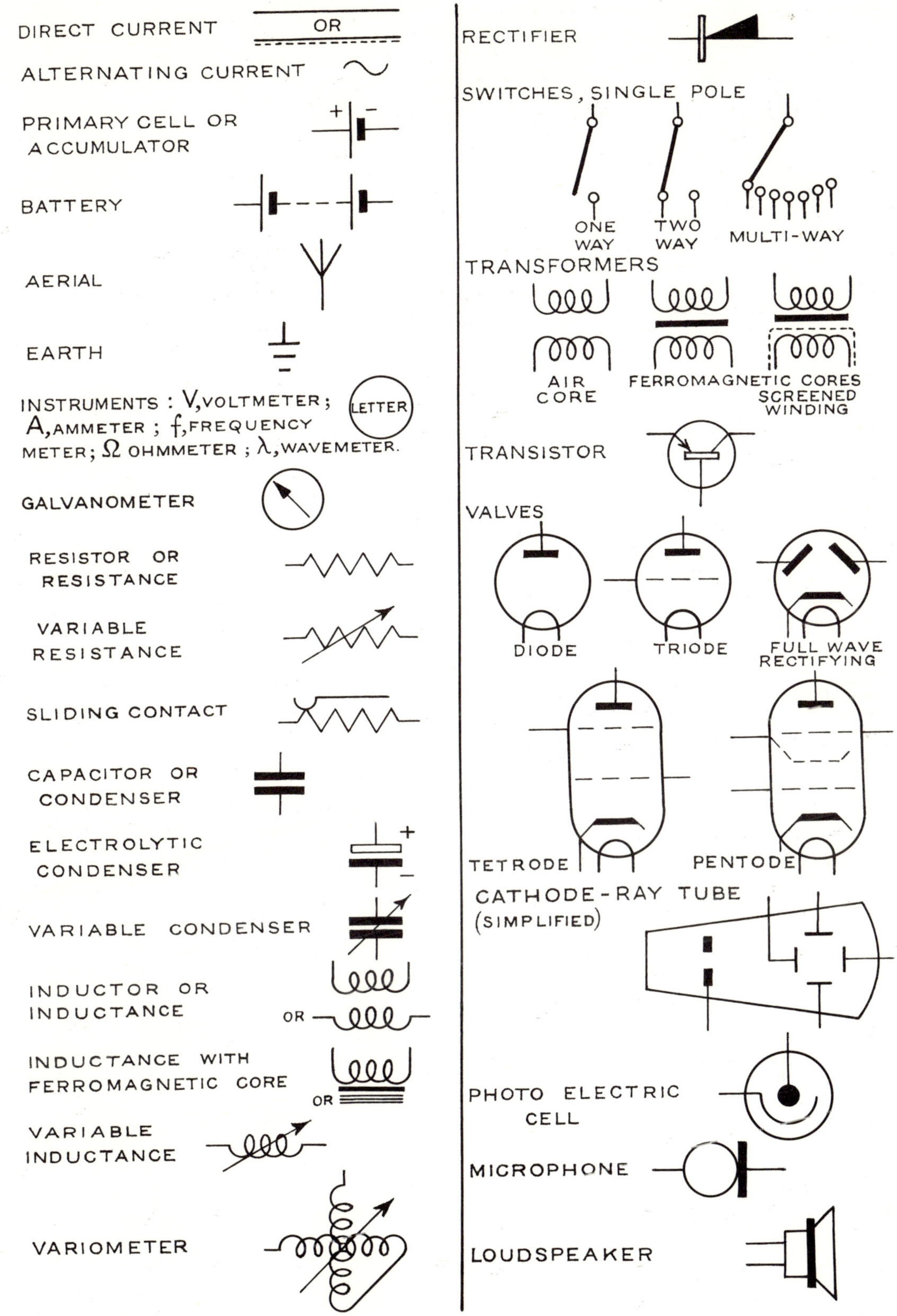

DIRECT CURRENT
OR
ALTERNATING CURRENT
PRIMARY CELL OR ACCUMULATOR
+
−
BATTERY
AERIAL
EARTH
INSTRUMENTS : V, VOLTMETER ; A, AMMETER ; f, FREQUENCY METER; Ω OHMMETER ; λ, WAVEMETER.
LETTER
GALVANOMETER
RESISTOR OR RESISTANCE
VARIABLE RESISTANCE
SLIDING CONTACT
CAPACITOR OR CONDENSER
ELECTROLYTIC CONDENSER
+
−
VARIABLE CONDENSER
INDUCTOR OR INDUCTANCE
OR
INDUCTANCE WITH FERROMAGNETIC CORE
OR
VARIABLE INDUCTANCE
VARIOMETER
RECTIFIER
SWITCHES, SINGLE POLE
ONE WAY
TWO WAY
MULTI-WAY
TRANSFORMERS
AIR CORE
FERROMAGNETIC CORES
SCREENED WINDING
TRANSISTOR
VALVES
DIODE
TRIODE
FULL WAVE RECTIFYING
TETRODE
PENTODE
CATHODE-RAY TUBE (SIMPLIFIED)
PHOTO ELECTRIC CELL
MICROPHONE
LOUDSPEAKER

Circuit Diagrams

Circuit diagrams show the *operation* of the arrangement of electrical components and their connections; they make use of standard symbols for the components, but they need not show these in their true relative positions. To be fully useful, circuit diagrams should give information about the components. Wiring diagrams differ from circuit diagrams in that they do show the components in their true relative positions and are designed to help the wireman connect up the parts correctly.

Wires passing and joining

Wires passing were formerly shown as in fig. 1*a*, and the method is still in use. But the cross-over loop is time-wasting and wires passing should be shown as in fig. 1*b*. Wires joining should be shown as in fig. 2*a*; the intention of the arrangement in fig. 2*b* is confirmed if shown as in fig. 2*c*, with a staggered junction.

Types of arrowhead

Three types of arrowhead are used: for instruments the head is closed, fig. 3*a*; for variability the head is half-closed, fig. 3*b*; and for direction the head is open, fig. 3*c*.

Arrangement of circuit diagrams

The sequence of effect in circuit diagrams should follow the sequence of the printed page: from left to right, and from top to bottom. Fig. 4 illustrates this: the aerial of the receiver is on the left at the top, the loudspeaker is on the right, and the earth is at the bottom. This circuit is of a simple receiver, employing a crystal rectifier and two transistors, which may be built up at home. The diagram is incomplete because information is lacking about the types and values of the components.

A more complete diagram is shown in fig. 5—which is a portion only of a larger circuit. Owing to limitations of space on the page opposite, fig. 5 has been drawn much too crowded and it would be better opened out; for the same reason the symbol for the fuse is smaller than the standard. The components are numbered and lettered: T for transformer, F for fuse, C for capacitor, L for laminated-core inductance, V for valve, SK for socket. Although values for capacitors are shown (μF meaning microfarad) a table would normally be provided with the circuit diagram giving a specification of each part.

Exercises

1. **Make a drawing of the circuit in fig. 4, using symbols of the correct size, and insert the name of each component.**
2. **Make a freehand sketch of the circuit in fig. 5 and from this sketch set out the circuit in a more spacious way.**
3. **Symbols for some of the components used in electrical installation work are shown at the lower part of the page opposite. Memorize these and identify the components which they represent in your home or at school or college. Set out the diagram of the installation in a small house.**

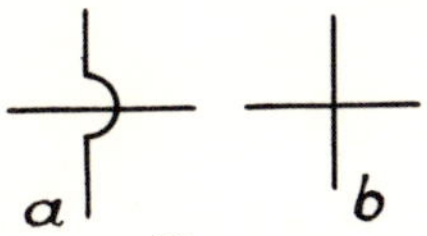

Fig 1

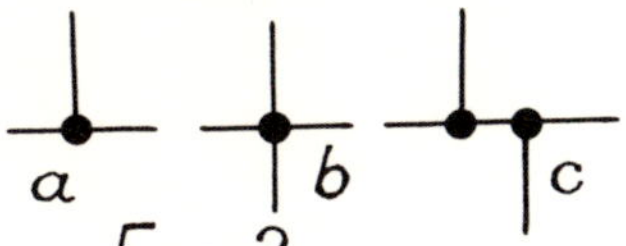

Fig 2

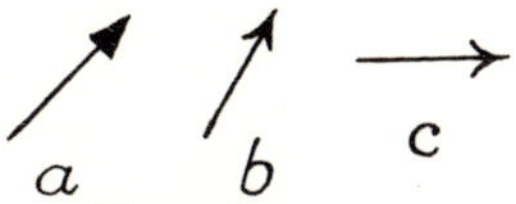

Fig 3

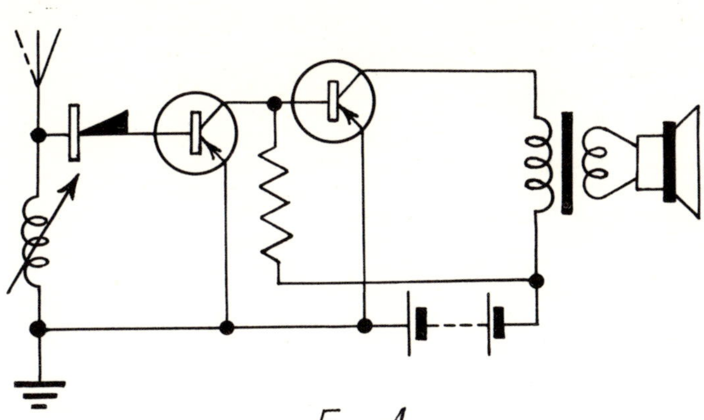

Fig 4

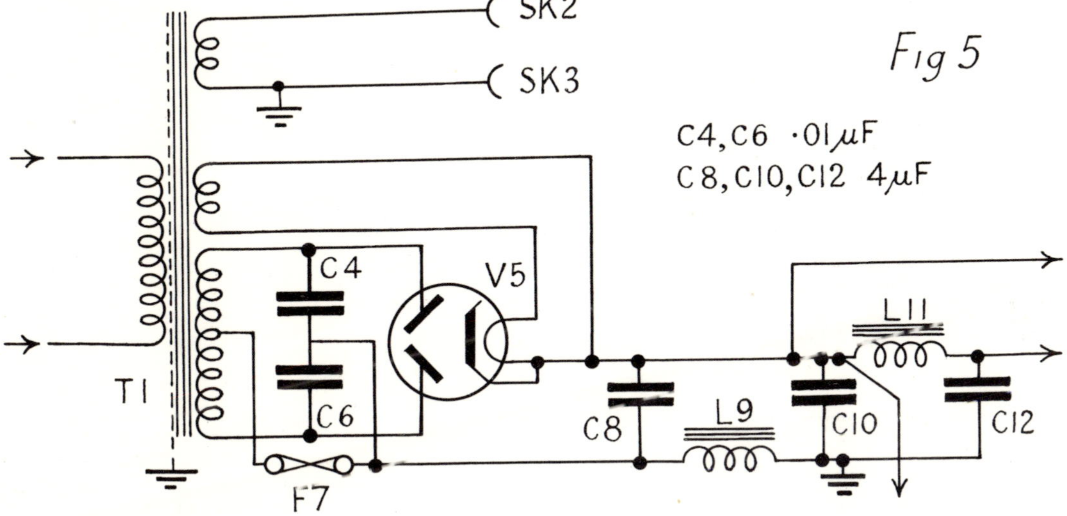

Fig 5

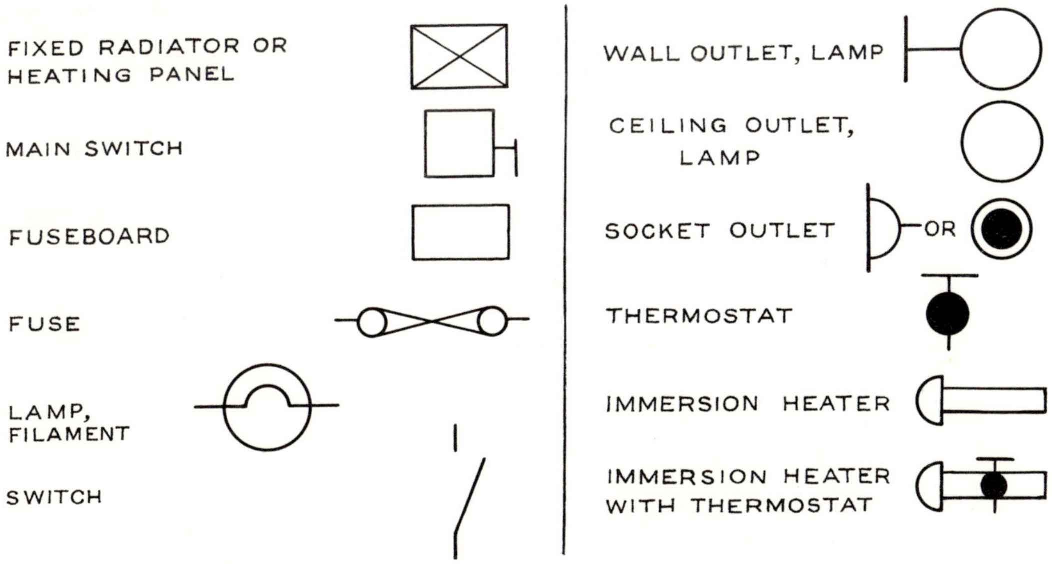

Handwheel

Exercise

The drawings show two undimensioned views of a handwheel for a machine tool. The wheel is of cast iron and has its rim offset from the boss; the four arms are elliptical in section; the handle is of steel and is screwed into the rim. The bore and ends of the wheel are machined; the outer semicircular surface of the rim is machined and plated; the handle is machined all over and, except for the screwed end, plated.
Measure the drawings given. Make full-size working drawings, separately, of (*a*) the wheel and (*b*) the handle. Show a keyway in the hub (refer to page 182). Insert all dimensions and add finish marks.

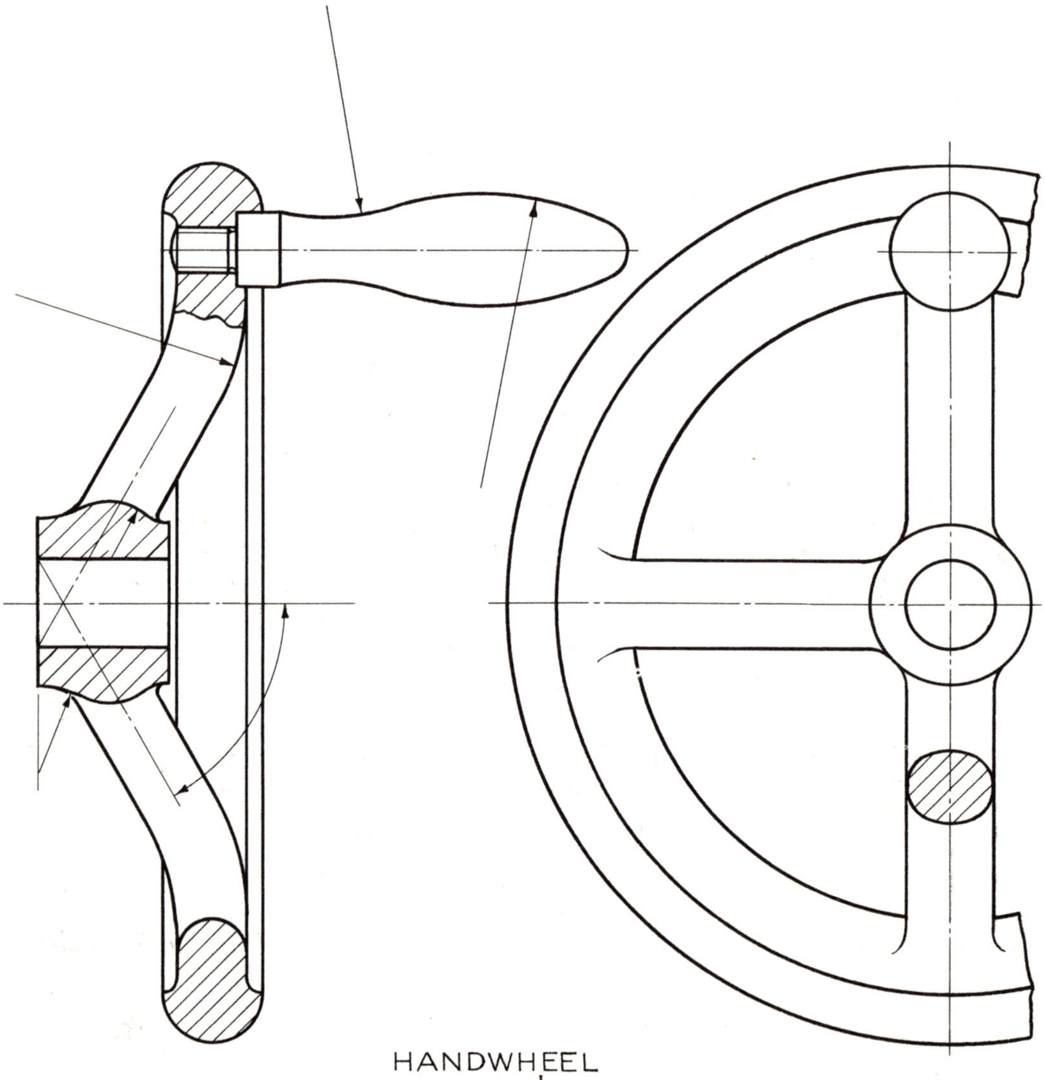

HANDWHEEL
SCALE $\frac{1}{2}$

Insulator

Exercise

The drawing shows one half of a central section of a circular insulator. This is built up of four parts cemented together: cast-iron base with pillar, cast-iron cap, and two porcelain elements. Certain surfaces are bitumen-coated and others sand-glazed.

Draw full size the section shown and add the complementary outside view on the right of the centre line. Project a half plan and a half underneath view. Insert only the principal dimensions.

Note.—Not all dimensions are given. The omission is intentional and the student should scale from the drawing and settle the missing dimensions.

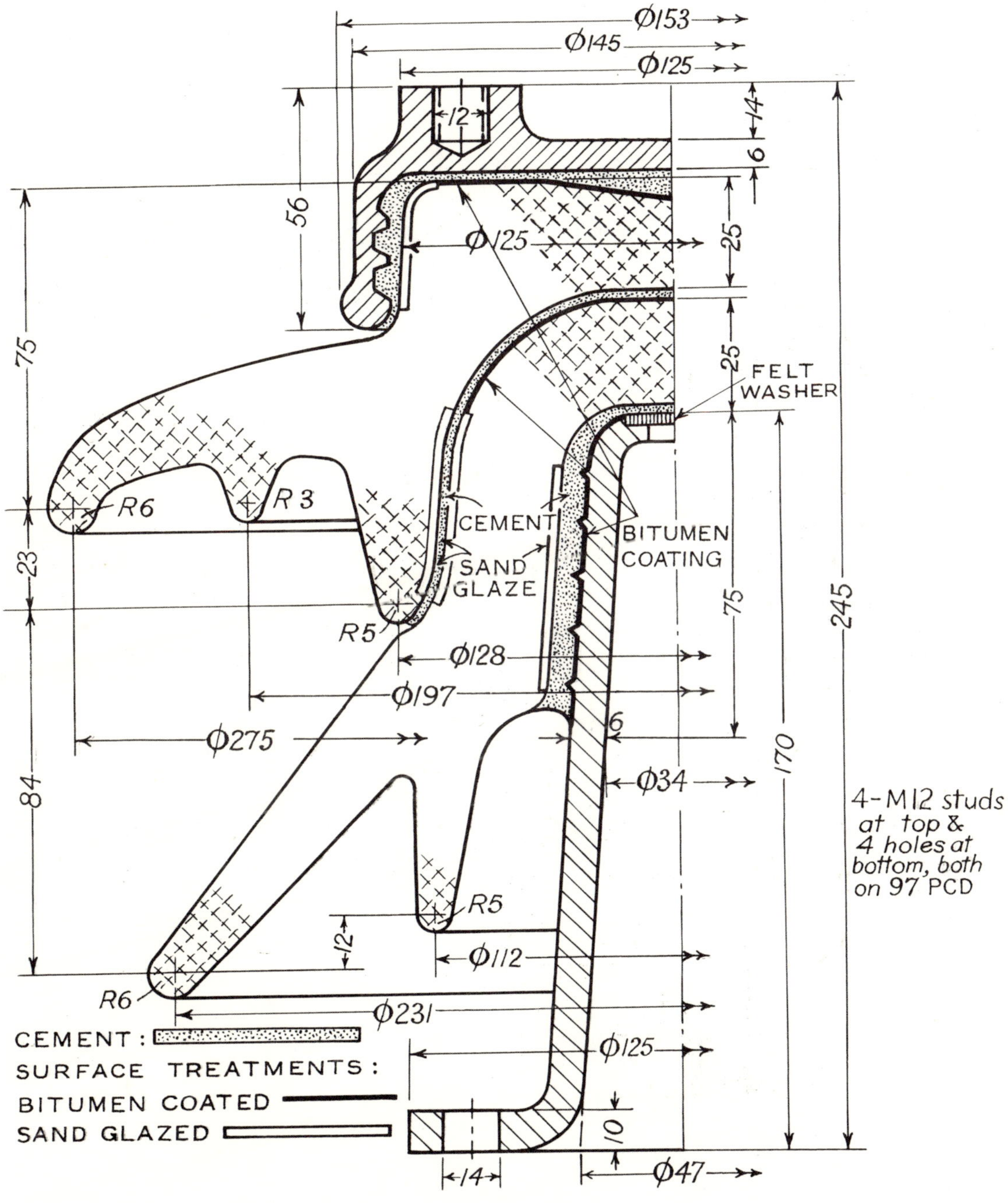

Vee Slide

Exercises

The drawing shows a cast-iron slide which is to have considerable movement over a cast-iron bed. The slide and the bed fit together over a vee bearing, one vee being a separate strip. The angle of vee is 45°.

It is necessary to compensate for wear and this is done by using a steel strip (on one side only) as shown in detail on the left. The strip is held against the slide on the bed by two 10 mm diameter tap bolts, which are locked in position by nuts. Two 10 mm diameter round-headed screws hold the strip to the slide; they are just slack when the adjustment is made by the tap bolts after which they are tightened. The holes for the screws must be made slightly larger than the screws and arranged to give the clearance space initially, necessary for the small take-up movement of the strip. The details of this compensating device are left for the student to settle.

1. Draw, full size, views of the slide in the directions A, B and C. The view C is to be in section through one of the adjusting screws. Insert the principal dimensions.
2. Draw, twice full size, completely dimensioned working drawings for : (*a*) a tap bolt and nut; (*b*) a cheese-headed screw; (*c*) the steel vee strip.

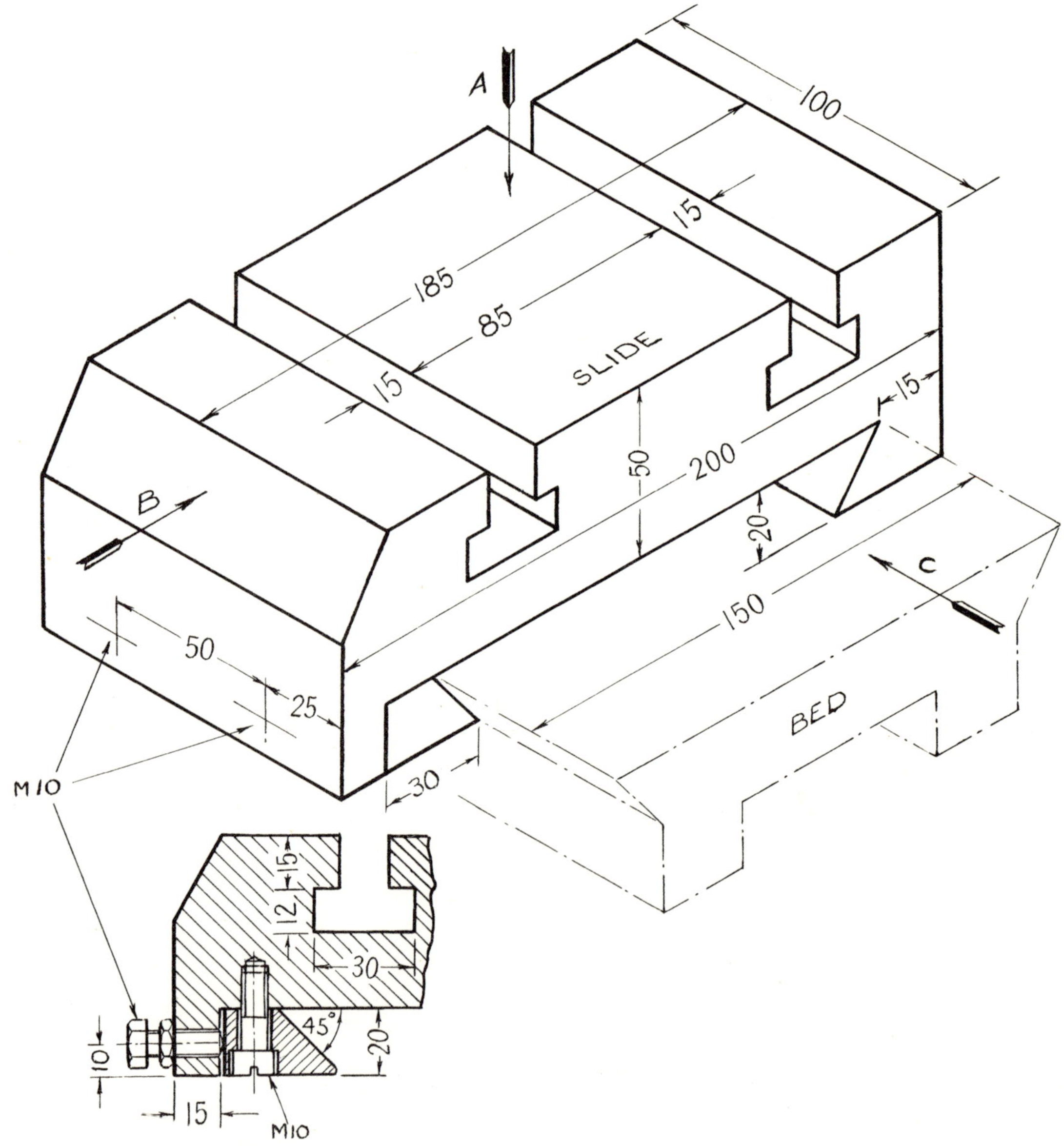

Guide Bracket

Exercises

The drawings show a cast-iron guide bracket. It has a barrel, through which passes a shaft 75 mm diameter. The bracket is bolted firmly in position and the shaft slides through the barrel.
It is necessary to grip the shaft in certain positions and a quick-locking device is required. That shown in the sectional view is composed of two gunmetal bushes, one on each side of the barrel, sliding in a cylinder and bearing on the shaft. A bolt passes through the two bushes, and the nut for the bolt has a hand lever. When this nut is tightened the two bushes are drawn together; they grip the shaft by the wedge action of their shaped ends.

Many minor dimensions have been omitted from the drawings and left to the judgment of the student. It will be seen that a peg is required in the head of the bolt to prevent it from turning when the nut is being tightened or unscrewed; this peg also is left to the student to provide.

1. Draw the two views given and add a third, this being an outside view looking along the shaft axis. Insert principal dimensions only.
2. Add, separately, full-size drawings, completely dimensioned, of the bolt, nut, and one bush. How would you shape the ends of the bushes?

PROJECTION

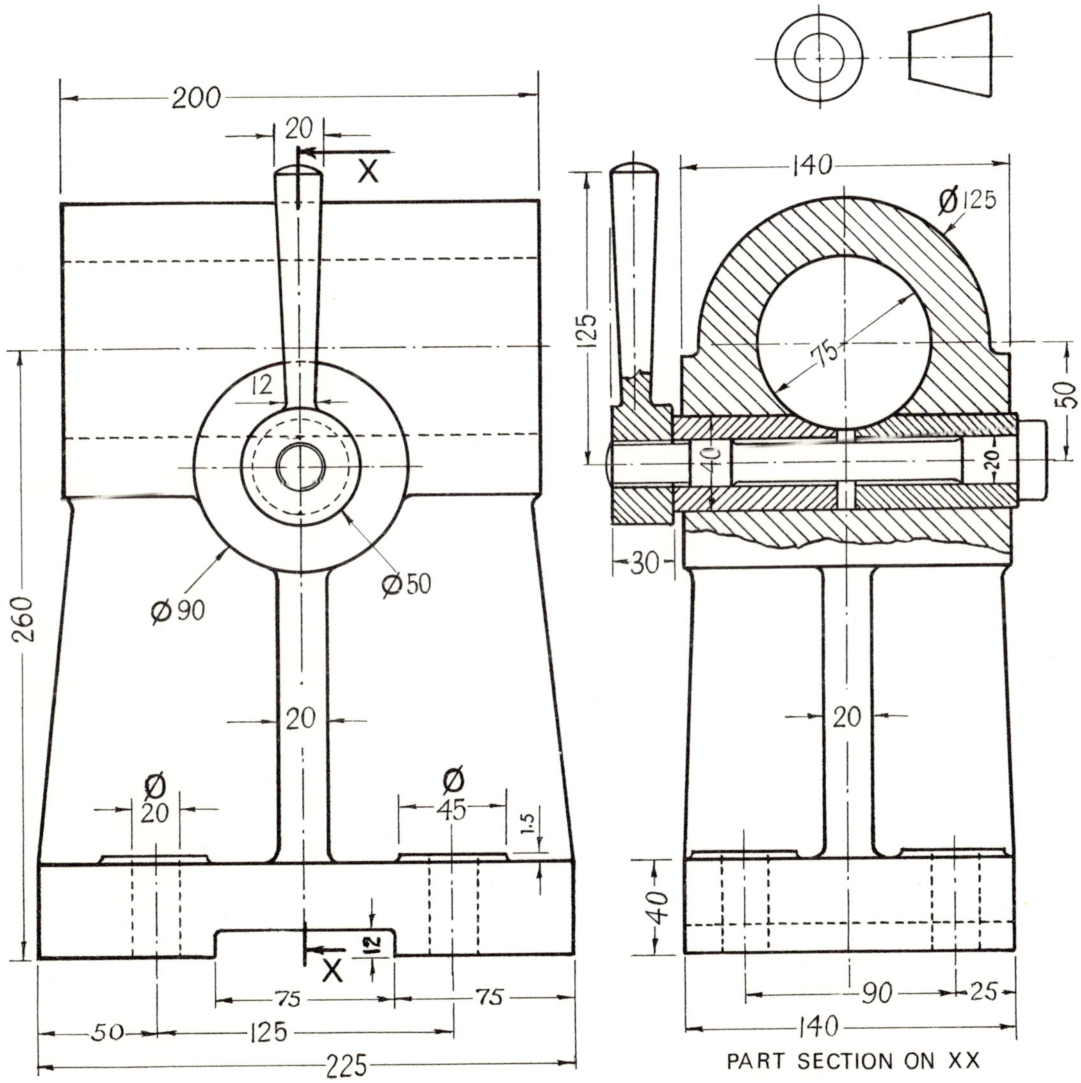

PART SECTION ON XX

Drawings in Ink

Many ordinary technical drawings are made in pencil on transparent paper. The paper is reasonably strong, and accurate drawings can be made upon it with, if necessary, very fine lines. By using a medium-hard pencil, lines can be made dense enough to give a good print from the drawing, the print being 'fixed' on sensitized paper and giving either a negative (white lines on blue or brown background) or a positive (black lines on white).

Records of drawings

The storage of drawings for future reference in the fire-proof room of a drawing office has in time past presented an ever-growing problem. Such drawings are now made in ink with lines a little wider than standard and photographed on to 35 mm film, which is then fixed to a record card and the original drawing destroyed. The film can be stored and enlarged prints taken at any future date. To ensure accuracy, scales are printed on the drawing paper both vertically and horizontally to check against slight distortion in the reproduction of the film.

Ink pens

The ink pen shown in fig. 1 is a tweezer-like arrangement, the gap between the blades being adjustable. Ink is inserted between the blades by a dipper. The tips must coincide when closed and must be smoothly rounded as in fig. 2, not too fine as in fig. 3 nor too broad as in fig. 4. A very fine oilstone slip is suitable for adjusting the points when they wear, fig. 5. Ink pens for compasses are of the same kind; when used for large radii the pen should be arranged to be at right angles to the paper, as in fig. 6.
Reservoir pens with round nibs giving a uniform thickness of line are useful. A range of sizes is available.

Use of instruments

Very satisfactory work can be done with ink instruments if certain precautions are taken. The pen should not be overcharged with ink or flooding may occur; it should be handled so that the ink is not transferred to the edge of the set square or rule; it should be held as in fig. 7, against a bevelled edge. Lines of various thicknesses may be drawn. When two lines meet it is better to let one of them dry before drawing the other.

Other kinds of drawings

In the aircraft industry it is common practice to make drawings full size on coated aluminium sheets using a fine metal stylus. The drawings are to scale and measurements may be taken from them. Copies for use in the shops are taken from the original by a photographic process, and are in fact full-size replicas on metal sheets.

Fig 1

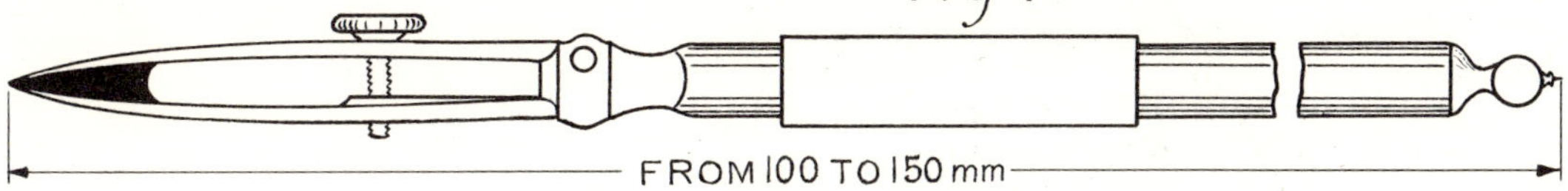

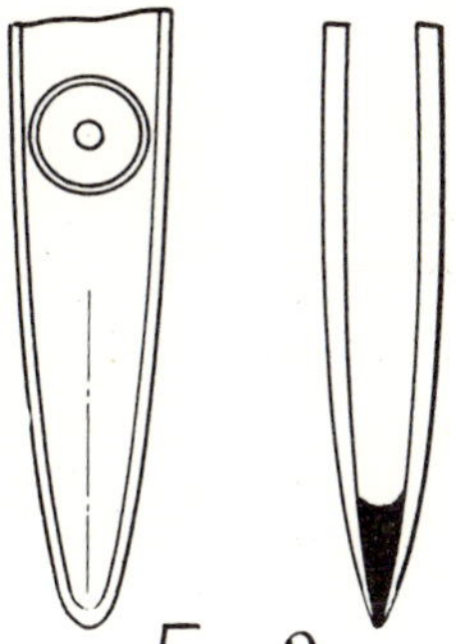

Fig 2

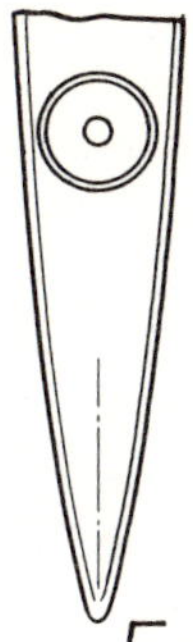

Fig 3

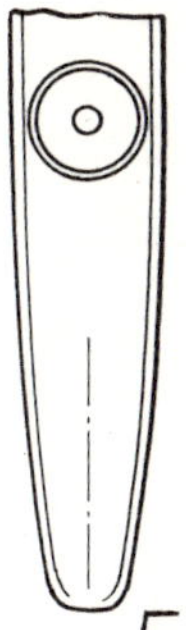

Fig 4

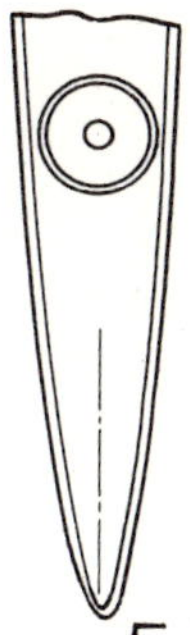

Fig 5

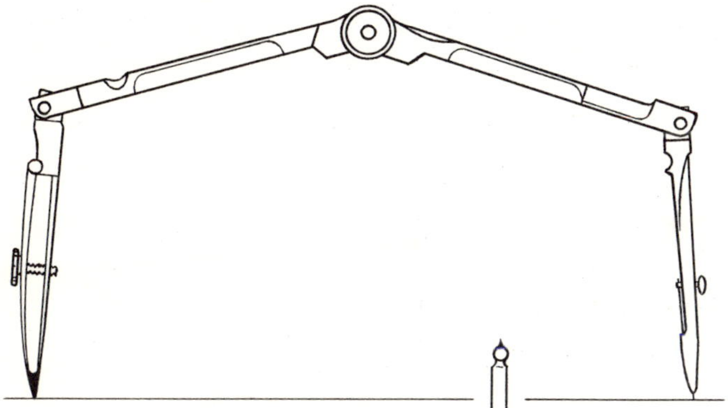

Fig 6

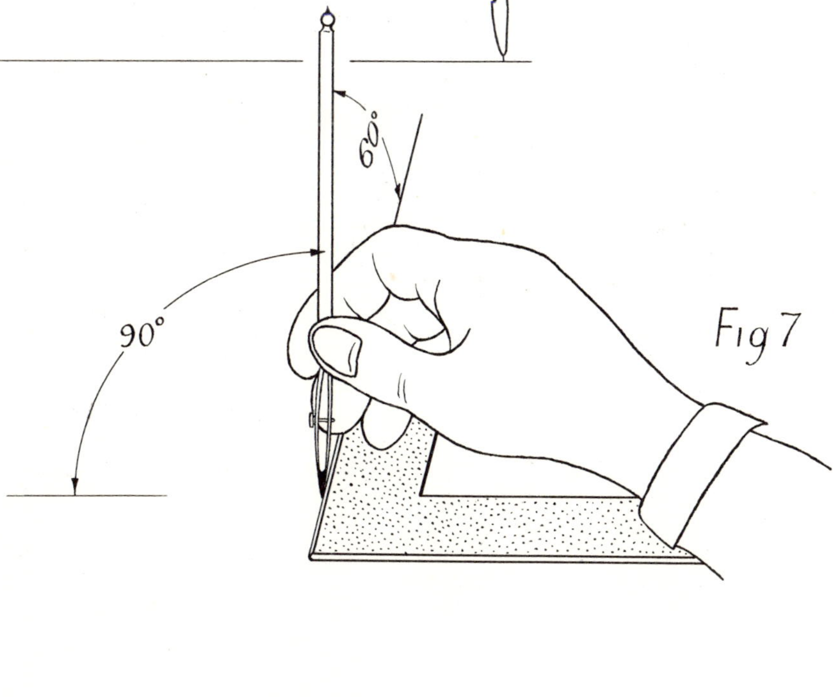

Fig 7

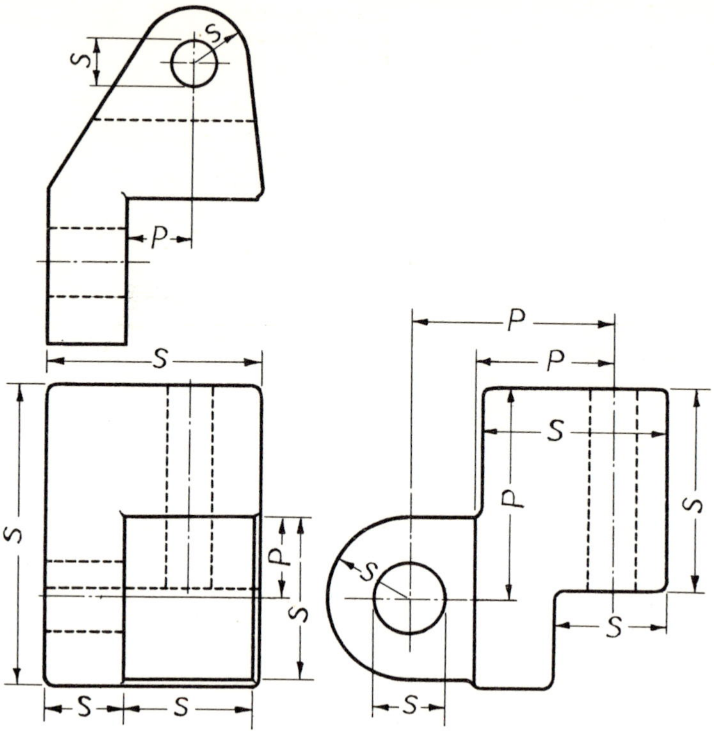
S
S
P
S
S
P
S
S
S
P
P
S
P
S
S
S
S

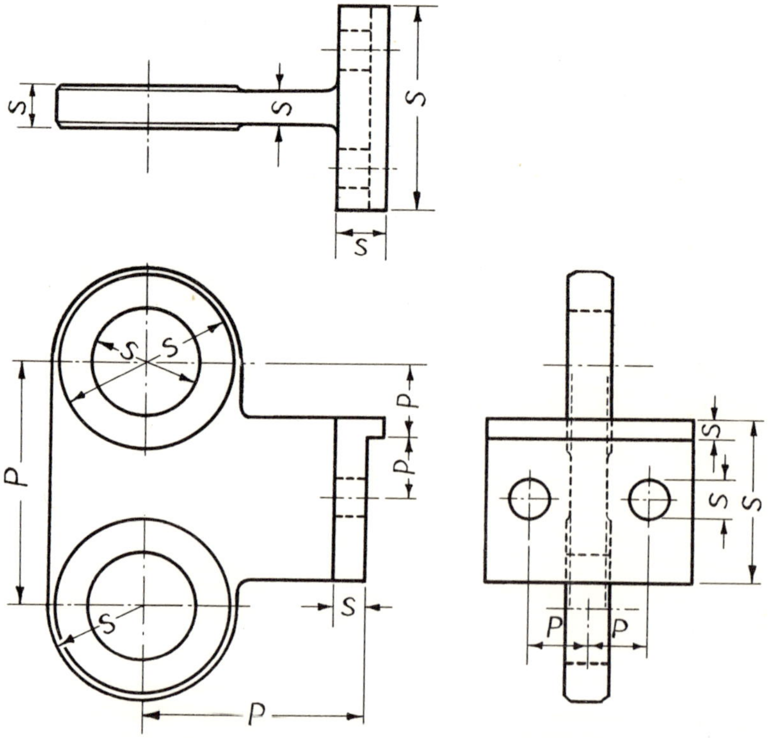
S
S
S
S
S
S
P
P
P
S
S
P
S
S
S
S
P
P

Index

Abbreviations 160
Accuracy in drawings 72
Angle bar 176
Angles, meaurement of 24
Approximate constructions 94
Archimedean spiral 100
Areas of irregular figures 102
Auxiliary projections:
cone 116
cube 112
cylinder 116
pyramid 114

Bearings 188
Bolts 168
Brackets 190 203
British Standards:
B.S. 108 196
B.S. 308 34 138
B.S. 499 180
B.S. 530 196
Butt joint 164

Cabinet projection 130
Cams 90
Cavalier projection 130
Channel bar 176
Checking constructions 72
Choice of view 140
Circles, for given conditions 84
Circuit diagrams 198
Compasses, use of 20
Cone 46
Conic sections 86 88 120
Conventions for nuts 168
Conventions for threads 166
Cottered joints 184
Couplings 182
Cube 28 112
Curves of intersection 122 124 144
Curves of intersection (approximations) 144
Cycloids 96
Cylinder 46

Development by triangulation 178
Development in sheet metal 178
Development of:
cone 126
cube 28
cylinder 126
prism, rectangular 30 36
pyramid 44
Dimensioning 62 154–158 162 194
Division of lines 22
Drawing board, size 16
Drawing errors 20
Drawing headings 146
Drawing procedure 142
Drawings in ink 204

Ellipse 86 88 120
Equal areas 82
Evolute 96
Extreme and mean ratio 74

Fabrication by welding 180
Feather key 182
First angle projection 106 138
Flanged coupling 182
Flap valve 193
Freehand sketching 64 172 174

Generator bracket 190
Geometrical mean 74
Gib and cotter 184
Gib headed key 182
Golden rule 76
Graphical symbols 196

Handwheel 200
Headings for drawings 146
Helical spring 100
Helix 100
Hexagonal prisms 40
Hidden lines 38
Hyperbola 86 120

Indexed plans 132
Insulator 201
Interpretation of drawings 66 68 70
Intersections of surfaces:
cones 124
cylinders 122
Involute 96 98
Involute gear teeth 98
Irregular figures 102
ISO metric thread 167
Isometric drawings 52–60
of circles 56 58
of curves 56
of rectangular solids 54
Isometric drawings, interpretation 66
Isometric projection 128
Isometric scale 128

Joints:
cottered 184
flanged 180
knuckle 186

Keys 182
Knuckle joint 186

Labelling of views 32 34
Lap joint 164
Letters and numbers 146
Limits of size 194
Line thicknesses 162
Line types 34
Loci, points on mechanisms 92
Locking devices 170

Index

Machining symbols 160
Mean proportional 74
Mechanisms, loci of points on 92

Non-return valve 192
Numbers 146
Nuts 168

Oblique projection 130
Orthogonal projection 104
Orthographic projection 104

Parabola 86 88 120
Parallel curves 90
Parthenon 76
Parts not sectioned 152
Plummer block bearing 189
Polygons 78 80 82
Polygons, similar 80
Position dimensions 154 156 204
Pressure angle, wheel teeth 98
Prism, hexagonal 40 42
Prism, rectangular 28 30
Projection:
  Cabinet 130
  Cavalier 130
  Horizontal 132
  First angle 106 138
  Oblique 130
  Orthographic 104
  Principles of 104 106
  Third angle 106 138
Pyramid 44

Ratios 74
Reading a drawing 140
Rectangle 76
Revolved sections 150
Rivets 164
Riveted joints 164
Rolled steel sections 176
Roulettes 96

Screws 166 170
Screw threads 166
Sectional views 48 50

121 SW MC 33½x41 10x8 3/16 685 8 on 10 20M

Sectional views, engineering 148 150 152
Sections of solids:
  cone 120
  cube 118
  pyramid 118
Set square 16
Shafts 182
Similar polygons 80
Size dimensions 154 156 206
Spheres, in isometric projection 128
Spiral 100
Square root 74
Square, set 16
Square spring 101
Square, tee 16
Steel sections 176
Straight line:
  inclinations 110
  projections 108
  true length 108
Straight lines:
  parallel 16 18
  perpendicular 16 18
Studs 168
Symbols for:
  electrical engineering 196 198
  machining 160
  welds 152 153 180

Tee square 16
Third angle projection 106 138
Threads 166
Tolerances 194
Toothed gearing 98
Trace of line, plane 104
Triangles 78
True shape of surfaces 42

Unified thread 166

Valves 192
Vee slide 202

Wall brackets 191
Welded flange 180
Welded link 180
Wheel teeth 98
Whitworth thread 166
Wiring diagrams 198